Physiology and Pathophysiology of Digestion

D. Neil Granger

James Morris

Peter R. Kvietys

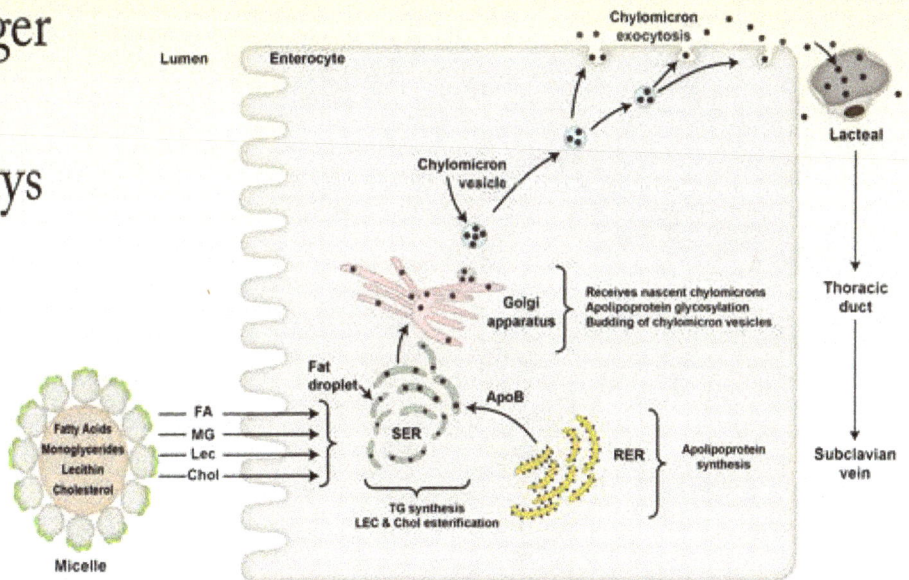

MORGAN & CLAYPOOL PUBLISHERS

Physiology and Pathophysiology of Digestion

Colloquium
Digital Library of Life Sciences

The *Colloquium Digital Library of Life Sciences* is an innovative information resource for researchers, instructors, and students in the biomedical life science community, including clinicians. Each PDF e-book available in the *Colloquium Digital Library* is an accessible overview of a fast-moving basic science research topic, authored by a prominent expert in the field. They are intended as time-saving pedagogical resources for scientists exploring new areas outside of their specialty. They are also excellent tools for keeping current with advances in related fields, as well as refreshing one's under-standing of core topics in biomedical science.

For the full list of available titles, please visit:
colloquium.morganclaypool.com

Each book is available on our website as a PDF download. Access is free for readers at institutions that license the *Colloquium Digital Library*.

Please e-mail info@morganclaypool.com for more information.

Colloquium Series on Integrated Systems Physiology: From Molecule to Function to Disease

Editors
D. Neil Granger, *Louisiana State University Health Sciences Center*
Joey P. Granger, *University of Mississippi Medical Center*

Physiology is a scientific discipline devoted to understanding the functions of the body. It addresses function at multiple levels, including molecular, cellular, organ, and system. An appreciation of the processes that occur at each level is necessary to understand function in health and the dysfunction associated with disease. Homeostasis and integration are fundamental principles of physiology that account for the relative constancy of organ processes and bodily function even in the face of substantial environmental changes. This constancy results from integrative, cooperative interactions of chemical and electrical signaling processes within and between cells, organs, and systems. This eBook series on the broad field of physiology covers the major organ systems from an integrative perspective that addresses the molecular and cellular processes that contribute to homeostasis. Material on pathophysiology is also included throughout the eBooks. The state-of the-art treatises were produced by leading experts in the field of physiology. Each eBook includes stand-alone information and is intended to be of value to students, scientists, and clinicians in the biomedical sciences. Because physiological concepts are an ever-changing work-in-progress, each contributor will have the opportunity to make periodic updates of the covered material.

Published titles
(for future titles, please see the website, www.morganclaypool.com/page/lifesci)

Copyright © 2018 by Morgan & Claypool Life Sciences

Physiology and Pathophysiology of Digestion
D. Neil Granger, PhD
James Morris, MD
Peter R. Kvietys, PhD
www.morganclaypool.com

ISBN: 9781615046966   paperback
ISBN: 9781615046973   ebook
ISBN: 9781615047901   Hardcover

A Publication in the

COLLOQUIUM SERIES ON INTEGRATED SYSTEMS PHYSIOLOGY: FROM MOLECULE TO FUNCTION TO DISEASE

Series Editors: D. Neil Granger, LSU Health Sciences Center, and Joey P. Granger, University of Mississippi Medical Center

Series ISSN

ISSN 2154-560X   print
ISSN 2154-5626   electronic

Physiology and Pathophysiology of Digestion

D. Neil Granger, PhD
Louisiana State University Health Sciences Center–Shreveport

James Morris, MD
Louisiana State University Health Sciences Center–Shreveport

Peter R. Kvietys, PhD
Alfaisal University, Riyadh, KSA

COLLOQUIUM SERIES ON INTEGRATED SYSTEMS PHYSIOLOGY: FROM MOLECULE TO FUNCTION TO DISEASE #82–84

MORGAN&CLAYPOOL LIFE SCIENCES

ABSTRACT

This collaboration of two physiologists and a gastroenterologist provides medical and graduate students, medical and surgical residents, and subspecialty fellows a comprehensive summary of digestive system physiology and addresses the pathophysiological processes that underlie some GI diseases. The textual approach proceeds by organ instead of the traditional organization followed by other GI textbooks. This approach lets the reader track the food bolus as it courses through the GI tract, learning on the way each organ's physiologic functions as the bolus directly or indirectly contacts it. The book is divided into three parts: (1) Chapters 1–3 include coverage of basic concepts that pertain to all (or most) organs of the digestive system, salivation, chewing, swallowing, and esophageal function, (2) Chapters 4–6 are focused on the major secretory organs (stomach, pancreas, liver) that assist in the assimilation of a meal, and (3) Chapters 7 and 8 address the motor, transport, and digestive functions of the small and large intestines. Each chapter includes its own pathophysiology and clinical correlation section that underscores the importance of the organ's normal function.

KEYWORDS

gastrointestinal physiology, digestion, motility, absorption, secretion, esophagus, stomach, pancreas, liver, biliary system, intestines, colon, gastrointestinal hormones, enteric nervous system

Contents

Preface

Over 35 years ago, we (Neil Granger, James Barrowman, Peter Kvietys) undertook the task of producing a textbook of medical physiology that focused exclusively on the digestive system. The resulting monograph, entitled *Clinical Gastrointestinal Physiology*, was published in 1985 by the W. B. Saunders Company. The textbook was produced for use in physiology courses offered to medical and graduate students, and as refresher material for medical or surgical residents, subspecialty fellows, and for the faculty assigned the task of teaching gastrointestinal physiology to these trainees.

When the initial version of this textbook was produced, GI physiology was commonly taught as part of a larger, comprehensive medical physiology course that covered the function of all body systems, and it devoted relatively little attention to pathophysiology and the clinical manifestations of GI dysfunction. A recent trend in medical education has resulted in the systems-based, integrated medical curriculum, wherein there is no longer an identifiable course in physiology. Consequently GI physiology is now often taught as part of a course that integrates basic science and clinical aspects of the digestive system, with a goal of bridging the gap between theory and practice, and helping take physiology from "bench to bedside."

The science of GI physiology has expanded quite remarkably over the past 35 years. Advancements in the field of biomedicine and the development of new high resolution imaging technology have greatly improved our understanding GI physiology at the molecular, cellular, and organ levels. These advances have also led to a more rational view of the molecular basis of the different pathological processes that affect the digestive system.

With the rapid expansion of knowledge on the digestive system and the emphasis placed on integration of basic and clinical material in the new medical curriculum structure, there is a clear need for a textbook that provides comprehensive and up-to-date coverage of digestive system physiology that highlights medical relevance of the physiological principles with discussions of pathophysiology and clinical correlation. Accordingly, we have extensively revised and updated this textbook material, which was first published in 1985. The underlying concepts of physiology presented in this text will help the clinical learner to develop a broader understanding of the digestive system to approach clinical challenges with the strength of a physiological approach to treatment.

We have retained the organizational feature of the original book that presents the physiologic concepts for the individual organs of the digestive system, rather than the more traditional organization in terms of the major GI functions, that is, motor, absorptive, and secretory functions. This approach allows the reader to follow a bolus of food down the GI tract and learn the physiologic functions that the bolus contacts, either directly or indirectly (e.g., pancreas, liver). It also serves to emphasize the sequential nature of the actions of each organ in the digestive system and places into perspective the contribution of each organ to the overall assimilation of a meal.

The textbook is divided into three parts. The first part (Chapters 1–3) includes coverage of basic concepts that pertain to all (or most) organs of the digestive system, salivation, chewing, swallowing, and esophageal function. The second part (Chapters 4–6) is focused on the major secretory organs (stomach, pancreas, liver) that assist in the assimilation of a meal, whereas the third (Chapters 7 and 8) addresses the motor, transport, and digestive functions of the small and large intestines.

The authors are very grateful to Janice Russell for her expert and timely production of the original illustrations that appear in this textbook.

D. Neil Granger, PhD
Boyd Professor Emeritus, Molecular & Cellular Physiology
Louisiana State University Health Sciences Center–Shreveport

James Morris, MD
Associate Professor of Medicine
Louisiana State University Health Sciences Center–Shreveport

Peter R. Kvietys, PhD
Professor
College of Medicine, Alfaisal University, Riyadh, KSA

Dedication

To James A. Barrowman, MD, PhD
(1936–1991)

CHAPTER 1

Basic Concepts

Introduction

Control of Gastrointestinal Function
The Enteroendocrine System
Gastrointestinal Hormones and Related Peptides
Intrinsic and Extrinsic Nerves
Intrinsic Innervation
Extrinsic Innervation
Parasympathetics
Sympathetics
Afferent Nerve Fibers
Sensory Transduction

Gastrointestinal Functions
Motility
Basal Electrical Rhythm and Interstitial Cells of Cajal
Excitation-Contraction Coupling
Functional Motor Patterns
Digestion
Absorption
Routes of Absorption
Mechanisms of Absorption
Passive Absorption
Active Absorption
Water Absorption
Secretion
Water and Electrolyte Secretion
Mucin Secretion

Circulation of the Digestive System

Pathophysiology and Clinical Correlations

INTRODUCTION

The energy required by all organisms for continued existence is derived from nutrients assimilated from the environment. Although single-celled organisms obtain these nutrients by simple processes, such as diffusion and phagocytosis, a more complex and specialized system exists for this purpose in multicellular organisms. In humans, this system takes the form of a tube, referred to as the gastrointestinal tract; the internal surface of which constitutes the largest interface between the external environment and the body's internal milieu. The gastrointestinal tract is richly supplied by blood vessels of the circulatory system, which distribute the assimilated nutrients to all tissues of the body.

The gastrointestinal tract (Figure 1-1) consists of a tube from mouth to anus (oropharynx, esophagus, stomach, and small and large intestine) and is associated with several secre-

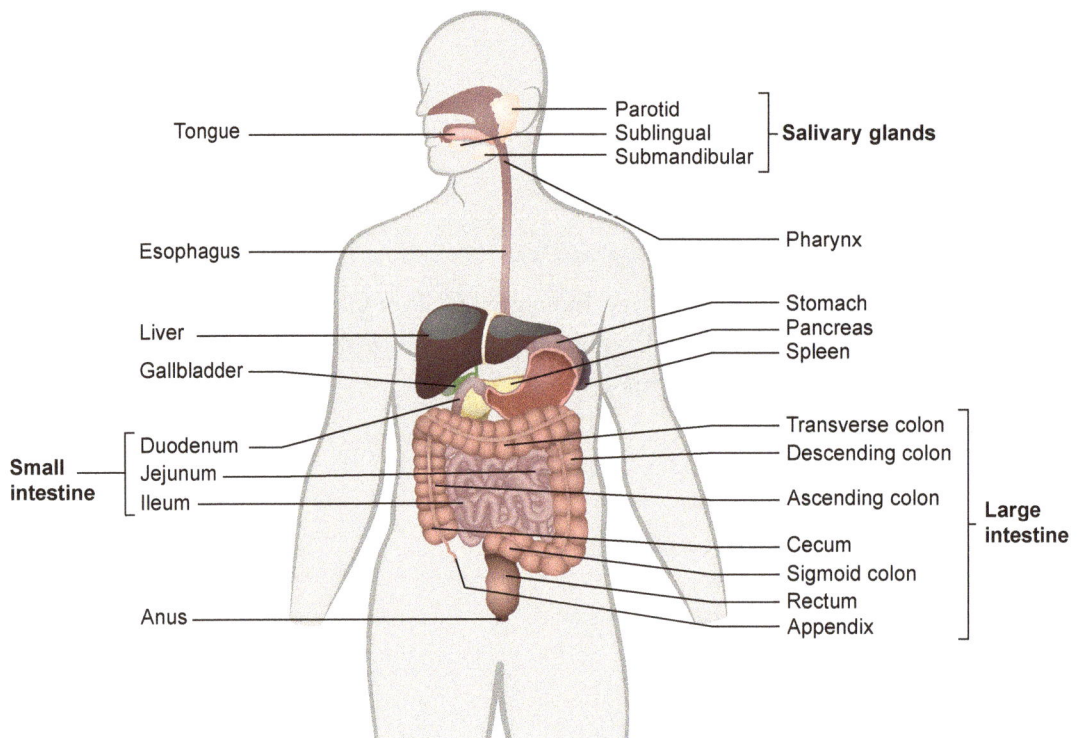

FIGURE 1-1: The digestive system includes the mouth, esophagus, gastrointestinal tract, and accessory organs, that is, the salivary glands, liver, gallbladder, the pancreas.

tory organs (salivary glands, liver, pancreas). Most of these components make a specialized contribution to the overall process of assimilation, that is, digestion and absorption, whereas other components simply store or propel food. The mouth and esophagus are primarily involved in food propulsion. The stomach and colon are mainly involved in the storage of food and food residues, respectively. The salivary glands, pancreas, and liver manufacture and deliver digestive juices, whereas the primary site of digestion and absorption of food is the small intestine.

The medical significance of the digestive system is verified by the fact that a large proportion of patients self-treat with over-the-counter products and consult their physicians concerning symptoms referable to the gastrointestinal tract. Common symptoms, such as indigestion, nausea, vomiting, heartburn, abdominal discomfort/pain, flatulence, constipation, and diarrhea, frequently reflect disorders of function. Nevertheless, a significant proportion of these symptoms may indicate underlying disease processes. Gastrointestinal disorders translate into high medical costs (direct and indirectly due to disability) and many hours of diagnostic effort on the part of the physician. It is essential that the normal functions of the digestive system be understood before the physician can effectively treat these symptoms in a rational manner.

Current research is leading to a rapid expansion of our knowledge of the digestive system. This book provides the medical and graduate student with the fundamentals of gastrointestinal (GI) physiology and can serve as a refresher for the medical or surgical resident, and subspecialty fellow. To illustrate the medical relevance of each area of gastrointestinal function, appropriate discussions of pathophysiology and clinical correlations are included. The material in this text is presented using an organ approach, so that the reader can follow a bolus of food down the GI tract and learn physiologic functions of the various organs which the bolus contacts, either directly or indirectly. To minimize repetition of material in subsequent chapters, a brief summary of basic concepts related to the three major functions of the digestive system, that is, motility, absorption, and secretion is presented here.

CONTROL OF GASTROINTESTINAL FUNCTION

The Enteroendocrine System

It is now well recognized that a large number of hormones and related peptides/amines are formed in tissues not generally regarded as part of the classic endocrine system (e.g., hypothalamic-pituitary-adrenal axis). The gut and kidney are examples of such tissues. Indeed, the first hormones to be discovered and so named were of gastrointestinal origin (secretin and gastrin). The endocrine cells of the gastrointestinal mucosa are scattered throughout the epithelium and referred to as "enteroendocrine cells" (EEC). Collectively, they form the largest endocrine system of the body. The peptides/amines produced by the EECs can function as hormones, which are secreted into the circulation to regulate target cells remote from their origin, either 1) within the gastrointestinal system (Table 1-1) or 2) at the level of the central nervous system (CNS), that is, hypothalamus (Table 1-2; the role of these peptides in the regulation of feeding/satiety is addressed in Chapter 2). Some of the listed substances do not behave like classic hormones in that they are released from cells and diffuse through the extracellular space to neighboring target cells. This type of local action is referred to as *paracrine* control.

Most EECs respond to feeding, specifically to the presence of hydrolytic products of food digestion in the lumen (Table 1-1). They act as "chemosensors" coordinating appropriate gastrointestinal responses to ingestion of food. How is a chemical/physical characteristic of luminal contents transduced by the EEC into a functional gastrointestinal response? Most EEC in the mucosa are of the "open type," that is, the microvilli on the apical surface of the cell are in contact with the lumen (Figure 1-2). The microvilli contain receptors for sampling luminal contents, and when these receptors are activated, the EEC discharge the content of their secretory granules into the interstitium. The secreted peptides/amine can then act either as a paracrine or a hormone. Paracrine targets include cells of the epithelium as well as afferent sensory neurons. For example, cytoplasmic protrusions of gastric D cells come in close apposition to G cells, allowing somatostatin to regulate gastrin secretion in a paracrine fashion. The I cells in the duodenal mucosa have basal protrusions that extend toward sensory neurons within the interstitium, facilitating the interaction of CCK with vagal afferents. Endocrine targets are generally located in accessory organs, such as pancreas and liver. Access of a hormone to the bloodstream is facilitated by the close proximity of the blood capillaries and the fact that the capillary fenestrae (pores) always face the basal aspect of the EEC. Some EECs are of the "closed" type that reside entirely in the interstitium and have no connection to the luminal surface. Although closed EEC do not directly respond to luminal contents, they can be regulated

TABLE 1-1: Products of Enteroendocrine Cells (ECC) that Regulate Gastrointestinal Function

EEC	PRODUCT	LOCATION	LUMINAL STIMULUS	PRINCIPAL EFFECT(S)
G cells	**Gastrin**	Stomach	protein hydrolysates, distension	↑ gastric acid secretion
I cells	**CCK***	Proximal Sm intestine	protein & lipid hydrolysates	↑ gall bladder contraction ↑ pancreatic enzyme secretion
S cells	**Secretin***	Proximal Sm intestine	acid	↑ pancreatic & biliary HCO_3- secretion
D cells	Somatostatin*	Stomach & Sm intestine	acid	↓ gastrin release
EC cells	5-HT*, ‡	Entire GIT	toxins, lipid hydrolysates	↑ intestinal motility & fluid secretion
ECL cells	Histamine*	Stomach	closed cell	↑ gastric acid secretion
M cells	**Motilin**	Sm intestine	fasting, nutrients inhibit	↑ migrating motility complex
N cells	Neurotensin	Distal Sm intestine & colon	lipid hydrolysates	↓ gastrointestinal motor activity
K cells	**GIP**	Proximal Sm intestine	carbohydrate & lipid hydrolysates	↑ insulin release
L cells	PYY*, GLP-1	Distal Sm intestine & colon	carbohydrate & lipid hydrolysates	↑insulin release ↓ gastrointestinal motor activity

Products in **bold** are peptides meeting all of the criteria for hormone status. * these products have paracrine actions. ‡ 5-HT can also be found in some I and L cells.

			LUMINAL	PRINCIPAL
CELL	PRODUCT	LOCATION	STIMULUS	EFFECTS
I cells	**CCK***	Proximal Sm intestine	lipid & protein hydrolysates	↓ food intake
A/X cells	**Ghrelin***	Stomach	fasting, nutrients inhibit	↑ food intake
P cells	leptin*	Stomach	feeding	↓ food intake
L cells	PYY*, GLP-1	Distal Sm & Lg intestine	lipid & carbohydrate hydrolysates	↓ food intake

TABLE 1-2: Products of Enteroendocrine Cells (EECs) with Effects on Hypothalamic Feeding/Satiety Centers

Products in **bold** are peptides meeting all of the criteria for hormone status. * these products have paracrine actions.

by paracrine, humoral, and/or neural inputs. For example, gastrin released by G cells can enter the circulation and stimulate enterochromaffin-like (ECL) cells, which reside entirely in the interstitium, to release histamine.

Gastrointestinal Hormones and Related Peptides

To date, there are six peptides generated by the EEC that meet all of the criteria for classification as a hormone (listed in bold in Tables 1-1 and 1-2). Some of the other peptides (e.g., somatostatin, PYY) are referred to as "candidate hormones" and, most likely, will attain full "hormone" status as more information becomes available. Conversely, some of the EEC-derived hormones influence local, rather than remote, target cells. For example, cholecystokinin (CCK)-induced pancreatic enzyme secretion is mediated in part via a neural reflex elicited by the paracrine activation of intramural afferent nerves. As information regarding the function of EEC-derived peptides accumulates, it is becoming apparent that some of these peptides have multiple modes of action, that is, endocrine, paracrine, and/or neurocrine. Thus, the EEC-

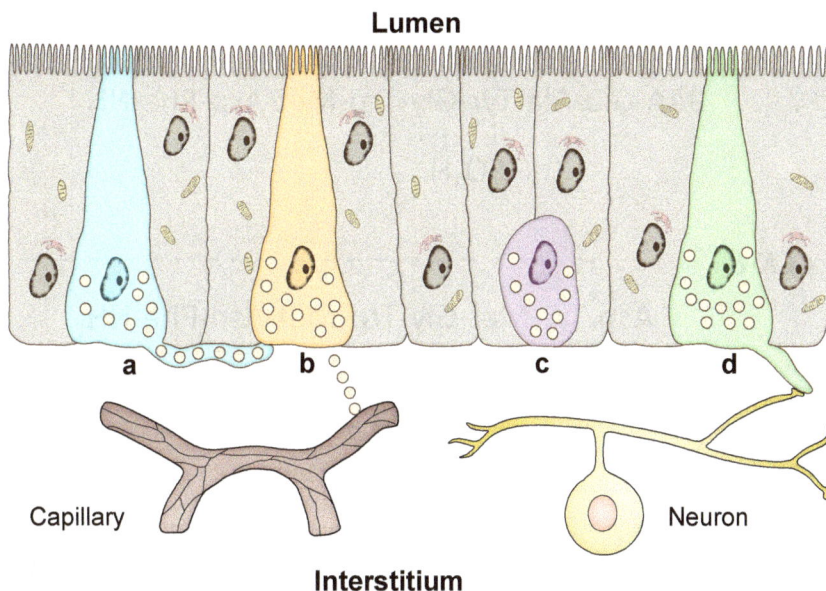

FIGURE 1-2: The enteroendocrine cell (EEC) types of the mucosal epithelium. The EEC labeled as "a, b and d" are of the "open" type, with apical microvilli that can detect and respond to specific luminal contents. "Closed" EEC, labeled "c." do not make contact with the lumen and cannot respond directly to luminal stimuli, instead they are activated by paracrine, endocrine, and/or neural mechanisms. EEC labeled "a and d" have specialized cell extensions that facilitate local, paracrine communication with an adjacent epithelial cell (**a**) or neuron (**d**), respectively. EEC labeled "b" represents a typical endocrine cell whose secreted product enters the circulation to reach its remote target (modified from Gribble FM and Reimann F. *Annu. Rev. Physiol.* 2016; 78: 277-299).

derived peptides are grouped into "families" based on the similarity of their amino acid sequence, rather than their mode of action on target cells.

Cholecystokinin and gastrin are members of the gastrin family. They share an identical sequence of five amino acids (Gly-Trp-Met-Asp-Phe) at the amidated carboxy terminal end of the peptide chain (Figure 1-3). In addition, they are defined by the sulfated tyrosyl residues neighboring the conserved active terminal sequence; position 6 for gastrin and position 7 for CCK. Gastrin and CCK are secreted in different biologically active forms and identified based on the amino acid chain length. The two forms of circulating gastrin are G-17 and G-34. CCK has at least four active forms (CCK-8, CCK-33, CCK-39, and CCK-58). In general, the

Gastrin-17

Pharmacophore CCK2

9AA-Glu-Ala-Tyr-Gly-Trp-Met-Asp-Phe-NH$_2$
|
SO$_3$H

CCK-8

Asp-Tyr-Met-Gly-Trp-Met-Asp-Phe-NH$_2$
|
SO$_3$H

Pharmacophore CCK1

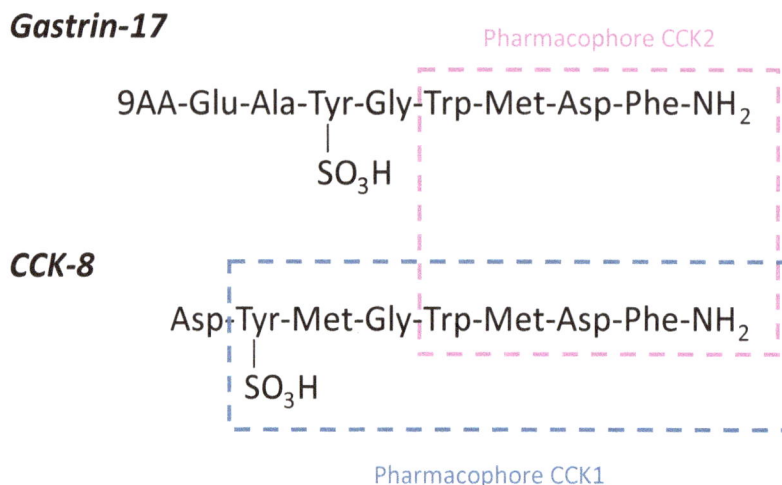

FIGURE 1-3: The amino acid sequences of human gastrin-17 and cholecystokinin-8 (CCK-8) peptides. The pharmacologically active (pharmacophoric) regions recognized by the CCK$_1$ and CCK$_2$ receptors are highlighted (from *Pharmacology & Therapeutics* 2008; 119: 83–95. Used with permission of Elsevier).

longer the amino acid chain length the longer the half-life in the circulation. The importance of the carboxy-terminal sequence in the primary structure of both gastrin and CCK is evidenced by the fact that both the synthetic carboxy pentapeptide (pentgastrin) and the carboxy octapeptide of CCK (CCK-8) are active and used clinically.

Because the two hormones share a common sequence of amino acids, they can interact with the same receptor and thus exhibit similar actions. Nonetheless, the relative potency of action of the two hormones differs among target organs. This results from subtle differences in structure and locations of the receptors, known as CCK$_1$ and CCK$_2$. Figure 1-3 shows the pharmacophores (molecular structure required for pharmacologic activity) of gastrin and CCK necessary for optimum activation of CCK$_1$ and CCK$_2$ receptors. Thus, although CCK can interact with the CCK$_1$ receptor, gastrin has little affinity for this receptor. Although CCK and gastrin have equal affinity for the CCK$_2$ receptor, gastrin exerts a dominant influence on the CCK$_2$ receptor in the gastric mucosa to induce acid secretion, which can be attributed to the 10-fold higher concentration of circulating gastrin (compared to CCK) under physiologic conditions. Ligation of the gastric CCK$_2$ receptor by CCK can be revealed under conditions of hypogastrinemia (low circulating gastrin). As expected, circulating CCK, but not gastrin, interacts with CCK_1 receptors in the gall bladder; contracting the bladder wall and propelling bile into the duodenum.

The second group of structurally related peptides, the secretin family, consists of peptides that circulate as single biologically active forms: secretin (27 aa), VIP (28 aa), and GIP (42 aa). Unlike the gastrin family, the structural homologies are internal, that is, identical amino acid sequences exist in segments within the peptide chain. Thus, the functional activity of the members of this family is more dependent on the tertiary, rather than the primary, structure of the peptides. The structural homologies within this group account for some shared spectrum of biologic activities. For example, secretin (a hormone) and VIP (a neurocrine) can stimulate fluid secretion from gastrointestinal organs, such as the duodenum and pancreas. On the other hand, glucose-dependent insulinotropic peptide (GIP) is an incretin (hormone that reduces blood glucose levels) that is secreted from the intestine in response to luminal glucose and stimulates insulin secretion by the pancreas.

Although the gastrin and secretin families of GI hormones are structurally dissimilar, members of one group may potentiate the effect of peptides from the other group in a given target cell. This is explained by the presence of receptors for each family on the same target cell. For example, in pancreatic acini, the interaction of CCK with the CCK_1 receptor leads to an increased intracellular Ca^{++} concentration, whereas the interaction of secretin with its receptor activates the cyclic AMP system (Figure 1-4). Both of these "second messengers" (Ca^{++}, cyclic AMP) work synergistically to promote near optimum enzyme secretion by the cell.

Motilin and ghrelin, the EEC-derived hormones of the ghrelin family, are unique in that they are both released during the interdigestive period (between meals) and inhibited by feeding. Motilin is released from the duodenum and jejunum, whereas ghrelin is produced by the stomach mucosa. Ghrelin is involved in the regulation of food intake by impacting on the feeding/satiety centers in the hypothalamus. Its transport across the blood brain barrier (and biologic activity) is dependent on a unique posttranslational modification, octanoylation (formation of ester with octanoic acid). The only proposed physiologic action of motilin is eliciting the interdigestive motility pattern, the migrating motility complex (MMC).

Intrinsic and Extrinsic Nerves

The gastrointestinal tract is extensively innervated by both the sympathetic and parasympathetic divisions of the autonomic nervous system. Efferent fibers in the vagus (parasympathetic) and

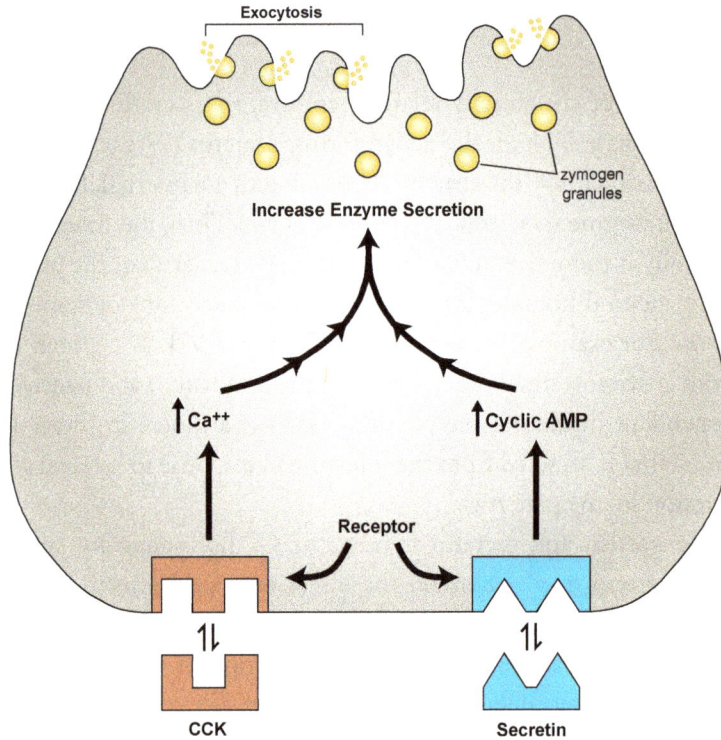

FIGURE 1-4: Mechanisms of action of the GI hormones CCK and secretin in pancreatic acinar cells. The interaction of CCK with its receptor (CCK_1) results in an increased intracellular Ca^{++} concentration, whereas the engagement of secretin with its receptor activates the cyclic AMP system. Both "second messengers" (Ca^{++}, cyclic AMP) work synergistically to promote the exocytosis of zymogen granules and release of pancreatic enzymes.

spinal (sympathetic) nerves allow the CNS to exert an influence on gastrointestinal function. Afferent fibers in these nerves carry information from GI organs to the brain. Furthermore, there is a complex intrinsic nervous system located within the wall of the stomach and gut. The intrinsic nervous system is perceived as an "enteric brain" which can function independently of the CNS to modulate the motor and secretory activities of the stomach and gut.

Intrinsic Innervation
The intrinsic nervous system, also referred to as the enteric nervous system (ENS), consists of two networks of ganglia and fibers located within the wall of the G.I. tract (Figure 1-5).

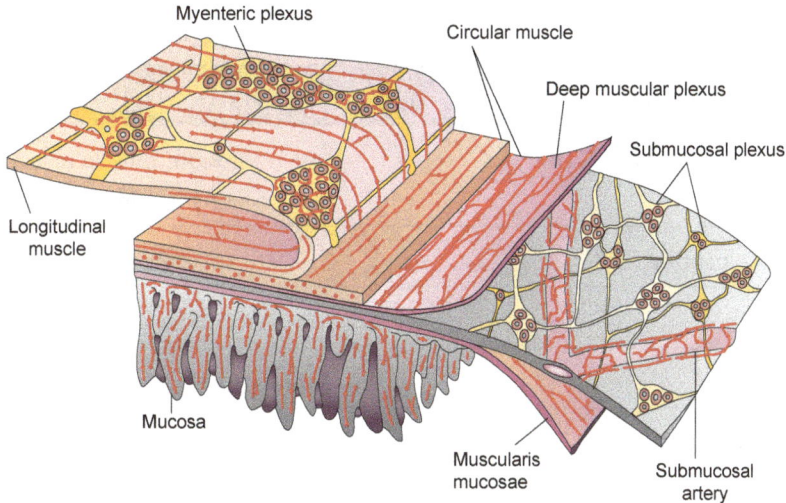

FIGURE 1-5: Organization of the enteric nervous system (ENS). The ENS has two major plexuses: the myenteric plexus, located between the longitudinal and circular muscle layers and the submucosal plexus (SMP). Nerve fibers connect the ganglia and also innervate the longitudinal muscle, circular muscle, muscularis mucosae, arterioles, and mucosal epithelium. **Circles** represent nerve cell bodies in ganglions; **Red lines** represent nerve fibers (modified with permission from *Nat Rev Gastroenterol Hepatol* 2012; 9:286–294, Figure 2, page 288).

One network, the myenteric plexus, is situated between the circular and longitudinal muscle layers and extends from the esophagus to the anal sphincter. The submucosal plexus separates the circular muscle and submucosal layers and is rather sparse or absent in esophagus, but continuous throughout the intestines. Efferent fibers from the myenteric plexus terminate on smooth muscle cells in both the circular and longitudinal layers; therefore, this plexus primarily influences motility (muscle tone and rhythm). The submucosal plexus efferents terminate on the mucosal epithelium (including endocrine cells), on the muscularis mucosa, and, to some extent, on the circular muscle layer. Therefore, this plexus primarily influences epithelial cell activity (e. g., secretion) and to a lesser extent, motor functions. There are many short neurons that link, and allow for communication between, the myenteric and submucosal plexuses.

Numerous transmitters have been identified in the ENS; some of the better characterized are listed in Table 1-3. In the motor neurons of the myenteric plexus, both acetylcholine (Ach) and substance P (SP) have been implicated in contraction of smooth muscles, whereas vasoactive intestinal polypeptide (VIP) and nitric oxide (NO) are considered the major mediators of

TABLE 1-3: Neurotransmitters of the Enteric Nervous System	
NEUROTRANSMITTER*	**PRINCIPAL EFFECTS**
Vasoactive intestinal peptide (VIP)	smooth muscle relaxation ↑ intestinal secretion
Gastrin releasing peptide (GRP)	↑ gastrin release
Substance P (SP)	smooth muscle contraction vasodilation
Nitric oxide (NO)	smooth muscle relaxation vasodilation
Acetylcholine (ACh)	smooth muscle contraction
Calcitonin gene-related peptide (CGRP)	vasodilation

* Neurotransmitters may be co-localized within to a given population of enteric neurons: VIP & nNOS (NO) or ACh & SP.

muscle relaxation. ACh and VIP release from the submucosal plexus have been implicated in the modulation of secretion as well as the vasodilation of submucosal arterioles.

Extrinsic Innervation

Parasympathetics

The parasympathetic nervous system has important secretomotor actions on the entire gastrointestinal tract. The organs of the digestive system are innervated by parasympathetic fibers originating from the medulla oblongata and the sacral part of the spinal cord (Figure 1-6). The medullary outflow comprises the seventh, ninth, and tenth (vagus) cranial nerves. The fibers of the seventh and ninth nerves supply the salivary glands. The vagi supply fibers to the esophagus, stomach, small intestine, proximal part of the large bowel, pancreas, liver, gallbladder, and bile ducts. The sacral parasympathetics originate from the second, third, and fourth sacral segments of the spinal cord and proceed to form the pelvic nerves (*nervi erigentes*). These fibers supply the distal portion of the large bowel, that is, sigmoid, rectum, and anus.

The preganglionic neurons of the parasympathetic supply terminate primarily on the ganglionic cells of the myenteric plexus in the stomach and gut and on the intraparenchymal ganglion cells of the pancreas, salivary glands, and liver. The preganglionic neurotransmitter is

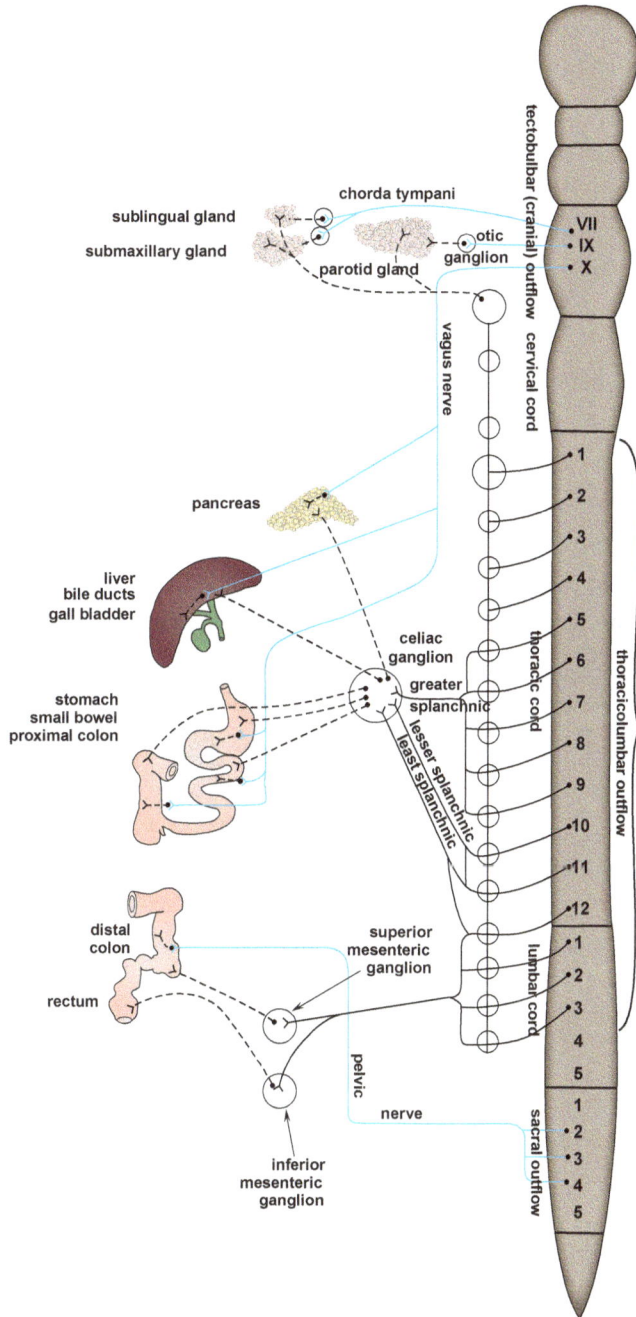

FIGURE 1-6: Extrinsic parasympathetic (blue lines) and sympathetic (black lines) innervation of the digestive system. Solid lines represent preganglionic fibers and broken lines represent postganglionic fibers (modified from Goodman, L.S. and Gilman, A. The Pharmacologic Basis of Therapeutics, 12th Edition, with permission of McGraw-Hill Professional).

acetylcholine, which interacts with *nicotinic* receptors on the postsynaptic membrane of ganglion cells. Acetylcholine also serves as a neurotransmitter at postganglionic nerve terminals, where it interacts with *muscarinic* receptors on effector cells (e.g., acinar cell of pancreas). A large number of other postganglionic neurotransmitters (or neurocrines) have been identified within the ENS of the stomach and gut that can exert either excitatory or inhibitory effects on GI function; some of the well-characterized neurocrines are listed in Table 1-3.

Sympathetics

Unlike the parasympathetic supply to the GI tract, which originates from the cranial and sacral segments of the cord, the sympathetic supply stems from the midportion of the cord, that is, the thoracic and lumbar segments (Figure 1-6). The preganglionic fibers enter the sympathetic chains after leaving the cord. The postganglionic fibers that terminate on the salivary glands arise directly from the superior cervical ganglia, which are located within the chains. All other digestive organs receive postganglionic fibers originating from outlying ganglia, such as the celiac, superior mesenteric, and inferior mesenteric ganglia, which in turn receive preganglionic fibers from the sympathetic chain. The celiac ganglia supply fibers to the esophagus, stomach, proximal duodenum, liver, and pancreas; the superior mesenteric ganglia supply fibers to the remainder of the small bowel and proximal colon; and the inferior mesenteric ganglia supply fibers to the distal colon and rectum. Some fibers make synaptic connections with the intrinsic nerve plexuses of the stomach and gut, whereas others end directly on blood vessels and, to a lesser extent, on other parenchymal structures. Synaptic transmission between preganglionic and postganglionic fibers of the sympathetic nervous system involves acetylcholine, whereas the postganglionic neurotransmitter is norepinephrine.

Afferent Nerve Fibers

Afferent fibers from the GI tract also travel in autonomic nerves (Figure 1-7). For example, 80% of vagal fibers and 50% of the sympathetic fibers are afferent. All of these fibers transmit sensory information to the central nervous system. Some of this information reach the conscious level, chiefly in the form of pain, whereas other information are a component of vegetative reflex arcs controlling GI function. Sensations of pain emanating from noxious stimuli in the GI tract are carried in spinal afferent fibers whose cell bodies are located in the dorsal root ganglia (DRG). Vagal afferent fibers involved in homeostatic regulation of GI functions (e.g., motility, secretion, satiety) have their cell bodies in the nodose ganglion (NG). In general, their

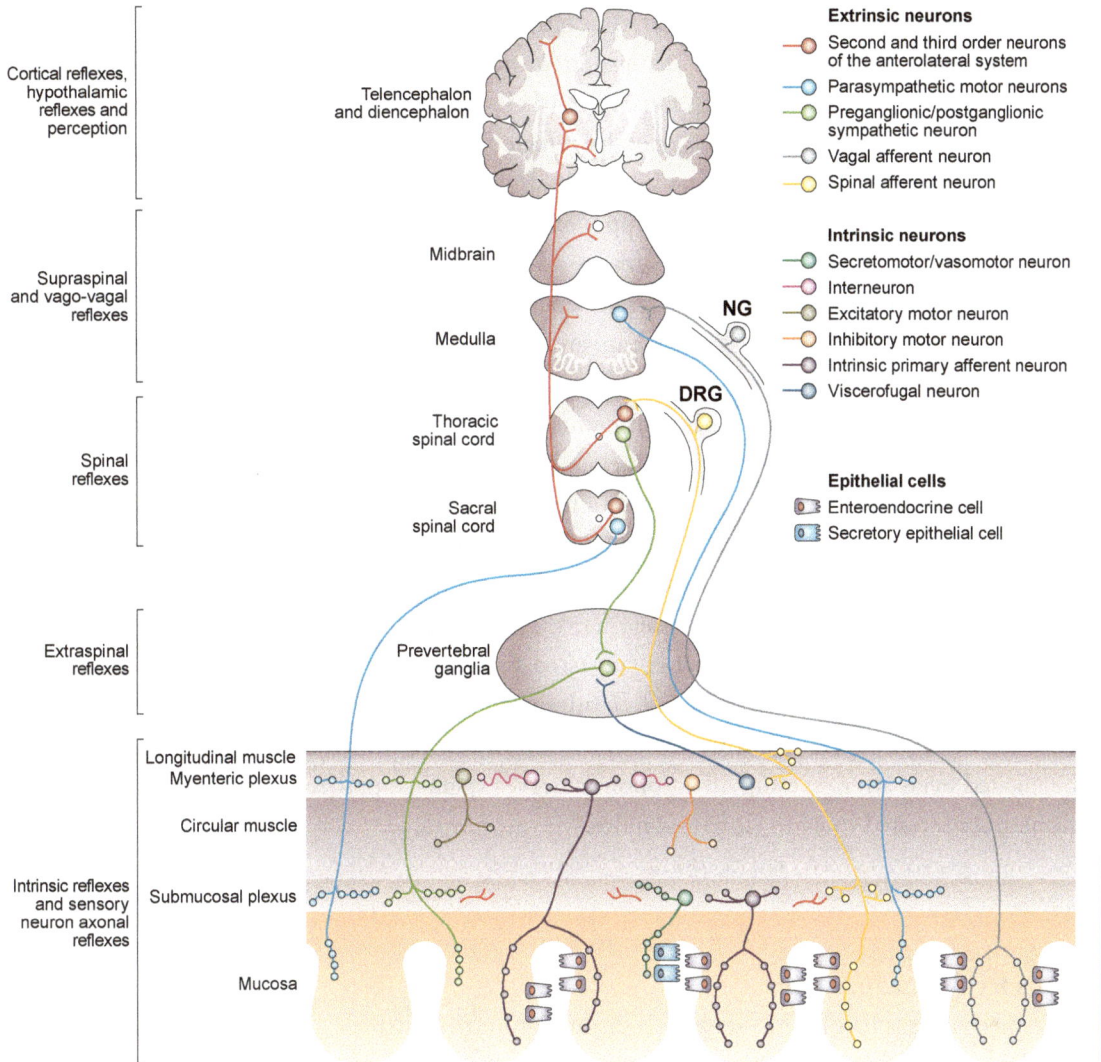

FIGURE 1-7: Multiple levels of reflex control of gastrointestinal function. Sensations of pain emanating from noxious stimuli in the GI tract are carried in spinal afferent fibers whose cell bodies are located in the dorsal root ganglia (DRG). Vagal afferent fibers involved in homeostatic regulation of GI functions (e.g., motility, secretion, satiety) have their cell bodies in the nodose ganglion (NG). Their efferent fibers impact on gastrointestinal smooth muscle, epithelial cells and blood vessels to complete various reflex arcs (reprinted with permission from *Nat Rev Gastroenterol Hepatol* 2014; 11: 611–627).

receptive fields and terminals are located in either the vicinity of the mucosa or muscularis to either receive chemical/irritant or distension/stretch input, respectively. Their efferent fibers impact on gastrointestinal smooth muscle, epithelial cells and blood vessels to complete various reflex arcs. These reflexes are generally referred to as "long reflexes" because they involve a response at a remote site from the initiating stimulus. Their specificity is denoted by a term combining the afferent and efferent components of the reflex arc. For example, the "enterogastric reflex" is initiated by activation of sensory afferents in the duodenum in response to mucosal irritation and terminates via vagal efferent fibers in the stomach to inhibit gastric emptying. Thus, the enterogastric reflex is a mechanism by which the intestine can regulate the delivery of gastric contents to its lumen.

There are also intrinsic sensory afferents whose cell bodies and effector arms are located within the wall of the gastrointestinal tract (Figure 1-6). By contrast to the extrinsic sensory afferents, the intrinsic afferent neurons do not transmit sensory information to the brain. Instead, they initiate reflexes through an interconnected series of neural networks to regulate motility, secretion, and blood flow. These reflex arcs are referred to as "short reflexes," because the reflex arc is usually confined to very short segments. For example, the "peristaltic reflex" is initiated by the presence of a bolus of chyme (ingested food) in the intestinal lumen. The intrinsic neurons initiate a local constriction of circular muscle proximal to the bolus and a relaxation distally, thereby propelling chyme aborally for only a few centimeters. The peristaltic reflex can be elicited even after severing all autonomic nerves (parasympathetic and sympathetic) and is sometimes referred to as an "intrinsic reflex."

Sensory Transduction

The mechanisms involved in sensory transduction in the extrinsic and intrinsic afferents is dependent on the stimulus. Distention/stretch of the musculature or mucosal deformation activates mechanoreceptors on afferent extrinsic or intrinsic neurons, respectively. It has been proposed that local afferents detecting mucosal deformation may monitor the movement of chyme over the mucosal surface. Moderate distension can elicit physiologic reflexes in the upper and lower gastrointestinal tract. Gastric distension can give rise to a sensation of "fullness" and impact food intake. Rectal distension usually results in the urge to defecate. Excessive distension, however, can elicit discomfort and pain, a nociceptor reflex probably mediated via the spinal afferents and dorsal root ganglion. In a similar manner, the presence of noxious stimuli (e.g., acid, bacterial toxins) in the gastric and/or intestinal lumen can result in activation of afferent sensory neurons and pain perception. However, because the afferent sensory fibers do not extend into the lumen (Figure 1-7), this latter function of sensory afferents depends on the noxious stimuli reaching the nerve terminals via transport across (or breech of) the epithelial lining.

The chemosensing of both innocuous (e.g., nutrients) and noxious (e.g., bacterial toxins) agents has been relegated to the various EEC of the mucosal epithelial lining. As mentioned above, the EEC can sample luminal contents and release a variety of mediators, some of which can activate adjacent sensory afferent neurons in the subepithelial space. The enterochromaffin cells (EC) are the most numerous of the EEC and their product serotonin (or 5-hydroxytryptamine; 5-HT) has received much attention as a primary chemosensory transducer. Serotonin has been implicated as a mediator in various GI functional activities, including motility, secretion, even the sensations of nausea and pain. For example, enterochromaffin cells detect the presence of chyme in the lumen and release serotonin, which initiates reflex intestinal contractions. Furthermore, the EC can detect bacterial toxins (e.g., *V. cholera* enterotoxin) and serotonin initiates the appropriate reflex secretory and motor activity to dilute and wash away the toxin. Negative feedback regulation of these responses is accomplished by uptake of serotonin by epithelial cells; most of which express the serotonin-selective reuptake transporter (SERT).

Other EEC have been implicated as chemosensory transducers in reflex responses of the gastrointestinal tract to nutrients in the lumen. The classic hormones CCK and secretin are now believed to exert at least some of their effects via activation of sensory afferent neurons. For example, the enterogastric reflex, which is initiated by irritants in the duodenal lumen, is likely mediated by CCK activation of afferent sensory neurons via the CCK_1 receptor. This is not entirely surprising considering that one of the major irritants initiating the reflex are long chain fatty acids, which are also the major stimuli for activating I cells (CCK-producing EEC in duodenum) to secrete CCK. The peptides released by EEC that have been implicated in feeding and satiety also appear to exert their effects, at least in part, via activation of sensory afferents (Table 1-2).

GASTROINTESTINAL FUNCTIONS

Motility

Efficient assimilation of nutrients by the gastrointestinal tract depends on an orderly intraluminal flow of ingested food at a rate that allows for optimum digestive and absorptive activity. Propulsion of food is accomplished by the coordinated motor activity of the gut; a function attributed to its smooth muscle coat. Motility is concerned not only with propulsive activity but also with the mixing of foodstuff with digestive secretions. One important digestive secretion, bile, is delivered into the lumen of the gut by contraction of the smooth muscle lining the gallbladder.

The smooth muscle coat of the gut is characteristically arranged in two layers, an inner circular and an outer longitudinal (Figure 1-5). In addition, a thin layer of smooth muscle (the muscularis mucosa), which may mediate mucosal (villus) motion, separates the mucosa from the submucosa. At the upper end of the gastrointestinal tract, striated muscle is found in the oropharynx and upper one third of the esophagus, and this striated muscle blends with the smooth muscle of the mid-esophagus. The remainder of the gastrointestinal tract is composed of smooth muscle except for the external anal sphincter, which consists of striated muscle. At various strategic points along the alimentary tract, specialized areas of smooth muscle serve to regulate the movement of chyme and secretions between adjacent luminal compartments (e.g., stomach and duodenum). The term sphincter is applied to these regions, some of which are characterized by a thickened band of circular smooth muscle, for example, the upper esophageal sphincter, the pylorus, and the sphincter of Oddi. Others are functional sphincters with no specialized structural features that, nonetheless, generate a zone of high intraluminal pressure. An example of these functional sphincters is the lower esophageal sphincter.

Basal Electrical Rhythm and Interstitial Cells of Cajal

Gastrointestinal smooth muscle is characterized by a resting membrane potential which can depolarize to initiate contraction. The resting membrane potential of gastrointestinal smooth muscle ranges from –55 mv to –80 mv, with a mean of about –60 mv. This potential is created by the distribution of Na^+, K^+, and Cl^- ions across the cell membrane. In contrast to most other excitable tissues, where the resting membrane potential remains fairly constant, the resting potential of gastrointestinal smooth muscle is characterized by rhythmic fluctuations (Figure 1-8). The oscillations of the potential have an amplitude of 15 to 20 mv, a duration of 1 to 5 seconds, and a frequency which varies along the gut, that is, 3/min in the stomach, 12/min in the duodenum, declining progressively to 8/min in the ileum. These rhythmic depolarizations are called slow waves or the *basal electrical rhythm* (BER).

Although isolated GI smooth muscle cells do not exhibit spontaneous depolarizations (slow waves), adjacent interstitial cells of Cajal (ICC) do. It is widely held that ICC act as "pacemakers" that elicit slow waves in smooth muscle. The oscillating slow waves generated in the ICC networks are transmitted across gap junctions to smooth muscle cells, which results in the BER. Within the GI tract, ICC are distributed in networks that lie in close proximity to the neural plexuses found within and between the muscle layers (Figure 1-9). These networks allow

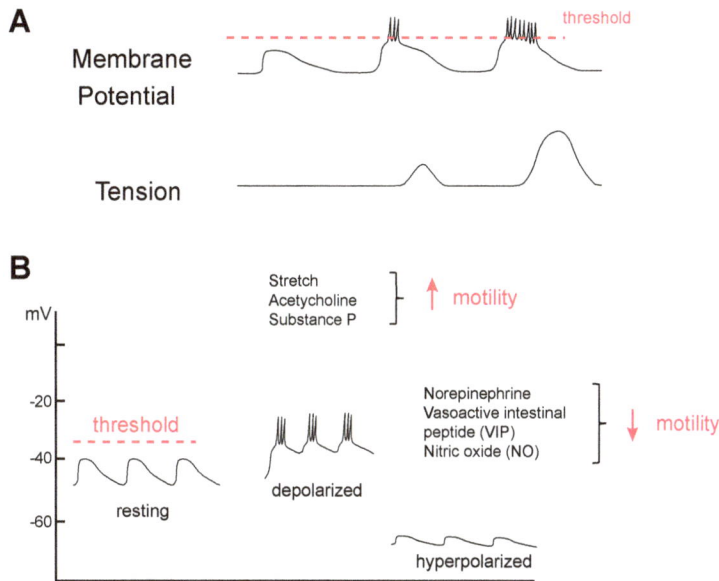

FIGURE 1-8: Relationships among slow waves, action potentials and contractile activity of intestinal smooth muscle. **A.** Slow waves are rhythmic cycles of membrane potential depolarizations and when a threshold potential is reached action potentials are generated. Smooth muscle contraction (increased tension) occurs only when action potentials are present on the slow waves; the strength of contraction is directly related to the number of action potentials. **B.** Stimuli either depolarize the membrane potential toward the threshold for generation of action potentials and increase smooth muscle contractile activity (motility) or hyperpolarize the membrane potential and inhibit motility.

for electrical coupling of the ICC with one another and with adjacent smooth muscle cells. For example, the ICC within the myenteric region (ICC-MY) are believed to function as the pacemakers that generate and transmit the slow waves in adjacent smooth muscle cells. The ICC network in the region of the deep muscular plexus (ICC-DMP) is in close apposition to nerve varicosities and is believed to play an important role in the transmission of nerve impulses to the smooth muscle.

Excitation-Contraction Coupling

Superimposed on the depolarization phase of the slow wave are small generator potentials, which, when they reach a critical threshold voltage, give rise to a spike or action potential (Figure 1-8). These spikes are responsible for initiating muscle contraction, the strength of which

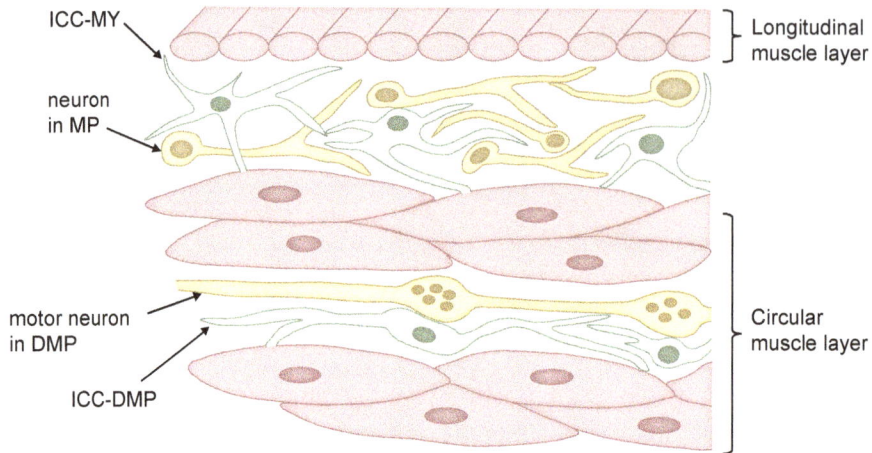

FIGURE 1-9: Interstitial cells of Cajal (ICC) between and within muscle layers of the intestine. Myenteric ICC (ICC-MY) are located in the region of the myenteric plexus (MP), between the circular and longitudinal muscle layers. This ICC network is responsible for generating the pacemaker slow waves that are transmitted to the smooth muscle cells. Intramuscular ICC (ICC-DMP) are located in the region of the deep muscular plexus (DMP) within the circular smooth muscle in close proximity to varicosities of motor neurons. This ICC network appears to relay neural information to smooth muscle cells and act as mechanosensors (adapted from Sanders KM, Kito Y, Hwang SJ, Ward SM. *Physiology* 31: 316–326, 2016).

is determined by the number of spikes generated (Figure 1-8A). The pre-potential and action potential involve an influx of Na^+ and Ca^{++} into the smooth muscle cell, owing to changes in membrane conductance for these ions. The Ca^{++} entering the cell is involved in the initiation of smooth muscle contraction. Stimuli that enhance motility result in smooth muscle depolarization (e.g., stretch, acetylcholine, substance P) and raise the membrane potential to the threshold for the generation of spike potentials, whereas factors that reduce motility (e.g., norepinephrine, VIP, nitric oxide) typically hyperpolarize the muscle and move the membrane potential further below the threshold potential (Figure 1-8B). This leads to a reduction in intracellular calcium and subsequent muscle relaxation.

The contractile activity of smooth muscle can be compared with and contrasted with that of striated muscle. Smooth muscle cells are much smaller and have a greater surface-to-volume ratio than skeletal muscle cells. Although both muscle types contain similar contractile proteins (actin and myosin filaments) that require Ca^{++} for activation, the excitation-contraction coupling is somewhat different in smooth and striated muscle cells. The rise in intracellular

Ca^{++} in smooth muscle is primarily a result of an influx of extracellular Ca^{++}, whereas mobilization of intracellular stores is more important in striated muscle. The high surface-to-volume ratio in smooth muscle cells facilitates the influx and intracellular diffusion of calcium. Smooth muscle cells also lack troponin, which is present in skeletal muscle cells. Therefore, Ca^{++} initiates smooth muscle contraction via calmodulin-mediated activation of myosin filaments. As in striated muscle, the energy for contraction is derived from degradation of adenosine triphosphate (ATP).

Smooth muscle contraction does not begin until 50 to 100 milliseconds after excitation and requires an additional 500 milliseconds for development of maximum tension. The entire cycle requires 1 to 3 seconds and is 30 times longer than a single twitch contraction of skeletal muscle. The latency to onset of contraction is due to the time required for Ca^{++} to enter and diffuse throughout the cell. The slow relaxation depends on the removal of Ca^{++} by intracellular depots.

How is electrical and contractile activity propagated longitudinally and perpendicularly in the two layers of the gastrointestinal tract? A fundamental feature of smooth muscle cells is the presence of *gap junctions* or *nexi*, which confer the properties of a functional syncytium. These gap junctions provide a low resistance pathway for the movement of ions and thereby facilitate electrical conduction from cell to cell. In addition, there is a prominent role for ICC in the propagation of electrical/contractile activity. ICC connections allow for propagation of electrical activity within the network and the connections of ICC to smooth muscle cells within the circular and longitudinal muscle layers elicits the activation of these cells. It is generally held that the coordinated mixing and propulsive contractile activity of the GI tract is dependent on a functional syncytium composed of both smooth muscle and ICC.

Contractile activity of the gastrointestinal tract is regulated by both extrinsic and intrinsic neural influences on the basal electrical rhythm (Figure 1-7B). For example, the postganglionic parasympathetic (vagus) neurotransmitter, acetylcholine, reduces the overall membrane potential while preserving the basal electrical rhythm, thereby increasing the tendency for spike discharge and contractile activity. Norepinephrine, on the other hand, tends to hyperpolarize or stabilize the membrane against other agents that cause depolarization. In general, sympathetic discharge tends to reduce gastrointestinal motor activity, whereas parasympathetic discharge enhances it. The myenteric plexus also contains both excitatory and inhibitory neurons that regulate motility in a manner similar to the autonomic nerves (Table 1-3). Excitatory neurons induce local contractions via neurotransmitters that elevate (depolarize) the membrane potential (e.g., acetylcholine, substance P). Inhibitory neurons produce local relaxations via neurotransmitters (e.g., VIP or NO) that lower (hyperpolarize) the membrane potential.

Motility pattern	Function	Site

Peristalsis

Propulsive
causes transport
non-propulsive
causes mixing

Esophagus
Stomach
Small intestine

Rhythmic segmentation

Mixing

Small and large
intestines

Tonic contraction

Blocking passage
Separation

Gastrointestinal
sphincters

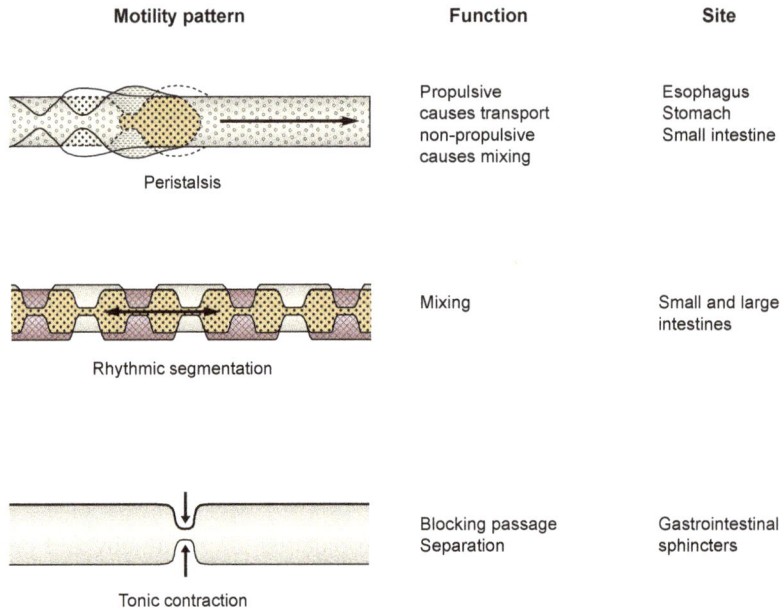

FIGURE 1-10: The three major patterns of GI motility. Peristalsis involves a local reflex that is elicited when the gut wall is stimulated by a food bolus. Peristaltic contractions consist of a wave of circular muscle contraction that is elicited upstream from the food bolus and a wave of relaxation in the adjacent downstream section of the bowel. Rhythmic segmentation involves alternating contractions and relaxations at different sites along the bowel, which function to mix ingested nutrients with digestive secretions. Tonic contraction, with intermittent relaxation, of the sphincters serves to regulate movement of luminal contents along (and secretions into) the GI tract.

Functional Motor Patterns

Electrical and contractile events in gastrointestinal smooth muscle lead to three major patterns of motility: rhythmic segmentation, peristalsis, and tonic contractions (Figure 1-10). Rhythmic segmentation, which is attributed to the activity of the circular muscle layer, mixes ingested nutrients with digestive secretions. In addition, segmentation tends to move intestinal contents anal-ward (downstream), because the frequency of these segmental contractions is higher in the upper than lower small intestine. Another factor that contributes to the forward propulsion of luminal content along the gastrointestinal tract is peristaltic activity. Peristalsis involves a local reflex mediated through the myenteric plexus that is elicited when the gut wall is stimulated by a food bolus. Peristaltic contractions consist of a wave of circular muscle contraction that is

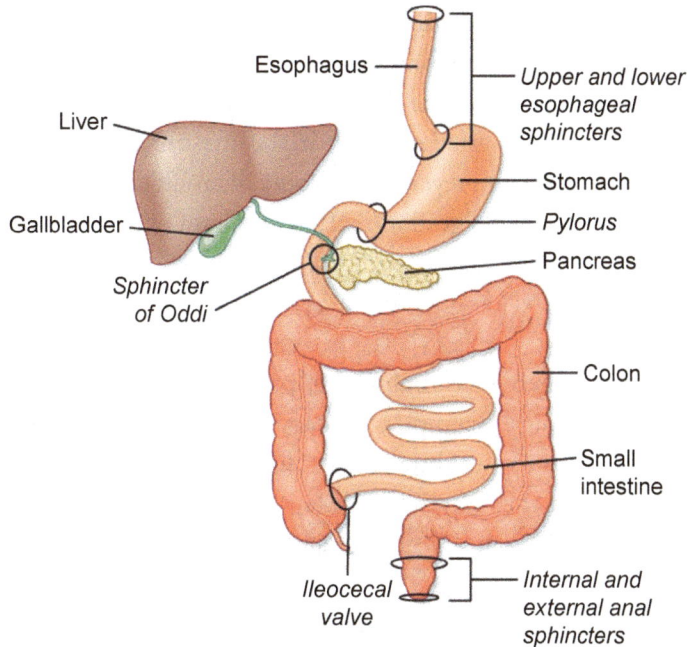

FIGURE 1-11: Sphincters control the flow of food/chyme within, as well as the movement of digestive secretions into, the GI tract. These areas of smooth muscle resistance include the upper and lower esophageal sphincters, pylorus, sphincter of Oddi, ileocecal valve, and the internal and external anal sphincters (modified with permission from Berne and Levi Physiology, 6th edition, 2008, Mosby, with permission of Elsevier).

elicited upstream from the food bolus via activation of excitatory neurons that release acetylcholine and substance P. The contractile wave is coupled to a wave of relaxation in the adjacent downstream section of the bowel that results from the reflex-mediated activation of inhibitory neurons that release nitric oxide and VIP. Although peristalsis can be conducted in both directions, the reflex relaxation confers the observed polarity of peristaltic activity, that is, movement in the oral to anal direction.

Tonic contraction, with intermittent relaxation, of the sphincters serves to regulate 1) movement of luminal contents along the gastrointestinal tract and 2) delivery of digestive secretions from the pancreas and liver. The upper esophageal and external anal sphincters are under somatic control, whereas the others are under control of the autonomic nervous system. The sphincters compartmentalize the different functional regions of the GI tract (Figure 1-11). For example, the lower esophageal and pyloric sphincters isolate the acid-producing stomach from adjacent esophageal and intestinal compartments, where acid-induced injury could result. These sphincters open intermittently to allow for the passage of ingested material to the next compartment. The sphincter of Oddi regulates the delivery of pancreatic enzymes and biliary secretions to coincide with the presence of ingested nutrients in the intestine to facilitate digestion/absorption. The ilealocecal valve serves to prevent the colonic microbiota from entering the intestine, where the bacteria can result in maldigestion/malabsorption of nutrients and even toxicity.

Digestion

The human diet consists largely of carbohydrate, protein, and fat derived from animal and vegetable sources. These chemically complex substances are reduced to smaller absorbable units by digestive enzymes. The sources of intraluminal digestive enzymes are the digestive glands, some of which are located in the GI tract itself (e.g., gastric glands), whereas others lie outside the GI tract (pancreas and salivary glands) and deliver their secretions to the gut lumen through ducts.

Digestive enzymes are proteins that are synthesized, stored, and secreted by specialized cells found in salivary glands, stomach, pancreas, and intestines. The secretion of proteins (i.e., enzymes) is an orderly process, beginning with synthesis of the peptide chain in the basal portion of the cell and culminating with extrusion of the proteins at the apical region. The secretory cycle can be divided into the following six intracellular events: synthesis, segregation, intracellular transport, concentration, storage, and discharge. The enzymes are synthesized by polysomes attached to the rough endoplasmic reticulum, and the elongating peptide chain is directed into the cavity (cisterna) of the endoplasmic reticulum. This form of synthesis effectively segregates the secretory proteins to a membrane-bound compartment, which assures appropriate channeling of the protein through a series of subcellular organelles to their ultimate site of secretion. The newly synthesized proteins move through the cisternae to "transitional elements" of the rough endoplasmic reticulum that are subsequently pinched off and transport the proteins to the concentrating vacuoles of the Golgi apparatus. Within these vacuoles, the nascent proteins are concentrated to produce mature storage granules (zymogen granules). After their formation in the Golgi complex, the secretory granules migrate to the apical portion of the cell and remain there until an appropriate stimulus (neural or humoral) triggers exocytosis (Figure 1-4). Exocytosis involves the orderly movement of the granule toward the apical cell membrane and fusion of the zymogen granule and cell membrane. Then, the membranes dissolve at the point of fusion, releasing the granule contents from the cell.

The digestive enzymes hydrolyze ingested carbohydrates, lipids, and proteins by attacking glycosidic, ester, and peptide bonds, respectively. Under physiologic conditions, the pancreatic enzymes are the most important and the small intestine the primary site of hydrolysis. Hydrolysis can occur within the gut lumen, at the apical membrane of the absorptive cell, or within the cytoplasm of this cell, depending on the nutrient and location of the digestive enzymes. Intraluminal hydrolysis, the initial digestive event, is accomplished primarily by enzymes secreted by the glands associated with the gastrointestinal tract, of which the pancreas is the most important. In general, all three classes of nutrients undergo preliminary hydrolysis in the lumen for effective absorption to occur. In the case of fat, this digestive step is facilitated by biliary secretions and yields monoglycerides and fatty acids, both of which are lipid soluble and thus require no further modification before absorption. Carbohydrates and proteins, on the other hand, undergo further processing. The complex glucose polymers (starch and glycogen) are hydrolyzed in the lumen to a mixture of monosaccharide and oligosaccharides (up to 4 to 5 glucose units). Because carbohydrates can only be absorbed as monosaccharides, the oligosaccharides are further hydrolyzed to yield glucose moieties by saccharidases located in the apical membrane of mucosal cells. Intraluminal hydrolysis of proteins also yields a mixture of smaller fragments consisting of amino acids and peptides of various sizes. Because only amino acids and small peptides (dipeptides and tripeptides) can be transported into enterocytes, larger peptides are further processed to a transportable size by peptidases in the microvilli. Intracellular dipeptides and tripeptides are further hydrolyzed to yield amino acids for final transport to the interstitium and blood stream.

Absorption

Absorption in the gastrointestinal tract involves the uptake of a heterogeneous group of substances (water, electrolytes, and nutrients) from an aqueous medium. These substances vary in several important aspects that determine their mode of absorption. Molecular size, charge (or lack thereof), and relative aqueous and lipid solubilities are major influences in this regard.

Routes of Absorption

The major site of absorption of the hydrolytic products of food digestion, electrolytes, and water is the small intestine. Solutes encounter various barriers and channels during their transit from the bulk aqueous phase of luminal content to the blood and lymphatic circulations. The first barrier is a mucus layer adherent to the luminal surface of the epithelium which can act as a sieve restricting molecules based on size and lipophilicity. The size restrictions of this mucus gel, although inhibiting indigestible particulate matter and pathogens, does not significantly hinder the passage of the small hydrophilic products of nutrient digestion. However, lipophilic solutes would be hindered significantly due to hydrophobic interactions with mucins. The mucus layer of the small intestine is estimated to be approximately 15 to 30 μm thick; the layer is thinnest or even absent in the upper small intestine (major site of absorption). Taken together, the physical and chemical characteristics of the adherent mucus gel do not present a substantially greater barrier to the absorption of nutrients than an unstirred water layer of similar thickness.

The next barrier to solute movement is the cell membrane and its associated structures. The cell membrane, a lipid bilayer containing phospholipids and cholesterol, has a mosaic pattern of proteins and glycoproteins associated with it. The proteins of the cell membrane can be classified as either integral or peripheral. Integral proteins are embedded in the phospholipid bilayer, bridging it from its extracellular to intracellular surface. The peripheral proteins are attached to either surface and are believed to subserve primarily enzymatic and signaling functions. The integral proteins behave as enzymes or carriers and provide structural channels through the lipid membrane. The large surface area offered by the exposed lipid portion of the membrane allows for easy passive transit of lipid-soluble molecules into the cell, regardless of their size. For water and water-soluble molecules, the lipid membrane provides a more formidable barrier to entry into the cell; consequently, they rely on the structural proteins to traverse the cell membrane. Aqueous channels formed by the integral proteins provide a route of access to the cell interior for small water-soluble substances (e.g., electrolytes) as well as water. The dimensions of these channels (or pores) in gastrointestinal epithelia average about 8 Å in diameter, with larger pores in the duodenum than in the ileum. Because these channels are created by charged proteins, electrolyte movement through them may be either facilitated or hindered, depending on the ion species and the pore charge. For example, the epithelium of the gallbladder has an excess of negatively charged pores, whereas in the stomach, cationic pores predominate. Some channels, called aquaporins, selectively conduct water molecules in and out of the cell, while preventing the passage of ions and other solutes.

Another route by which electrolytes and water cross the mucosal epithelia is the tight junctions that connect adjacent absorptive cells. These paracellular or intercellular channels (Figure 1-10) behave as if they have different degrees of porosity in different tissues. For example, much more fluid and electrolytes cross the intercellular junctions of the gallbladder epithelium than cross the intestinal epithelium.

Neither the tight junctions nor the protein channels of the cell membrane offer suitable routes for the absorption of large water-soluble nutrients such as glucose and amino acids. For the uptake of these substances, mechanisms exist that involve specialized "carrier" proteins. These proteins can "shuttle" molecules across the cell membrane. For large molecules, the possibility also exists that pinocytosis may account for cellular uptake from the intestinal lumen.

The remaining barriers in the absorptive process are the basolateral membrane of the enterocyte, its basement membrane, the interstitium of the lamina propria, and the endothelial cell wall of capillaries and lymphatics. Relatively little is known about these barriers. Located in the basolateral cell membrane are energy-dependent ion pumps that promote the exit of electrolytes and water. Larger water-soluble molecules, such as monosaccharides and amino acids, cross the basal membrane by facilitated diffusion using carrier proteins. Chylomicrons (750 to 6000 Å diameter), which are formed in the cell during fat absorption, leave by exocytosis through the basolateral cell membrane. The lateral intercellular space between enterocytes is the final common pathway for a heterogeneous collection of water, solutes, and lipid particles which then cross the basement membrane, traverse the interstitium, and reach the blood and lymphatic circulations. The water-soluble substances enter blood capillaries and chylomicrons enter the initial lymphatics.

Mechanisms of Absorption

Substances are primarily absorbed by two basic processes, diffusion and active transport. Diffusion can be defined as the movement of molecules along an electrical (difference in charge) or chemical (difference in concentration) gradient or both (electrochemical). Diffusion is a passive process governed by simple physical laws and requires no energy input. Active transport, on the other hand, involves the movement of molecules uphill against a concentration, electrical, or pressure gradient and therefore requires energy.

Passive Absorption Lipid-soluble substances cross the lipid portion of the cell membrane in accordance with their concentration gradients. Water-soluble substances can also enter the cell by passive diffusion. Small solutes (electrolytes), whose dimensions are smaller than the aqueous pores, can diffuse through these channels. Larger water-soluble molecules are obliged to use an alternative route involving carrier proteins in the membrane (Figure 1-12). The molecule to be transported, such as the monosaccharide fructose, couples with its carrier on the external side of the membrane and the complex either diffuses or rotates within the membrane. At the inner face of the membrane, the transported molecule dissociates from its carrier. This carrier is analogous to the aqueous pores in that the laws governing simple diffusion still operate,

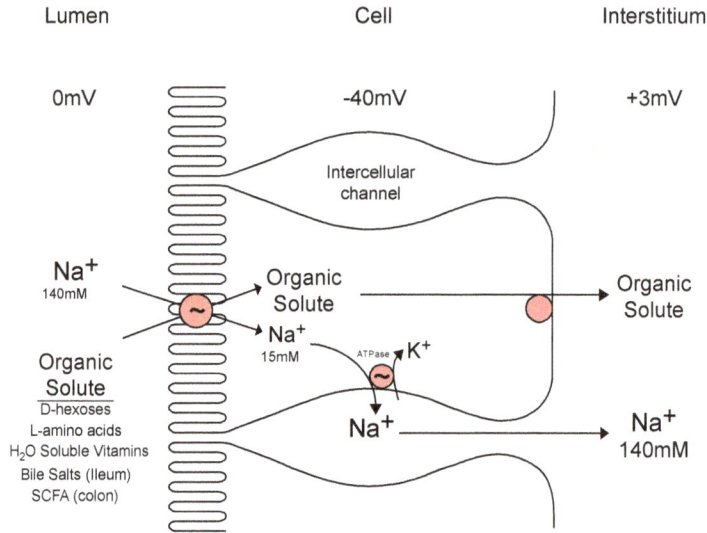

FIGURE 1-12: Secondary, or electrolyte-coupled transport, is used by intestinal epithelial cells to absorb monosaccharides, amino acids, and other water-soluble organic solutes. A Na^+-solute (e.g., glucose) carrier protein on the brush border membrane has binding sites for both Na^+ and the organic solute. Both molecules move into the enterocyte by facilitated diffusion down Na+ concentration ($140 \rightarrow 15$ mM) and electrical ($0 \rightarrow -40$ mv) gradients created by the energy-dependent Na^+-K^+ pump located in the basolateral membrane.

and this process is termed *facilitated diffusion*. However, an important difference exists between simple diffusion and facilitated diffusion. Because facilitated diffusion depends on a fixed number of carrier protein sites, this process is saturable, unlike simple diffusion.

Active Absorption   Water-soluble substances can also be actively transported across the cell membrane. Actively transported substances include ions such as Na^+, K^+, Cl^-, H^+, Ca^{++}, and Fe^{++}, as well as certain monosaccharides, amino acids, and peptides. Active transport processes can be divided into two main types, primary (ATP-coupled) and secondary (electrolyte-coupled).

An example of a primary active transport process is the exchange of Na^+ and K^+ across the cell membrane as illustrated in Figure 1-12. The active transport of Na^+ and K^+ is a characteristic of all cells in the body and is responsible for the extracellular and intracellular distribution of these ions. Both ions are moved against their concentration gradients by means of a carrier protein possessing enzymatic activity (Na^+, K^+-ATPase). The enzymatic activity of this carrier provides the energy required for transport by virtue of its intrinsic ability to hydrolyze ATP. The hydrolysis of ATP occurs on the inner portion of the membrane, and the energy so derived enables the carrier to transport sodium out of the cell and potassium into the cell. The transport of these ions is coupled in such a manner that three Na^+ ions are extruded from the cell for every two K^+ ions taken in. In the absorptive epithelia of both the intestines and gallbladder, the Na^+-K^+ pump is located in the basolateral portion of the cells.

Secondary, or electrolyte-coupled transport, is a very important mechanism for the active absorption of monosaccharides, amino acids, and peptides. An example of electrolyte-coupled transport is the absorption of glucose by the enterocyte (Figure 1-12). As mentioned above, the sodium-potassium pump is located in the basolateral portion of the epithelial cell, and by depleting intracellular sodium, it creates a favorable electrochemical gradient for the diffusion of Na^+ from the lumen into the mucosal cell. The apical portion of the cell membrane contains a Na^+-solute (e.g., glucose) carrier protein. The carrier has binding sites for both glucose and Na^+, and it will not traverse the membrane unless both Na^+ and glucose are attached. Glucose and Na^+ are moved into the apical portion of the enterocyte by facilitated diffusion down Na^+ concentration ($140 \rightarrow 15$ mM) and electrical ($0 \rightarrow -40$ mv) gradients created by the energy-dependent Na^+-K^+ pump in the basolateral membrane. This mechanism can transport glucose against its concentration gradient when necessary, that is, when intraluminal concentrations of glucose are low compared with intracellular levels. The glucose then leaves the basolateral portion of the cell and moves down its concentration gradient by facilitated diffusion. Thus, the overall movement of glucose from the intestinal lumen to the circulation is an active process deriving its energy from the Na^+-K^+ pump. A similar electrolyte-coupled transport mechanism is also important for the active absorption of other water-soluble nutrients in the small intestine (e.g., amino acids) and colon (short chain fatty acids).

Water Absorption

The absorption of water occupies a special position in gastrointestinal physiology because up to 10 liters of water are normally transported from the lumen to the blood every 24 hours. Water absorption is accomplished by osmotic forces generated by transepithelial movement of solutes from lumen-to-blood (absorption). The active transport of ions (e.g., Na^+) and organic solutes (e.g., glucose) by GI epithelium leads to the creation of an osmotic gradient between the gut lumen and intercellular spaces (Figure 1-13). The accumulation of solutes and consequent increase in osmolality within the intercellular spaces draws water across the intercellular junctions and through water channels (aquaporins) on the apical and basolateral membranes of the epithelium. The flow (and transient accumulation) of water in the intercellular compartment

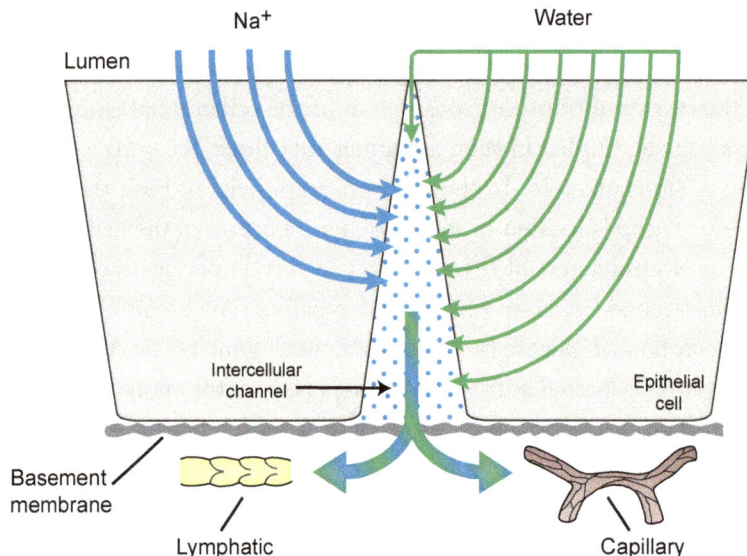

FIGURE 1-13: The standing osmotic gradient mechanism of trans-epithelial water movement. The active transport of ions (e.g., Na^+) by GI epithelium creates an osmotic gradient between the gut lumen and intercellular spaces. The increased osmolality within the intercellular spaces draws water across the intercellular junctions and through water channels (aquaporins) on the apical and basolateral membranes of the epithelium. The accumulation of water in the intercellular compartment leads to a rise in hydrostatic pressure, which drives the absorbed fluid into blood and lymphatic vessels that normally lie in close proximity to the epithelial layer.

leads to a rise in hydrostatic pressure, which in turn drives the fluid into blood and lymphatic vessels within the lamina propria that normally lie in close proximity to the epithelial layer. The osmotic pressure differences required to drive water absorption are smaller than 3 to 5 mOsm/l, which allows for water absorption by both transcellular (through cells) and paracellular (between cells) routes. It is estimated that over 50% of fluid absorption occurs through the intercellular junctions. When large volumes of water are moved through the paracellular pathway by the osmotic gradients developed during solute absorption, the phenomenon of solvent drag (convection) is observed, that is, solutes are transported due to bulk fluid motion. This passive process can mediate the movement of solutes against their electrical and chemical gradients.

Secretion

In addition to the fluid delivered to the intestines by gastric, pancreatic, and biliary secretions to facilitate digestion and absorption of nutrients, the intestines also secrete water and mucus. Intestinal electrolyte-driven water secretion is derived from crypt epithelial cells, whereas mucus secretion emanates from goblet cells. Rather than subserving digestive/absorptive functions, the primary objective of these secretions is to protect the epithelium from inadvertently ingested luminal toxins. The presence of noxious material in the intestines stimulates the secretions of both water and mucus to remove the offending substances from the surface of the mucosal epithelium.

Water and Electrolyte Secretion

Secretion of electrolytes and water by the GI tract and associated glands involves mechanisms comparable to those described earlier for absorption, the major difference being the direction of solute transport, that is, lumen-to-blood movement with absorption and blood-to-lumen movement with secretion. As with absorption, an osmotic gradient is established by electrolyte secretion; therefore, water movement also occurs in the direction of blood to lumen. Parietal cells in the stomach secrete H^+ into the lumen to generate acid (HCl) for efficient protein digestion. On the other hand, pancreatic duct cells secrete HCO_3^- in the lumen to neutralize any acid entering the duodenum from the stomach. Although a major function of the intestines is to absorb the fluid load from the stomach, pancreas, and liver, the intestine also secretes electrolytes and water. Although electrolyte/water absorption normally occurs on the most luminal aspect of the epithelium (e.g., villus tips), the epithelial cells of the crypt region are the source of electrolyte/water secretion. The osmotic gradient generated by crypt cells for blood-to-lumen fluid movement is established primarily by Cl^- transport. An important physiologic benefit of water secretion from the crypts to the lumen is to ensure the fluidity of the luminal contents at the villus tips for optimal hydrolysis and absorption of ingested nutrients. In addition, the presence of noxious material in the lumen elicits an intrinsic neural reflex that increases crypt cell fluid secretion, presumably to wash away the toxins. Noxious stimuli are detected by enterochromaffin cells, which release serotonin into the interstitium. Serotonin, in turn, activates intramural afferent sensory fibers, which pass through the submucosal plexus and activate efferent secretory nerves, which release VIP at the base of the crypt cells to induce secretion.

An example for Cl^- secretion by the intestinal crypt cells is depicted in Figure 1-14. The major pathway for Cl^- secretion is the cystic fibrosis conductance regulator (CFTR) channel located at the apical membrane of intestinal crypt cells. Cl^- is taken up into the cell by a specific co-transporter in the basolateral membrane that allows for the uptake of Cl^-, K^+, and Na^+. Because this co-transporter moves 2 Cl^- (along with a Na^+ and K^+) into the cell, it provides adequate amounts of Cl^- for the secretory process. As noted above for solute-coupled Na^+ absorption (Figure 1-10), the energy required for this Na^+-K^+-2Cl^- co-transporter is derived

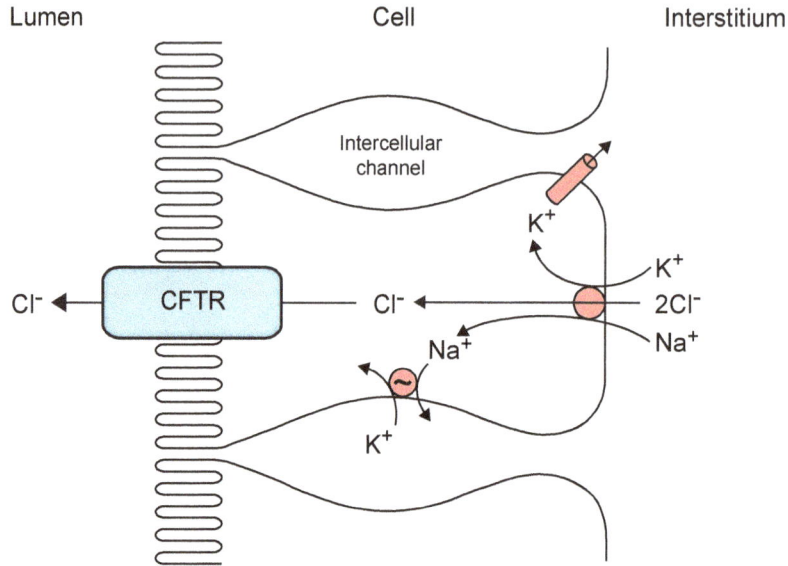

FIGURE 1-14: Mechanism of chloride secretion from the intestinal crypt cells. Cl– enters the cell via a specific co-transporter in the basolateral membrane that allows for the uptake of Cl⁻, K⁺, and Na+. This Na^+-K^+-$2Cl^-$ co-transporter moves 2 Cl– (along with a Na+ and K+) into the cell using energy derived from the Na+ gradient created by the Na^+-K^+ pump in the basolateral membrane. The accumulating intracellular Cl^- then moves out of the cell via the CFTR (cystic fibrosis transmembrane conductance regulator) channel.

from the Na^+ gradient created by the Na^+-K^+ pump. The accumulating intracellular Cl^- then moves out of the cell via the CFTR channel. Of note, the CFTR channel is an anion channel that can transport either Cl^- or HCO_3^-. In addition to its role in intestinal crypt cells, the CFTR channel plays an important role in electrolyte and water secretion at other sites within the gastrointestinal tract (e.g., stomach, pancreas) as well as in extra-gastrointestinal organs (e.g., lung). A dysfunctional CFTR channel is the basis for the genetic disorder, cystic fibrosis, which leads to multiorgan dysfunction that is associated with a lack of water secretion.

Mucin Secretion

Mucus is secreted by specialized epithelial cells, the gastric mucous cells and intestinal goblet cells. The major constituents of mucus that are responsible for its viscosity are the mucins; heavily glycosylated proteins. The mucin glycoproteins tend to repel each other due to the negative charges of the carbohydrate moieties. The mucin glycoproteins are stored in a compact state within gran-

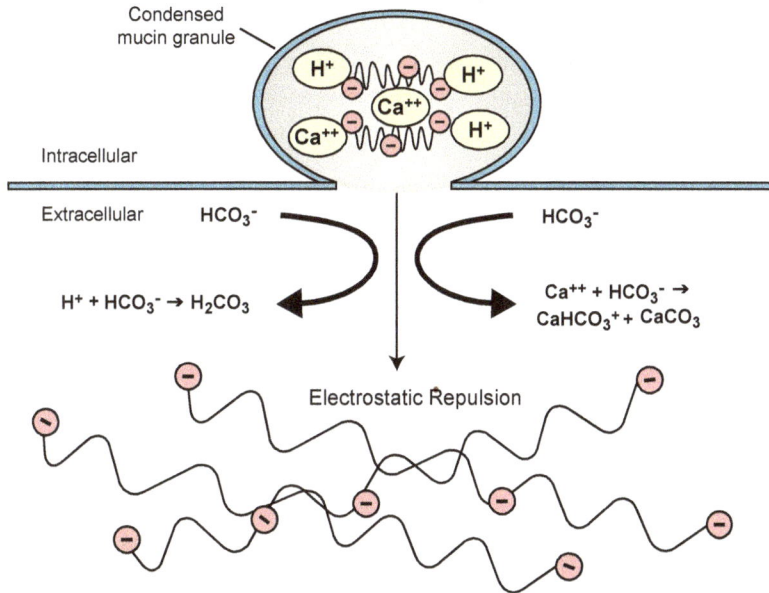

FIGURE 1-15: Schematic model of the role of HCO_3- in the formation of the mucus layer of the gastrointestinal tract. The mucin stored in goblet cells is maintained in compact form by Ca^{++} and H^+ ions that shield the negative charges. When mucin is released into the lumen, the negative charges repel each other and the mucin molecule expands; a process dependent on HCO_3- co-secretion via the CFTR channel. (Modified from *Am J Physiol Cell Physiol* 2010; 299: C1222–1233).

ules (secretory vesicles); the condensed packaging of mucins is a result of intracellular acidity and excess Ca^{++} that neutralizes the negative charges (Figure 1-15). When the granules release the mucins into the lumen, they expand dramatically in volume, spread out and organize into a sheet covering the epithelium. This process is facilitated by the co-secretion of HCO_3^- to neutralize the acid and bind the Ca^{++}, allowing the mucins to expand by electrostatic repulsion. The osmotically driven water secretion that accompanies HCO_3^- transport is quickly imbibed by the expanding mucin network. A critical anion channel integral to secretion of HCO_3^- is CFTR. Genetic deletion of CFTR results in mucoviscidosis (thick mucus) within the intestine.

The mucus layer can be subdivided into two sub-layers: an outer loose layer that can be easily removed by the passage of luminal contents and an inner layer that remains firmly adherent to the epithelium. The outer loose layer is believed to be derived from the deeper adherent layer by loosening of the network via surface digestion by luminal or bacterial enzymes. The mucus layers are not static entities, but dynamic. Removal of the outer loose layer, by motility or the shear stress resulting from the movement of chyme, leads in its replenishment by additional secretion of mucus. The mucus layers are thick in the stomach and colon, providing a lubricating cushion that prevents damage to the gastric and colonic epithelium during the mixing and propulsion of coarse lumen contents. As mentioned above, the mucus layers in the small intestine tend to be very thin or absent, thereby allowing for absorption of nutrients from the fluid-rich chyme.

CIRCULATION OF THE DIGESTIVE SYSTEM

The digestive organs receive the largest fraction of cardiac output, that is, 25% to 30%. This high rate of perfusion is presumably required to meet the metabolic needs of this large mass of tissue. Following a meal, blood flow to all digestive organs increases to meet the enhanced demand for oxygen imposed by motility, absorption, and secretion. Inasmuch as the processes of absorption and secretion involve the transport of large volumes of fluid and solutes between the blood and gut lumen, the circulation also plays an important role in these processes.

The organization of the blood supply to the gastrointestinal tract can be characterized in terms of parallel and series coupled circuits. The three major arteries supplying the digestive organs are the celiac, superior mesenteric, and inferior mesenteric arteries (Figure 1-16). These vessels compose the parallel vascular circuit. The venous drainage from the stomach, pancreas, and intestines empties into the portal vein which, in turn, perfuses the liver and constitutes the series component of this circuit. The parallel arrangement of the splanchnic circulation allows for independent regulation of blood flow to individual organs in the GI tract, whereas the series arrangement of the portal venous system ensures that the liver is first exposed to all absorbed substances.

There are several characteristic features of the microcirculation of digestive organs that optimize the ability of these tissues to move large amounts of fluid and electrolytes between the blood and transporting epithelia. In comparison to other tissues (e.g., skeletal muscle), the digestive organs have a high capillary density and consequently a large capillary surface area for secretion or absorption. The capillaries in the digestive organs are generally of the fenestrated type. These fenestrations provide an enormous pore area for water and solute exchange. Furthermore, the fenestrations usually face the basal aspect of the transporting cell, thereby minimizing the distance fluid must travel between the blood and epithelia. Capillaries of digestive organs are highly permeable to small solutes, yet they are relatively impermeable to macromol-

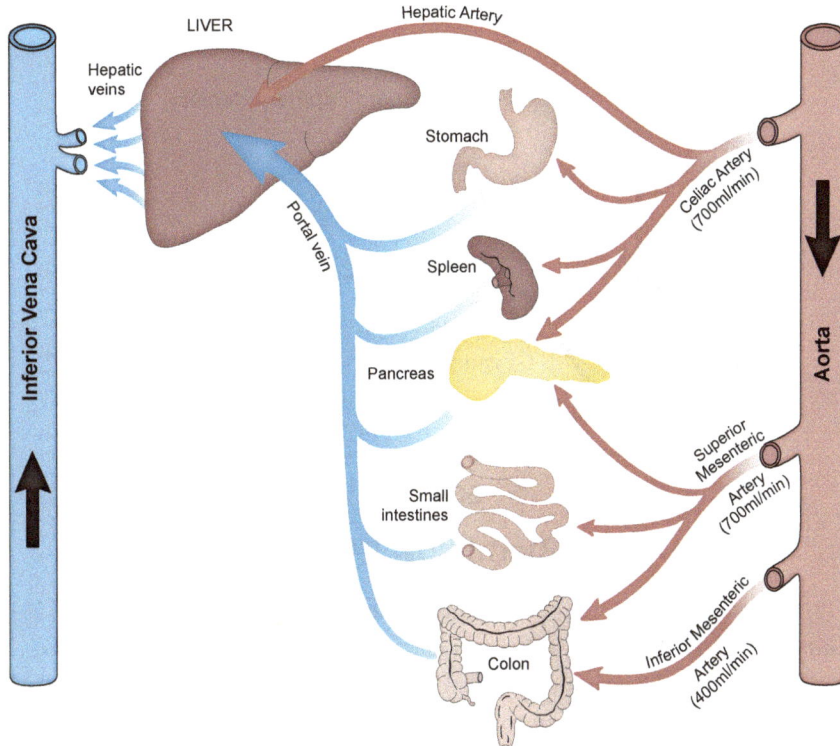

FIGURE 1-16: The splanchnic circulation. The celiac, superior mesenteric, and inferior mesenteric arteries supply blood to the organs that comprise the digestive system. The venous drainage from the spleen, stomach, pancreas, and intestines empties into the portal vein which, in turn, perfuses the liver, along with the hepatic artery (a branch of the celiac artery) (modified from *Anesthesiology* 2 2004, Vol. 100, 434–439, with permission of Wolters Kluwer Health).

ecules. This means that molecules the size of glucose readily gain access to the blood stream, whereas plasma proteins are highly restricted by the capillary wall.

Another feature of the GI microcirculation is its ability to adjust both blood flow and capillary surface area (density) to meet the oxygen requirements of digestive organs. The rate of oxygen uptake (or consumption) by the GI tract is a function of both the rate of delivery of O_2 via the blood and the capillary surface area available for O_2 diffusion. Each GI organ can regulate both blood flow and capillary density to meet its specific moment-to-moment O_2 requirements. An example of this intrinsic regulatory mechanism is illustrated in Figure 1-17, where changes in capillary density are depicted as a function of blood flow. At a normal blood

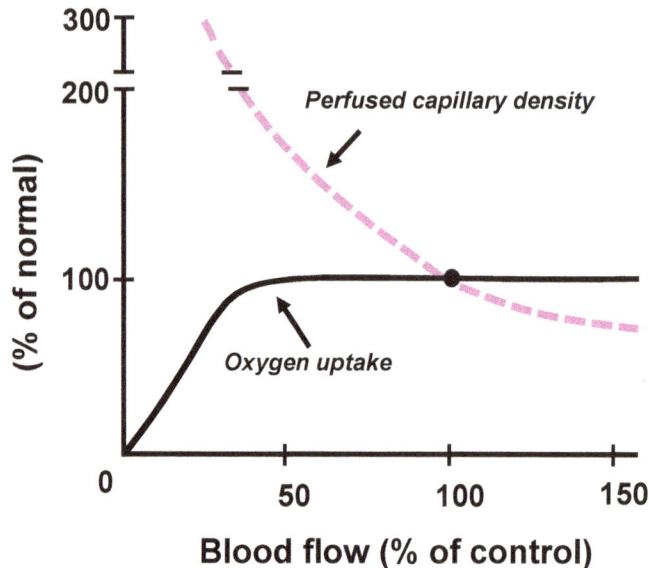

FIGURE 1-17: Gastrointestinal oxygen uptake is maintained during moderate reductions in blood flow, because the recruitment (opening) of more capillaries facilitates O_2 extraction by the tissue (increase in arteriovenous O_2 difference) when blood flow is reduced. Oxygen uptake is calculated as the product of blood flow and arteriovenous O_2 difference (modified from *Gastrointestinal Mucosal Defense System*: https://doi.org/10.4199/C00119ED1V01Y201409ISP058).

flow, approximately 25% of the capillaries are open to perfusion. If blood flow is moderately decreased, oxygen consumption remains stable because the regulatory mechanisms increase capillary density. The increase in perfused capillary density yields an increase in capillary surface area and a consequent reduction in the diffusion distance for O_2 exchange between the capillary and parenchymal cell, thereby enhancing the ability of the tissue to extract oxygen. However, with more drastic reductions in blood flow (>50%), the ability of the tissue to maintain oxygen uptake is compromised because the regulatory increase in capillary density (and O_2 extraction) is insufficient to compensate for the decrease in O_2 delivery.

Both capillary blood flow and the number of perfused capillaries in the gut increase following ingestion of a meal. These vascular responses ensure that there is an enhanced delivery of oxygen to the more metabolically active cells of the gastrointestinal tract and associated accessory organs. Furthermore, such microvascular adjustments serve to supply the water that is secreted by transporting epithelial cells of the salivary gland, pancreas, and stomach. Without these postprandial vascular changes, the critical motor, secretory and absorptive roles of the gastrointestinal system cannot be sustained with normal digestive loads.

The lymphatic system also plays an important role in the transport functions of digestive organs. Lymphatic vessels are particularly prominent in the small bowel. Although blood flow is about 1000 times greater than lymph flow in digestive organs, approximately 1 to 2 l of lymph derived from the GI tract enters the thoracic duct each day in man. Lymph is the major route for the transport of absorbed fat into the circulation. Lymphatics also contribute to the removal of absorbed water from the small and large bowels.

PATHOPHYSIOLOGY AND CLINICAL CORRELATIONS

Figure 1-18 illustrates the relationships among GI oxygen consumption (uptake), epithelial solute transport (absorptive or secretory), and tissue injury, relative to blood flow. Moderate (<50%) reductions in blood flow do not compromise tissue oxygen uptake and do not affect absorptive or secretory activity in the GI tract. However, more severe reductions in blood flow that lead to a reduction in tissue oxygen uptake can have an adverse effect on GI function. For example, gastric acid secretion or intestinal glucose absorption are compromised once blood flow is reduced to levels that limit oxygen uptake, that is, solute transport become limited by blood flow and oxygen delivery. A similar relationship is noted with gastrointestinal motor activity; motility becomes dependent on blood flow only when blood flow is reduced to levels that compromise oxygen uptake. Tissue injury does not become apparent until oxygen consumption is reduced by ≥50%.

Acute mesenteric ischemia is a syndrome resulting from inadequate blood flow in the mesenteric vessels. From a pathophysiologic standpoint, the bowel is resilient to reductions in blood flow. Ischemia results when blood flow is not adequate to provide critical levels of oxygen. Collaterals will develop shortly after vessel obstruction develops distal to the obstructive site. With prolonged ischemia, vasoconstriction develops (due to angiotensin II, vasopressin, and prostaglandins) that eventually leads to irreversible ischemia. With the restoration of blood flow (reperfusion), further tissue injury can result, due in part from the generation of reactive oxygen species and the recruitment and activation of inflammatory cells. Mesenteric ischemia presents with abdominal pain that typically results from embolic occlusion or thrombosis of the superior mesenteric artery. The pain may be focal or present diffusely. Infections and tissue necrosis may develop if emergent therapy, using interventional radiologic or surgical approaches, is not initiated to restore blood flow to the affected organ. In *chronic mesenteric ischemia*, postprandial

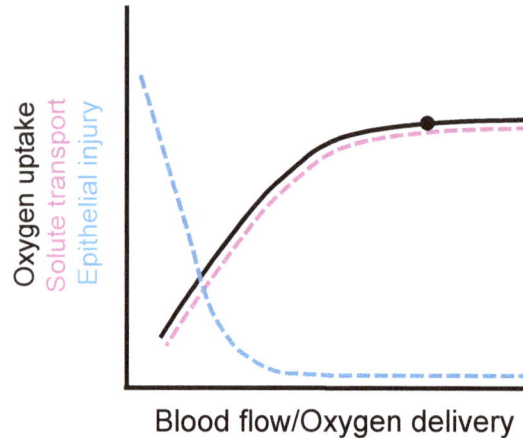

FIGURE 1-18: Reductions in GI blood flow yield parallel reductions in epithelial solute transport and tissue oxygen uptake. However, epithelial injury is not manifested until blood flow is reduced to levels yielding >50% decline in oxygen uptake (modified from *Gastrointestinal Mucosal Defense System*: https://doi.org/10.4199/C00 119ED1V01Y201409ISP058).

abdominal pain or "intestinal angina" may develop shortly (less than half an hour) after a meal, resolving in 1 to 3 hours. A more extensive collateral circulation develops with atherosclerotic narrowing of the major vessels, which occurs predominantly in the elderly. Treatment can involve surgical revascularization or endovascular recanalization with mesenteric angioplasty.

Ischemic colitis is the most common form of ischemic injury in the GI tract. This condition is often seen in the elderly with underlying cardiovascular disease. Low blood flow in the inferior mesenteric artery or colonic branches of the superior mesenteric artery can lead to edema, submucosal hemorrhage and an inflammatory response in the affected bowel segment. The predominant symptom is left lower quadrant abdominal pain associated with abdominal distension and need to defecate; diarrhea and hematochezia (passage of fresh blood through the anus) may also develop. In a patient with right-sided colitis, the pain in focused in the lower abdomen and may not be associated with bloody diarrhea. Diagnosis is made based on history, physical examination findings and endoscopic evaluation. The colonoscope is used to identify the location of the ischemic segment, and this procedure is completed with minimal air or carbon dioxide insufflation. Biopsies may be taken for confirmation histologically, unless the colitis is severe, which would increase risk for bowel perforation. Treatment is mostly supportive with intravenous fluids and broad-spectrum antibiotics due to the potential for spontaneous recovery within 1 to 3 weeks. Occasionally, ischemic strictures, resulting in obstruction, or gangrene (peritonitis) may develop that requires surgical intervention.

REFERENCES

Barrett KE. Gastrointestinal physiology. 2nd Edition, McGraw Hill Education, 2014.

Feldmanna M, Friedman MD, Brandt LJ (editors). Sleisenger and Fordtran's Gastrointestinal and Liver Disease: Pathophysiology, Diagnosis, Management. 2 Volume set, 10th Edition, Elsevier, 2016.

Johnson LR. Gastrointestinal physiology. 8th Edition, Elsevier Mosby, 2014.

Podolsky DK, Camilleri M, Fitz JG, Kalloo AN, Shanahan F, Wang TC (editors). Yamada's Textbook of Gastroenterology, 2 Volume Set, 6th Edition, John Wiley & Sons, 2015.

Reinus RF, Simon D. Gastrointestinal anatomy and physiology. The essentials. Wiley Blackwell, 2014.

Said HM, Gishan FK, Kaunitz JD, Merchant JL, Wood JD. Physiology of the gastrointestinal tract. 2 Volume Set, 6th Edition, Academic Press, 2018.

CHAPTER 2

Eating: Salivation, Chewing, and Swallowing

INTRODUCTION

The amount of food and water ingested by a person is determined principally by the drives of hunger and thirst. These drives assure that the intake of food and water is appropriate for the needs of the individual. After food and fluid ingestion, initial events in the mouth, such as mastication and salivation, prepare the food bolus for swallowing. This chapter describes these early events in food assimilation, which are collectively referred to as eating.

PHYSIOLOGY OF HUNGER AND THIRST

Hunger

Hunger is defined as a craving for food. It is associated with a series of objective sensations, such as hunger pangs. Accompanying phenomena include salivation and increased food-searching behavior. The term appetite is often used in the same sense as hunger; however, it more aptly describes the desire to eat food and attitudes toward different types of food. Appetite may persist even when hunger has been appeased. Satiety is the lack of a desire to eat that occurs after ingestion of food, that is, the sensation of feeling full. Anorexia, on the other hand, is an aversion to eating despite an existing stimulus for hunger.

Regulation of food intake occurs primarily within hypothalamic centers (Figure 2-1). Hypothalamic regulation of feeding and the perception of satiety involve the interaction and coordination of signals between different neuronal centers (nuclei), some of which have gained the designation of the "satiety center" and the "feeding center." The satiety center is localized in the ventromedial nucleus, whereas the feeding center is situated in the ventrolateral nucleus. Stimulation of the ventrolateral nucleus evokes eating behavior in conscious animals, and its destruction causes severe fatal anorexia. On the other hand, stimulation of the ventromedial nucleus leads to cessation of eating, whereas lesions in this region lead to excessive eating (hyperphagia). Other neuronal centers, such as the paraventricular, dorsomedial, and acuate nuclei, integrate chemical and neuronal signals from the periphery that influence the feeding and satiety centers. These centers produce neurotransmitters that either stimulate (orexigenic) or inhibit (anorexigenic) feeding. For example, the arcuate nucleus contains both orexigenic neurons that produce neuropeptide Y and anorexigenic neurons that produce α-melanocyte stimulating hormone (α-MSH).

Chemical and neural signals from peripheral organs exert a major influence on the hypothalamic centers that control food intake (Figure 2-1). These include blood levels of the pri-

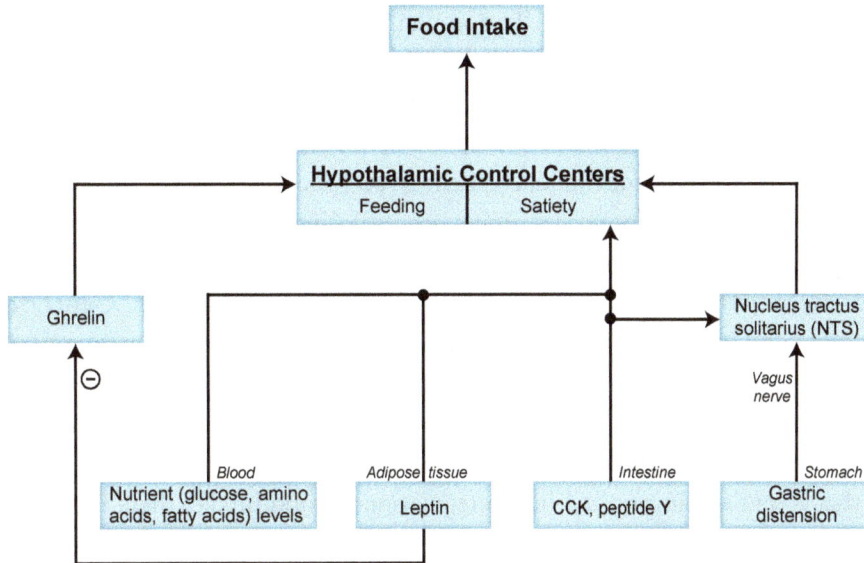

FIGURE 2-1: Neurohumoral regulation of food intake. Neuronal centers in the hypothalamus regulate feeding and the perception of satiety. Chemical and neuronal signals from peripheral tissues, including the GI tract, influence the activity of these hypothalamic centers.

mary nutrients (glucose, fats, amino acids), afferent (vagal) nerve activity from the stomach, and hormones/peptides released from the GI tract (CCK, peptide YY, ghrelin), pancreas (insulin) and adipose tissue (leptin). The peripheral sensing mechanisms that involve the GI tract are thought to play an important role in the short-term (meal-to-meal) regulation of food intake. Gastric distension after ingestion of a meal elicits inhibitory signals (via the vagus nerve) to the feeding center that reduces the desire for food intake. This response is reinforced by the actions of several GI hormones (CCK) and peptides (peptide YY) that are released after a meal. These appetite-suppressing peptides bind to receptors on vagal afferents in the bowel wall to modulate signaling of the hypothalamus by the vagus nerve. Ghrelin, an appetite-enhancing peptide, is released by the stomach during fasting, achieves a peak concentration in plasma just before eating, and falls rapidly after a meal.

Changes in plasma nutrient levels and body energy stores have been implicated in intermediate- to long-term regulation of food intake. An example of nutrient control of food intake is the "glucostat theory," which states that the glucose receptors present in the ventromedial nucleus of the hypothalamus are sensitive to changes in blood glucose levels. The hyperglycemia associated with feeding activates the satiety center and consequently inhibits feeding.

Hypoglycemia has the opposite effect. The similar effects of blood amino acid and fatty acid concentrations on food intake have also lead to the proposal of "aminostatic" and "lipostatic" theories of feeding regulation. Leptin, a hormone that is constitutively secreted by fat cells, suppresses food intake by engaging with its receptors in the arcuate nucleus of the hypothalamus. Leptin is considered a key factor that enables the hypothalamus to detect energy storage status in the body. As the mass of adipose tissue increases (signaling an excess in energy storage), more leptin is produced and released into blood, which delivers it to the brain to suppress food intake. Leptin is also secreted by the stomach after feeding and in response to CCK. Genetic deficiency of leptin in mice and in humans is associated with morbid obesity. However, in most obese humans, leptin production is not impaired and plasma leptin increases in proportion to increasing adiposity. This has led to the assertion that obesity is linked to resistance of leptin receptors to activation. With the rapidly increasing prevalence of obesity worldwide and the expanding list of factors that have been implicated in the regulation of food intake, there is a growing need (and hope) for the development of novel clinical strategies for weight loss therapy.

Thirst

Fluid intake, although largely a result of habit, is also controlled by a specific region of the brain called the thirst center (Figure 2-2). This center is located in the anterolateral region of the hypothalamus. Stimulation of this area results in drinking. The thirst center is activated by an increased extracellular fluid osmolality and by reductions in extracellular fluid volume and blood pressure. Of these stimuli, osmolality is the most potent. Only a 1% to 2% increase in plasma osmolality will elicit the perception of thirst, whereas ≥15% reductions in blood volume and arterial pressure are required to produce the same response. Injection of a hypertonic solution into the thirst center causes cell shrinkage that, in turn, stimulates thirst. This observation is consistent with the concept that cells of the thirst center act as osmoreceptors, which control thirst and drinking. Another population of osmoreceptors, situated in the mouth and pharynx, has also been implicated in the enhanced desire for water ingestion that accompanies a dry mouth and throat. These oropharyngeal osmoreceptors may also explain why the perception of thirst is terminated shortly after the amount of water needed to correct a change in plasma osmolality is consumed, even though it takes longer for the intestines to absorb the ingested water and plasma osmolality is restored to a normal level. This fact is further demonstrated by the immediate relief of thirst after fluid ingestion in patients with esophageal fistulae, in whom the fluid does not reach the G.I. tract and is therefore not absorbed. Furthermore, simple distension of the stomach with a balloon affords transient relief of thirst. These mechanisms likely

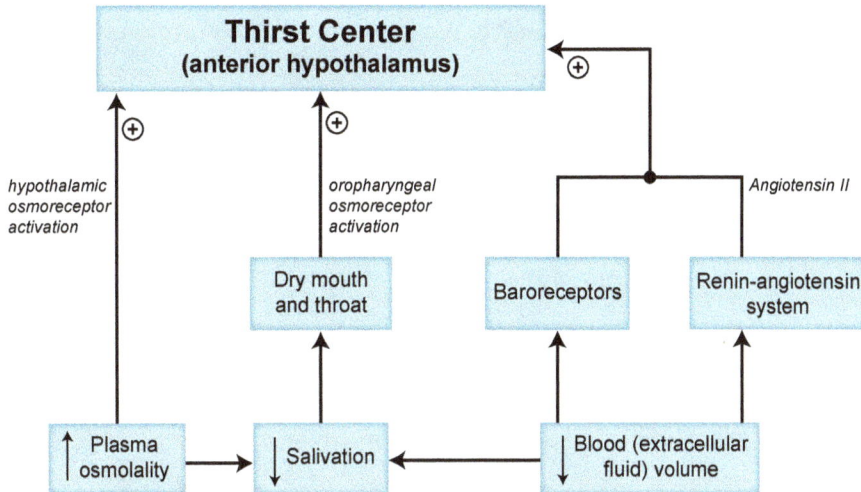

FIGURE 2-2: Factors that promote drinking via activation of the thirst center in the anterior hypothalamus. Osmoreceptors located within the anterior hypothalamus and mouth/pharynx activate the thirst center in response to changes in plasma osmolality and oropharyngeal dryness, respectively. A reduction in blood volume stimulates the thirst center via activation of baroreceptors and the renin-angiotensin system.

serve to prevent excessive ingestion of fluid during the period required for absorbed fluid to correct the disturbances of extracellular osmolality and volume.

Reductions in blood (and extracellular fluid) volume and blood pressure stimulate thirst through mechanisms independent of the osmoreceptor system. Low blood volume and pressure are sensed by the thirst center via input from arterial baroreceptors in the carotid sinuses and aortic arch as well as stretch receptors in the cardiac atria. The renin-angiotensin system has been implicated in this response. Hypovolemia results in increased renin secretion, thereby producing an elevated circulating level of angiotensin II. The angiotensin II acts on specific areas of the hypothalamus that, in turn, activate the thirst center. Thus, hemorrhage increases the desire for water intake despite an unchanged extracellular fluid osmolality.

SALIVARY GLANDS

Saliva is important to the hygiene and comfort of the teeth and mouth, and it contributes to the normal digestion of food. Saliva is the product of a heterogeneous group of exocrine glands

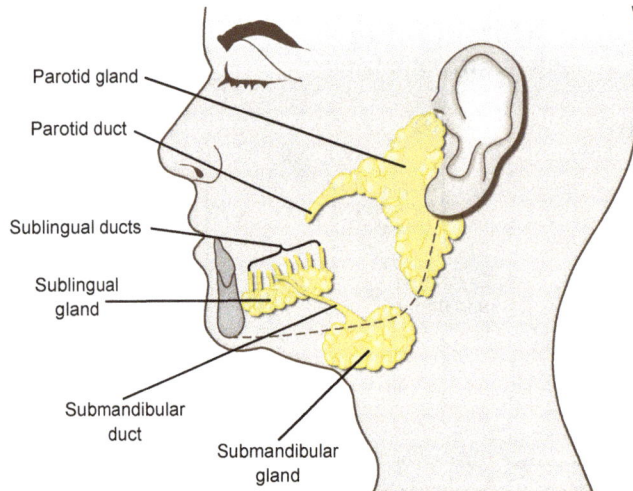

FIGURE 2-3: Location of the three major of salivary glands that empty (via their respective ducts) into the oral cavity.

that drain into the mouth (Figure 2-3). The three main pairs of salivary glands are the parotid, submandibular (submaxillary), and sublingual glands. In addition to these glands, there are numerous smaller glands located in the lips (labial glands), palate (palatine glands), tongue (lingual glands), and cheeks (buccal glands). The salivary glands are characterized by the nature of their secretions. The largest of the glands, the parotids, as well as the lingual glands, secrete a watery non-viscous solution (serous) containing primarily water and electrolytes. All of the other glands secrete a more viscid solution, with the viscosity resulting from the presence of mucins in the serous secretion. The glands producing this type of saliva are commonly referred to as mixed mucus and serous glands. The contributions of the three main salivary glands to total output of saliva (~0.5 to 1.0 l/d) vary with the extent of stimulation. Between meals (resting), the parotid, submandibular, and sublingual glands contribute 20%, 60%, and 20% to total salivary flow, respectively. During the enhanced stimulation elicited by ingestion of food, the contribution of the parotid increases to more than 50% of total flow.

The microscopic structure of the salivary glands is illustrated in Figure 2-4. The functional secretory unit of the salivary gland (sometimes referred to as the salivon) consists of the acinus, the intercalated duct, the striated duct, and the excretory (collecting) ducts. The acinus, the primary secretory unit, is a blind-end sac lined by large pyramidal cells. These cells are of either the serous or the mucous type. The serous cell is characterized by secretory granules

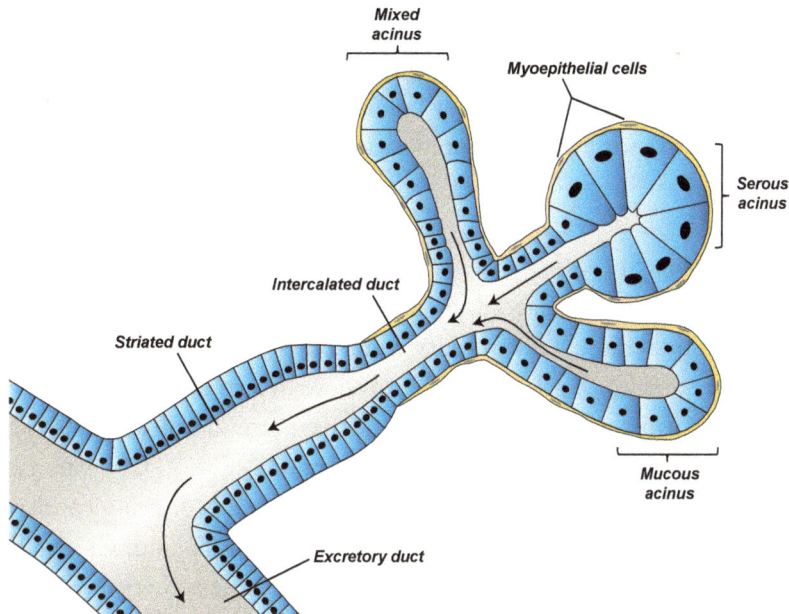

FIGURE 2-4: Microscopic structure of a mixed salivary gland. The functional secretory unit of the gland consists of both serous- and mucous-type acini, an intercalated duct, striated duct, and an excretory (collecting) duct. Myoepithelial cells surrounding the acini and striated duct can contract to enhance the movement of saliva.

and a pronounced rough endoplasmic reticulum (RER), features common to protein-secreting cells. In contrast, the mucous cells contain histochemically demonstrable mucous droplets and a sparse RER. Lying between the secretory cells and their basement membrane are stellate-shaped myoepithelial cells that contain contractile elements. When these cells contract, the acini are compressed and their contents (the primary secretion) are propelled into the intercalated duct.

The intercalated ducts, so-called because they lie between the acinus and the striated duct, are lined by low cuboidal epithelium. They are also surrounded by myoepithelial cells that, upon contraction, further facilitate the movement of secreted fluid between the acinus and striated duct. Whether the acinar secretions are modified within the intercalated duct remains uncertain.

The striated ducts are lined by simple columnar epithelial cells. The striated appearance of the ducts is attributed to the characteristic in-foldings of the basal cell membrane and the columns of mitochondria contained within these folds. The epithelial cells of the striated duct modify the ionic composition and osmolality of the primary secretions produced by the acini. The larger excretory ducts are also lined by columnar epithelium, and there is evidence that these structures may further modify the composition of saliva.

Water and Electrolyte Secretion

The acini are the primary sites of elaboration of saliva (Figure 2-5). An isotonic solution is produced by the acinar epithelium. This primary secretion is plasma-like in composition, that is, the Na^+, K^+, HCO_3^-, and Cl^- levels are similar to those detected in plasma. The fluid secreted into the acinar lumen is primarily linked to the active secretion of Cl^- ions. The accumulation of Cl^- in the acinar cells results from the activity of $Na^+/K^+/2Cl^-$ cotransporter on the basolateral membrane. This Cl^- movement into the acinar cell is driven secondarily by the low intracellular Na^+ concentration created by the Na-K exchange pump. Cl^- secretion into the acinar lumen occurs through apical membrane Cl^- channels, the activity of which is modulated (e.g., during cholinergic stimulation) by changes in intracellular calcium. Cl^- movement into the lumen creates a more electronegative environment that drives Na^+ into the lumen via the tight junctions. Water follows via these paracellular junctions and through aquaporin (AQP5) channels expressed on the apical membrane. K^+ enters the acinar lumen via calcium-activated K^+ channels. HCO_3^-, generated within the cell by the carbonic acid catalyzed reaction of carbon dioxide with H_2O, enters the acinar lumen via the same electrogenic channels used by Cl^-. The net result of these ionic transport processes is an isotonic primary secretion containing Na^+, K^+, Cl^-, and HCO_3^-.

The primary secretion elaborated by the salivary acini is significantly modified as it moves through the striated and excretory ducts This modification largely reflects the ability of ductal epithelial cells to absorb sodium and chloride ions, while secreting K^+ and HCO_3^- into the duct lumen (Figure 2-5). The low permeability of these epithelial cells also impedes the flow of water across the tight junctions; consequently more water remains in the lumen as Na^+ and Cl^- are absorbed, which yields hypotonic saliva.

The capacity of the ductal epithelium to absorb sodium is evidenced by observations that the concentration of sodium in the primary secretion leaving the acinus is 100 to 130 mEq/l, whereas the sodium concentration in saliva obtained at the excretory duct opening is 10 to 30 mEq/l. The absorption of Na^+ in the ducts results from a two-step process wherein Na^+ enters the cell from the lumen via channels (abbreviated as ENaC) in the apical membrane that

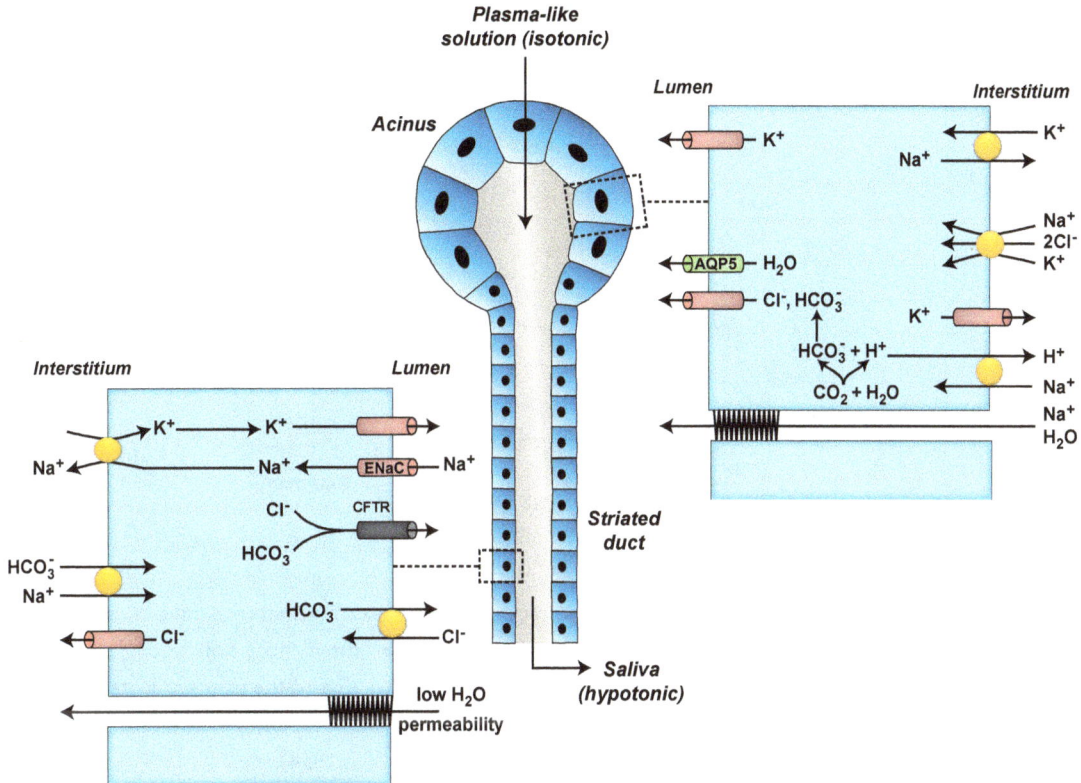

FIGURE 2-5: Electrolyte transport mechanisms that underlie the production of saliva by the acinus and the modification of this primary secretion as it courses through the striated duct. The fluid secreted into the acinar lumen is primarily linked to the active secretion of Cl^- ions that is driven by a $Na^+/K^+/2Cl^-$ cotransporter on the basolateral membrane and Cl^- channels on the apical membrane (right inset). Ductal cells modify the acinar secretion by absorbing Na^+ and Cl^- from, while secreting K^+ and HCO_3- into the duct lumen (left inset). The low water permeability of the ductal epithelium yields hypotonic saliva.

are selectively permeable to sodium ions. The driving force and route for cellular extrusion of this sodium is provided by the Na-K pump in the basolateral membrane. The K^+ entering the cell via the Na-K pump is secreted into the duct lumen via K^+ selective channels in the apical membrane (Figure 2-5). Chloride ions move out of the duct lumen in exchange for bicarbonate and by passive diffusion along an electrochemical gradient (accompanying sodium). Chloride channels in the basolateral membrane allow for the exit of absorbed Cl^- from the cell and entry into plasma. A Na^+-HCO_3- transporter, also situated in the basolateral membrane, allows for

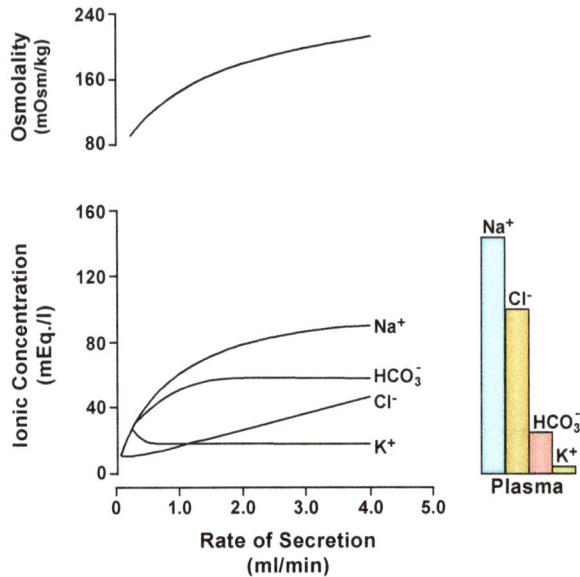

FIGURE 2-6: Changes in osmolality and electrolyte composition of saliva at different rates of salivary secretion. As the rate of secretion increases and the transit time of the primary secretion through the ducts diminishes, the electrolyte composition and tonicity of saliva approaches that of the primary acinar secretion (redrawn based on data from Thaysen JH, Thorn NA, and Schwartz IL: Excretion of sodium, potassium, chloride and carbon dioxide in human parotid gland. *Am J Physiol* 178:155, 1954).

the intracellular accumulation of HCO_3^-, which is then secreted into the duct lumen via the Cl^--HCO_3^- exchanger. The cyclic-AMP-activated cystic fibrosis transmembrane conductance regulator (CFTR) channel in the apical membrane can also contribute to the movement of both Cl^- and HCO_3^- between the cell interior and ductal lumen. The chloride ions that enter the duct lumen via the open CFTR channels fuel the adjacent Cl^--HCO_3^- pumps to secrete HCO_3^- into the duct lumen in exchange for Cl^- because the rate of HCO_3^- secretion by the Cl^--HCO_3^- pump depends on the availability of luminal Cl^-. When luminal Cl^- concentration is low, the selectivity of the CFTR channels is shifted towards conducting HCO_3^-, rather than Cl^-. A result of HCO_3^- secretion into the duct lumen is a slightly alkaline saliva, with a pH of approximately 8.0 as it enters the mouth.

The degree of hypotonicity and the electrolyte composition of saliva are dependent upon the rate of salivary secretion (Figure 2-6). At low secretion rates (0.5 ml/min), the movement of the primary secretion within the ducts is slow enough to allow the ductal transport processes to decrease the osmolality of the primary secretion by 70%, that is, to 88 mOsm/l. This fluid contains approximately 10 mEq/l of chloride and 26 mEq/l of sodium, bicarbonate, and potassium. As secretion rate is increased, the reduced time of exposure of the primary secretion to the ductal epithelium limits the amount of electrolytes that can be removed (sodium) or added (potassium). Thus, at higher secretion rates, the tonicity of the saliva approaches that of the primary secretion. For example, at a secretion rate of 4 ml/min (approximately 10 times the resting value), the osmolality of the saliva is approximately 212 mOsm/l with the following ionic composition: sodium, 90 mEq/l; bicarbonate, 58 mEq/l; chloride, 46 mEq/l; potassium, 18 mEq/l. Although the dependence of ionic composition on secretion rate varies from one salivary gland to another, a common feature of all secretions is that at low rates of secretion, the saliva is hypotonic. Relative to their plasma concentrations, potassium and bicarbonate concentrations are high, whereas the sodium and chloride levels are low (Figure 2-6).

Substances that influence the secretory activity of salivary epithelium, such as neurotransmitters and hormones, do so by interacting with specific membrane receptors. The cellular events that link receptor activation and saliva formation for acetylcholine and norepinephrine, the two most important physiological stimulants of acinar cell secretion, are summarized in Figure 2-7. Acetylcholine, the primary parasympathetic neurotransmitter, stimulates the acinar cell by engaging with muscarinic receptors expressed on the cell surface. Engagement of these receptors results in the formation of inositol triphosphate (IP_3) and the subsequent accumulation of intracellular Ca^{++} that is derived from intracellular stores and an influx of extracellular Ca^{++}. The major action of this Ca^{++}-mobilizing signaling pathway is to stimulate the secretion of electrolytes and water into the acinar lumen. Substance P and VIP are other neurotransmitters released in the salivary gland that stimulate acinar cell secretion via Ca^{++} mobilization. Norepinephrine, the primary sympathetic neurotransmitter, exerts its effects on the acinar cell by engaging with beta-adrenergic receptors to increase the intracellular level of cyclic AMP (cAMP). The primary response of the acinar cell to cAMP-mediated signaling is the mobilization and exocytosis of granules containing amylase, mucins, and other proteins (Figure 2-7). cAMP signaling also results in an increased intracellular Ca^{++}, which is necessary for the mobilization and exocytosis of enzyme and mucin rich granules. Hence, although the Ca^{++}- and cAMP-dependent signaling mechanisms are capable of independently stimulating salivary secretion, the former mechanism exerts a dominant influence on fluid and electrolyte (volume) secretion, whereas the latter primarily regulates the release of enzymes and mucins into saliva. Because of "cross-talk" between the two signaling pathways, simultaneous

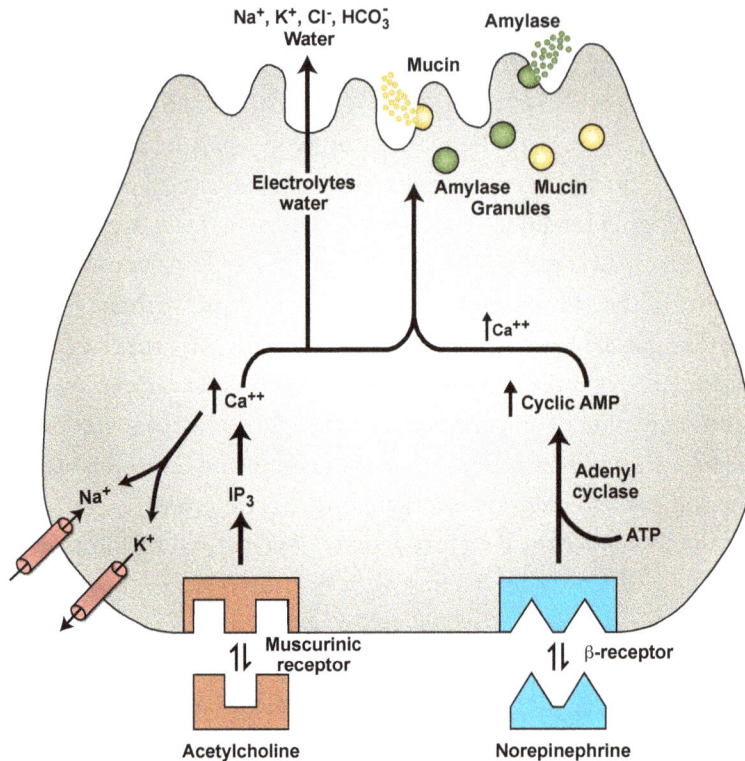

FIGURE 2-7: Cellular events that link acetylcholine- and norepinephrine-mediated receptor activation to saliva formation by acinar cells. Acetylcholine primarily stimulates the secretion of electrolytes and water via an IP_3-Ca^{++} signaling mechanism, whereas the major action of norepinephrine is the mobilization and exocytosis of granules containing amylase, mucins, and other proteins. Norepinephrine exerts its effects on the cell by engaging with beta-adrenergic receptors to increase the intracellular level of cyclic AMP (cAMP), resulting in an increased intracellular Ca^{++} that elicits the exocytosis of enzyme and mucin rich granules.

engagement of muscarinic and beta-adrenergic receptors is associated with a potentiated salivation response.

Composition and Functions of Saliva

Saliva contains several thousand unique proteins that yield a total protein concentration that ranges between 100 and 300 mg/dl. Although most of these salivary proteins are stored and released from secretory granules in acinar cells, some proteins (e.g., lysozyme, ribonuclease, epidermal growth factor) enter saliva from ductal epithelium. The wide variety of proteins found in saliva exhibit diverse biologic functions that generally fall under four broad categories: 1) enzymes, 2) antimicrobial agents, 3) mucoproteins (mucins), and 4) miscellaneous.

Alpha-amylase and lingual lipase represent two of the major enzymes found in saliva. Salivary amylase initiates starch digestion, via the cleavage of alpha-1,4-glycosidic bonds, when food enters the mouth. The enzyme has a pH optimum near 7.0 and is rapidly denatured when pH falls below 4.0. Because ingested food can remain unmixed with gastric acid in the upper stomach for an extended period, significant carbohydrate digestion (estimates range between 30% and 75%) may occur in a well-chewed meal before the salivary amylase is inactivated by gastric acid. However, in the absence of salivary amylase, the large amounts of pancreatic amylase secreted into the bowel after food ingestion is more than sufficient to ensure complete carbohydrate digestion in normal individuals.

Lingual lipase plays a role in the initiation of lipid digestion. This enzyme has a pH optimum of 4.0, and therefore, it can continue its digestive activity in the stomach. Lingual lipase hydrolyzes triglycerides to fatty acids and monoglycerides and diglycerides. Unlike pancreatic lipase, salivary lipase is not influenced by surface-active detergents, such as bile salts and lecithin. Although the actions of lingual lipase accounts for a small proportion of overall fat digestion in adults, it makes a larger contribution to the digestion of milk fat in newborn infants, which exhibit low pancreatic lipase activity and a low pH throughout the GI tract.

The human mouth is populated by a wide spectrum of pathogenic bacteria and the maintenance of proper oral hygiene is facilitated by production and secretion of antimicrobial agents by the salivary glands. The antibacterial action of saliva can be attributed to a variety of proteins, including lactoferrin, lysozyme, lactoperoxidase, and the binding glycoprotein for immunoglobulin IgA. The iron-binding protein, lactoferrin, which is found in a number of exocrine secretions, including milk and pancreatic juice, exerts its action by depriving microorganisms of nutrient iron. Lysozyme (muramidase) is a glycoprotein that hydrolyzes the muramic acid constituents of bacterial cell wall polysaccharides, thus destroying the microorganism. Lactoperoxidase catalyzes the oxidation of a number of inorganic (iodide, thiocyanate) and organic substrates by hydrogen peroxide, thereby imparting potent bactericidal activity to these substrates. The combination of IgA and its glycoprotein-binding protein, both of which are secreted into saliva, leads to the formation of secretory IgA, which is immunologically active against bacteria by binding to antigenic components of the bacterial cell wall.

Mucin represents one of the major proteins in saliva produced by the sublingual and submandibular glands. The steps (storage, exocytosis), signaling pathways, and stimuli involved in the release of mucin from mucous-type acinar cells are similar to those described for amylase (Figure 2-7). Mucin is the primary determinant of salivary viscosity. The viscosity of whole saliva is 2.9 centipoises, yet the viscosities of the secretions of the sublingual, submandibular, and parotid glands are 13.4, 3.4, and 1.5, respectively. Mucin strongly adheres to other proteins, forming a thin film or coating on tissue surfaces and on ingested food. Mucins lubricate both hard and soft tissue surfaces, and allow particles to slide along these surfaces with ease and at low resistance. The strong adhesivity of mucin also enables it to form stable bridges between bacteria and other salivary proteins, such as antimicrobial agents.

Other salivary proteins that are known to exert important physiologic actions in the GI tract are haptocorrin and growth factors. Haptocorrin (also known as "R-protein") binds to and forms a stable, acid-resistant complex with dietary vitamin B_{12} (cobalamin). The complex prevents acid degradation of the vitamin in the stomach. Once the complex enters the less acidic environment of the duodenum, pancreatic enzymes release the cobalamin, allowing the vitamin to bind to intrinsic factor, which facilitates B_{12} absorption in the ileum. A number of growth factors, including epidermal growth factor (EGF), nerve growth factor (NGF) and transforming growth factor (TGF), are synthesized by salivary glands and secreted into saliva. These proteins stimulate epithelial cell proliferation and differentiation, and promote wound healing. Consequently, it has been proposed that the growth factors produced by the salivary glands may contribute to the growth and repair of epithelial cells lining the oral cavity and the mucosa of more distal regions (esophagus, stomach) of the GI tract.

Saliva serves several important physiologic functions that can be broadly grouped into protective and digestive. The benefits derived from saliva production are evidenced both between meals and during the ingestion of food (Table 2-1). The protective functions of saliva are largely due to the presence of antibacterial agents, bicarbonate, and mucins. Bicarbonate, a major ionic constituent of saliva, plays an important role in the neutralization of acid in the mouth. The two major sources of acid in the mouth are ingested materials and the products of bacterial metabolism. Ingestion of acidic foods results in an increased secretion of saliva, which is rich in bicarbonate. This bicarbonate serves to partially neutralize the ingested acid. Acid released by bacteria plays an important role in the formation of dental caries (by dissolving enamel and dentine). In the absence of salivary bicarbonate, there is an increased incidence of dental caries. The buffering capacity of salivary bicarbonate is also important in the neutralization of gastric acid, which is refluxed into the lower esophagus. Saliva also protects oral structures between meals by providing a continual cleansing action in the mouth and by moistening the oral mucosa.

TABLE 2-1: Physiological functions of saliva
BETWEEN MEALS:
1) Maintains hydration & prevents abrasion of oral mucosa
2) Prevents demineralization of enamel (prevents caries)
3) Antimicrobial actions (enhances oral hygiene)
4) Facilitates speech
DURING MEALS:
1) Facilitates mastication and swallowing (lubrication)
2) Buffers ingested acids and acid reflux from stomach
3) Dilutes hot solutions (e.g., soup) and foul tasting substances
4) Enhances taste bud sensitivity by solubilizing food
5) Starch digestion (α-amylase, pH optimum ~7.0)
6) Lipid digestion (lingual lipase, pH optimum ~4.0)

The contributions of saliva to digestion can be divided into lubricating properties and hydrolytic activities. The presence of mucins in saliva facilitates mastication and deglutition by diminishing the frictional interaction between the food bolus and the oral and esophageal mucosa. These functions are demonstrated by the observation that patients with inadequate salivary flows have difficulty in swallowing dry foods, even when they are taken with large amounts of water.

Other important functions of saliva include facilitation of speech, oral comfort, and taste, and modification of the temperature of ingested food. The moistening and lubricative properties of saliva are essential for oral comfort and clear speech. Inasmuch as gustatory function is enhanced by solubilization of foodstuffs, saliva facilitates taste sensation. The free flow of saliva dilutes and lowers the temperature of the ingested hot fluid, thereby preventing scalding.

Regulation of Salivary Secretion
Neural Control

In most organs/tissues in the digestive system, parasympathetic and sympathetic nerve stimulation results in opposite physiologic responses, with parasympathetic stimulation yielding an excitatory response, whereas sympathetic stimulation results in an inhibitory response. By contrast, in the salivary glands, both arms of the autonomic nervous system stimulate secretion when activated. However, parasympathetic nerve activity is recognized as the most important physiologic regulator of salivary secretion, exerting a stronger and longer-lasting response than sympathetic nerve activation (Table 2-2).

Parasympathetics

All salivary glands appear to receive parasympathetic nerve endings, which originate from the salivary nuclei of the medulla and the facial (seventh) and glossopharyngeal (ninth) nerves (see Figure 1-6 in Chapter 1). Fibers destined for the parotid gland are contained in the ninth cranial nerve and synapse in the otic ganglion, from which postganglionic fibers in the auriculotemporal branch of the trigeminal nerve terminate on acinar and ductal elements of the gland. Preganglionic fibers for the submandibular and sublingual glands, the chorda tympani, course with the seventh cranial nerve and synapse in the submaxillary ganglion, from which ganglionic fibers innervate the gland and blood vessels. Both acetylcholine and vasoactive intestinal polypeptide are found in vesicles in the cholinergic nerve terminals. However, the influence of

TABLE 2-2: Comparison of Sympathetic and Parasympathetic Influences on Salivary Secretion		
RESPONSE	**SYMPATHETIC**	**PARASYMPATHETIC**
Saliva output	Scant	Copious
Time-course	Transient	Sustained
Composition	Protein rich High K^+ & HCO_3-	Protein poor Lower K^+ & HCO_3-
Denervation response	Minimal	Decreased secretion Glandular atrophy

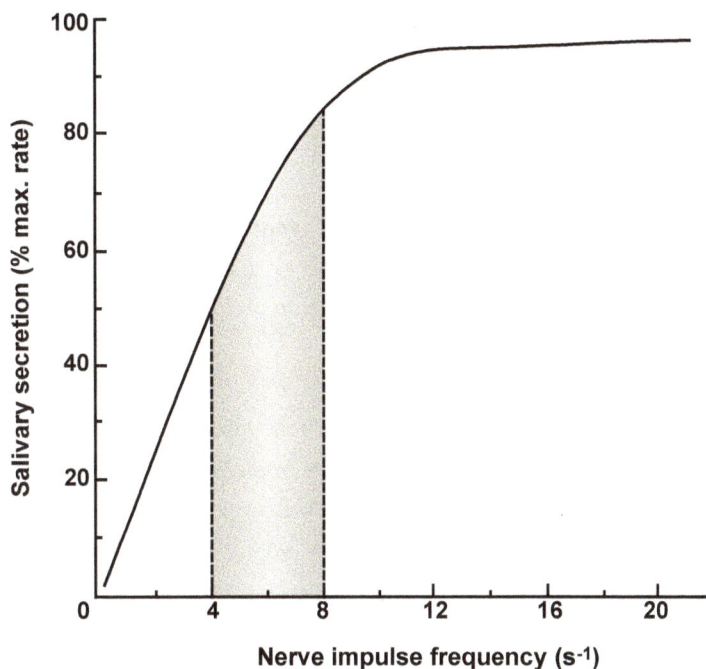

FIGURE 2-8: Relationship between parasympathetic nerve stimulation frequency and the rate of salivary secretion. The vertical bar shows the range of secretory rates achieved by the glands when maximally activated by feeding (redrawn based on data from Emmelin N and Holmberg J: Impulse frequency in secretory nerves of salivary glands. *J Physiol* (Lond) 191:205, 1967).

parasympathetic activation on salivary glands largely reflects different responses of the gland to muscarinic (m₃) receptor activation by acetylcholine. These responses include increased synthesis and secretion of amylase and mucin, enhanced electrolyte transport by ductal epithelium, and increased glandular metabolism, blood flow, and growth.

Events associated with the ingestion of a meal (e.g., smell, taste) excite the salivatory nuclei that, in turn, increase the impulse frequency of the parasympathetic nerves supplying the gland. As shown in Figure 2-8, the rate of saliva formation is dramatically increased by stimulation of parasympathetic nerves over a range of 1 to 10 pulses per second, with maximal secretion rates obtained at frequencies greater than 10 pulses per second. Under resting conditions, the impulse frequency is usually less than 1 pulse per second and it can increase to 4 to 8 pulses per second during feeding. Thus, as a result of the parasympathetic activation associated with feeding, the rate of salivary secretion increases by 4 to 8 times. Noxious stimuli (such as a sore in the mouth) can produce parasympathetic nerve impulse frequencies exceeding 8 pulses per second, thereby resulting in maximal salivary secretion.

The importance of the parasympathetics in controlling salivation is further demonstrated by the observation that severing all parasympathetic nerves to the parotid gland abolishes the secretory response to citric acid placed in the mouth. Parasympathetic denervation also results in atrophy of the salivary glands. Anticholinergic agents, such as atropine effectively inhibit salivary secretion, whereas drugs (such as pilocarpine) that inhibit the degradation of acetylcholine by cholinesterase enhance the rate of salivation.

Sympathetics

The salivary glands are also innervated by sympathetic fibers (see Figure 1-6 in Chapter 1). These fibers are derived from the first two thoracic segments via the superior cervical ganglion. The postganglionic fibers accompany the blood vessels into the glands, where they supply elements of the salivon (acini and ducts) and blood vessels. In contrast to the copious salivation elicited by parasympathetic stimulation, sympathetic stimulation produces a scant secretion. The maximal secretory rates produced by sympathetic stimulation are approximately one fifth the values observed with parasympathetic stimulation. The influence of sympathetic nerves on salivary secretion results primarily from activation of β-adrenergic receptors. Alpha-adrenergic receptors play a more important role in mediating the myoepithelial cell contraction associated with sympathetic stimulation.

Although both sympathetic and parasympathetic nerves are stimulants of salivary secretion, there are differences between the responses of the glands to the two neural influences (Table 2-2). Unlike parasympathetic stimulation, which produces a sustained copious flow of saliva, sympathetic stimulation produces a secretory response that tends to diminish or even cease despite continued stimulation. The composition of saliva is also dependent upon which division of the autonomic nervous system is stimulated. Although the electrolyte composition of the acinar secretion is similar for the two neural influences, the saliva obtained from the main duct has a higher K^+ and HCO^-_3 concentration, presumably indicating an effect of the sympathetics on the ductal epithelium. Another characteristic feature of sympathetically mediated secretion is a high protein concentration in the saliva. The myoepithelial cells are innervated by both divisions of the autonomic nervous system; however, the sympathetics appear to play a more important role in the contraction of acini and ducts, a process mediated by alpha-adrenergic receptors. Prolonged stimulation of either sympathetics or parasympathetics leads to salivary gland growth; however, parasympathetic denervation results in glandular atrophy, whereas sympathetic denervation does not. This fact presumably reflects a greater influence of the parasympathetics on basal salivary function.

Hormonal Regulation

In contrast to neural influences, which affect all elements of the salivon, some hormones, such as aldosterone and antidiuretic hormone, exert their effects only on the ductal epithelium. Hormonal influences do not alter salivary secretion rate, yet they do modify the ionic composition of saliva. Aldosterone acts directly on the ducts to increase both sodium absorption and potassium secretion. Adrenocortical insufficiency is characterized by an increased salivary sodium concentration, whereas primary hyperaldosteronism and pregnancy (which is associated with high levels of mineralocorticoids) are characterized by a low-sodium concentration in saliva. Antidiuretic hormone reduces the sodium concentration of saliva, presumably by enhancing sodium absorption in the ducts. The physiologic role of hormones in the regulation of salivary secretion appears to be far less important than that of neural influences.

Stimuli for Salivary Secretion

The most important stimuli for salivary secretion relate to events occurring before and during food ingestion. Even before ingestion, anticipation of food initiates salivation through signals from higher centers to the salivatory nuclei. This cephalic phase is reinforced by conditioned reflexes related to olfactory (smell) and visual stimuli. Salivation is not sustained unless food is eaten, which activates proprioceptive (touch) and gustatory (taste) reflexes. Proprioceptive reflexes can be initiated by placing tasteless objects in the mouth, with smooth objects being more effective stimulants than rough ones. Of the gustatory stimuli, sour-tasting or acidic material are the most potent.

Factors unrelated to food ingestion that decrease salivary secretion include sleep, anxiety, fear, mental effort, and dehydration. Nausea is associated with increased salivation, and the symptom of "water brash," which is the filling of the mouth with copious amounts of saliva, can occur in the presence of certain diseases of the upper GI tract, notably reflux esophagitis, and duodenal ulcer.

Salivary Blood Flow

The maximally stimulated salivary gland of the dog can secrete its own weight in saliva every two minutes (in man this requires 10 minutes). These high secretory rates create the need for high blood flows, which provide the oxygen, nutrients, electrolytes, and water required for salivation. Blood flow to the salivary gland in man is provided by arterial vessels originating from the external carotid arteries. Blood vessels enter at the hilus of the glands and ramify within the parenchyma in accordance with the lobular structure, forming capillary plexuses around the

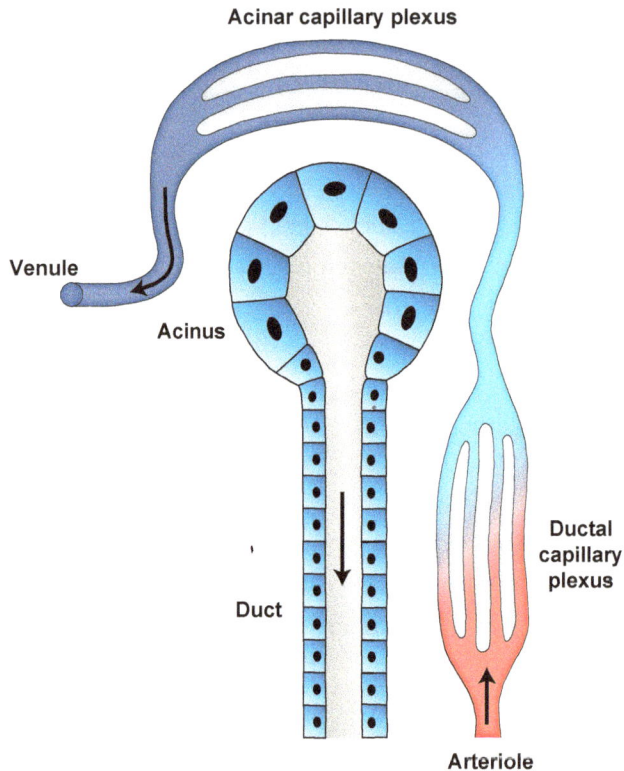

FIGURE 2-9: Arrangement of blood vessels within the salivary glands. The vasculature represents a series-coupled portal network in which arterial inflow first supplies the ductular capillary network and then drains (via venules) into the capillary network supplying the acini. The flow of blood in the microvasculature is countercurrent to the flow of saliva in the acini and ducts.

acini and ducts. The plexuses of fenestrated capillaries surrounding the ducts are much denser than those surrounding the acini. Veins draining the salivary capillaries course with the arteries and exit the tissue at the hilus to finally drain into the external jugular vein. The classical view of the vascular arrangement within the salivary glands is that of a series-coupled portal network in which blood flows countercurrent to salivary flow. In this model, arterial inflow first supplies the ductular capillary network, which in turn drains via venules into the acinar capillary network, forming a portal system whose flow is countercurrent to salivary flow (Figure 2-9). Such an arrangement would allow substances produced and released by ductal epithelium (such as epidermal growth factor) to influence the activity and/or function of acinar cells.

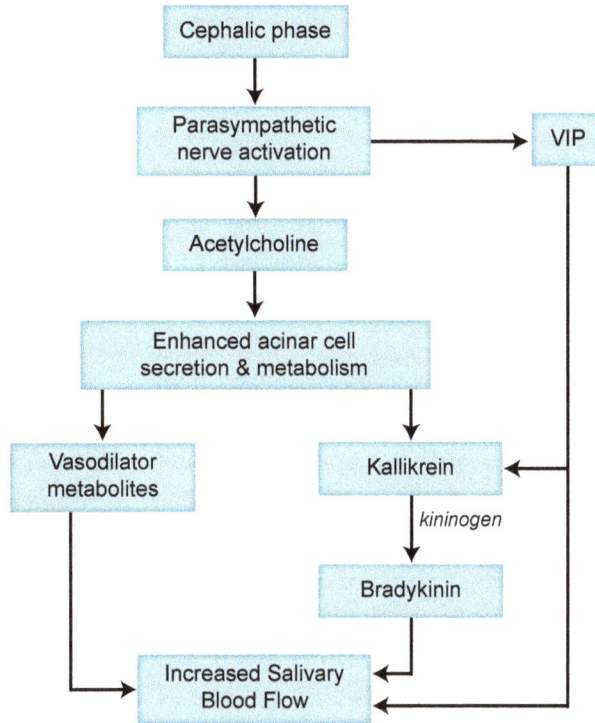

FIGURE 2-10: Mechanisms involved in the intense vasodilation (and increased blood flow) that occurs in the salivary gland during the cephalic phase of digestion. Bradykinin, a major mediator of the vasodilatory response, is generated in response to the release of acetylcholine and VIP from parasympathetic nerve terminals. VIP and products of tissue metabolism (e.g., adenosine) may also act directly on the vasculature to elicit vasodilation.

Resting blood flow in the salivary glands is comparable to that of other gastrointestinal organs (i.e., about 50 ml per min per 100 g) when expressed per unit mass of tissue. However, blood flow to the salivary gland can increase ten-fold in response to enhanced functional activity (secretion). This intense hyperemia compares with the 30% to 130% increase in intestinal blood flow observed after a meal. The functional hyperemia in the salivary gland is associated with an increase in the number of capillaries open to perfusion, which facilitates the exchange of oxygen, nutrients, and water between the blood stream and transporting epithelium of the acini and ducts.

The possible mechanisms involved in the functional vasodilation of the salivary gland are illustrated in Figure 2-10. Although acetylcholine is clearly responsible for the secretory response to parasympathetic stimulation, the observation that atropine (anticholinergic agent) blocks the secretory response (but not the vasodilation produced by parasympathetic stimulation) indicates that acetylcholine does not mediate the functional hyperemia. The most widely accepted mediator of this response is bradykinin, a potent vasodilator that is derived from the proteolytic action of glandular kallikrein on plasma kininogen. The fact that atropine only partially blocks the vasodilation induced by parasympathetic stimulation suggests that another neurotransmitter (e.g., VIP) is involved in the release of kallikrein. Vasoactive intestinal polypeptide and other noncholinergic neurotransmitters (e.g., substance P) may also act directly on the vasculature to elicit vasodilation. Metabolite accumulation has also been invoked to explain the vasodilation associated with enhanced salivary secretion. Such vasodilators might include adenosine, potassium ions, osmolality, and lactic acid.

The increased blood flow and perfused capillary density (capillary recruitment) associated with enhanced salivary secretion are essential for the maintenance of high rates of salivation. Arteriolar dilation leads to a rise in capillary hydrostatic pressure, which, coupled to the increased surface area for fluid exchange provided by capillary recruitment, allows for a greatly enhanced rate of capillary filtration. Thus, the fluid necessary for salivary secretion is derived from the blood circulation. The importance of the circulation in salivary secretion is demonstrated by the fact that, at high rates of secretion, the salivary venous hematocrit can increase from a value of 45% to 55%, owing to extraction of fluid from the blood. Furthermore, reductions in blood flow during maximal salivary stimulation lead to a reduction of secretion rate.

CHEWING (MASTICATION)

Although chewing is not essential for the assimilation and digestion of foodstuffs, it does greatly facilitate these processes. Chewing involves the tearing, grinding, and cutting of solid food. It is accomplished by rhythmic muscular contractions, which move the jaws to facilitate the frictional interaction between the food bolus and teeth. The tongue also takes part in this coordinated activity, moving the food along the surfaces of the teeth. The incisors provide a cutting and tearing action, whereas the molars grind food. The molars are capable of developing 3 to 4 times more pressure on food particles than are the incisors.

Mechanisms
Chewing can be considered both a voluntary and an involuntary (reflex) activity. Stimulation of centers in the hindbrain can elicit and maintain chewing. Fibers from these centers run along the fifth cranial nerve to supply the muscles involved in mastication. The two groups of muscles

involved in chewing can be divided into those that open the mouth and those that close it. The masseter, medial pterygoid, and temporal muscles are involved in closing the mouth, whereas the lateral pterygoid and digastric muscles open the mouth. When a meal is ingested, the mouth, which is tonically closed, opens voluntarily. When the mouth is closed over a bolus of food, the pressure of food against the tongue, teeth, gums, and hard palate stimulates receptors that initiate the chewing reflex. The first event in the chewing cycle is relaxation of the jaw-closing muscles and contraction of jaw-opening muscles. This reduces the pressure exerted on the receptors and results in rebound relaxation of jaw-opening muscles and contraction of jaw-closing muscles. The basic rhythmic activity of the jaw-opening and jaw-closing muscles is likely evoked by pattern generating neurons located in the brain stem. Characteristics of food, like hardness, and water and fat content, are sensed during chewing and consequently influence both the force generated during mastication as well as the number of chewing cycles before swallowing. In man, the duration of the chewing cycle is approximately one second.

Functions

Mastication serves several important functions related to food ingestion. The crushing and grinding of food during chewing not only stimulates salivary flow but also enhances the lubrication of food by mixing it with saliva. The carbohydrate component of many plant materials is encased in an indigestible fibrous (e.g., cellulose) coat. Chewing mechanically disrupts this coat, thus making the nutrient content available for digestion and absorption. Reduction of particle size of digestible foods by chewing facilitates the action of salivary and gastric enzymes by increasing the surface area available to these enzymes. Chewing and lubrication also protect the oral and esophageal mucosa from physical injury (abrasion) by hard and rough food particles.

SWALLOWING (DEGLUTITION)

Swallowing usually begins as a voluntary act, but once initiated it proceeds through an involuntary reflex involving the coordinated action of the oral and pharyngeal musculatures (Figure 2-11). The first step in this process (the voluntary stage) involves the separation of a portion of the food bolus in the mouth by lifting it with the tongue against the hard palate. The pressure exerted by the tongue then propels the bolus backward toward the pharynx. When the bolus exerts pressure on sensory receptors near the opening of the pharynx (tonsillar pillars, soft palate, tongue base), impulses are transmitted to the swallowing center in the medulla via various

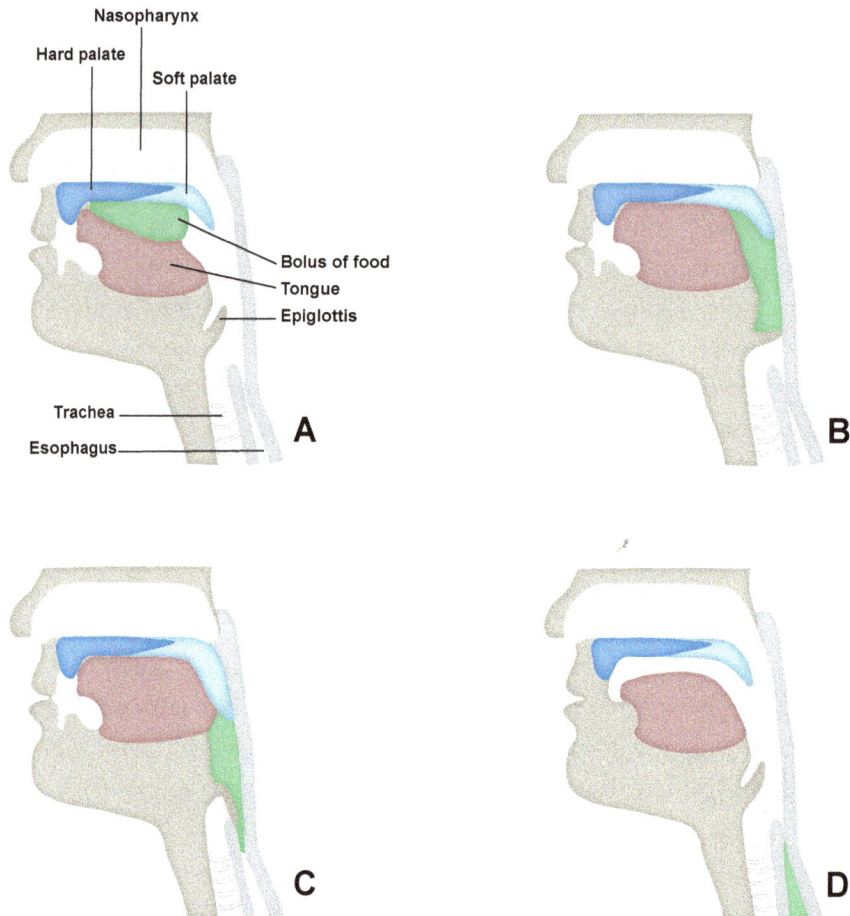

FIGURE 2-11: Events associated with swallowing that move a food bolus from the mouth into the esophagus. In the voluntary stage, a portion of the food bolus in the mouth is lifted by the tongue against the hard palate (Panel A). The pressure exerted by the tongue then propels the bolus backwards toward the pharynx. When the bolus exerts pressure on sensory receptors near the opening of the pharynx, impulses are transmitted to the swallowing center, which results in a peristaltic wave that propels the bolus toward the upper esophagus. During the pharyngeal phase (Panel B), the soft palate rises and the palatopharyngeal folds move inward, preventing nasopharyngeal reflux and creating a channel for the passage of the food bolus, respectively. Then, the vocal chords are pulled together, and the epiglottis closes over the chords (Panel C), which inhibits respiration and prevents food from entering the trachea. The entire larynx is then pulled upward and forward, thus stretching the entrance to the esophagus. The pharyngeal phase ends and the esophageal phase begins (Panel D) with the contraction of muscles in the anterior pharynx and opening of the upper esophageal sphincter, which allows the food bolus to enter the esophagus via pharyngeal peristalsis. The entire process occurs in approximately 1 second.

nerves (trigeminal, glossopharyngeal, and vagal). Stimulation of the swallowing center results in the coordinated contraction of a series of small pharyngeal muscles, resulting in a peristaltic wave, which propels the bolus toward the upper esophagus. This motor activity involves various cranial nerves (fifth, seventh, ninth, tenth, and twelfth).

The passage of food through the pharynx (pharyngeal phase) involves the following sequence of events: the soft palate rises and the palatopharyngeal folds move inward, preventing nasopharyngeal reflux and creating a channel for the passage of the food bolus, respectively. Then the vocal chords are pulled together, and the epiglottis closes like a lid over the chords. At this point, respiration is inhibited and food is prevented from entering the trachea. The entire larynx is then pulled upward and forward, thus stretching the entrance to the esophagus. The pharyngeal phase ends with the contraction of muscles in the anterior pharynx and opening of the upper esophageal sphincter, thus allowing the food bolus to enter the esophagus via pharyngeal peristalsis.

To summarize the mechanics of the pharyngeal stage of swallowing: the trachea is closed, the esophagus is opened, and a fast peristaltic wave originating in the pharynx then forces the bolus of food into the upper esophagus, the entire process occurring in approximately 1 second.

PAIN IN THE MOUTH AND OROPHARYNX

Unlike pain arising from the rest of the gastrointestinal tract, which involves autonomic nerves, pain from the mouth and oropharynx is conveyed by somatic nerves. The somatic nerves account for the fact that oral pain is not referred, but is localized to the affected area. The most common causes of pain in this region are dental and associated gingival diseases, ulcers of the oral mucosa, tonsillitis, and pharyngitis.

PATHOPHYSIOLOGY AND CLINICAL CORRELATIONS

Salivary Glands

A number of pathological processes and some medications are known to affect the salivary glands. These conditions are generally associated with diminished saliva production that results from an impaired function and/or injury of acinar and ductal cells. The clinical manifestations of these disease- and drug-induced salivary disorders are generally expected from hyposalivation given the known functions of saliva and can include dry mouth (xerostomia), difficulty tasting (dysgeusia) and swallowing (dysphagia) food, tooth decay (caries), and ulceration of the oral mucosa. Sialadenitis, inflammation of the salivary glands, generally results from a viral or bacterial infection. Previously, the most common form of salivary gland inflammation was *mumps*, a viral infection, which affects several tissues but has a predilection for the parotid gland. The inflammatory process, which produces vascular congestion and edema of the gland, results in marked swelling and pain. Mumps is caused by the paramyxovirus (RNA) virus that is spread by airborne droplets from salivary, nasal, and urinary secretions. The virus incubates for 2 to 3 weeks in the upper respiratory tract before symptoms develop. Before antibody tests were developed, lemon juice, a potent stimulant for salivary gland secretion, placed on the tongue was used to identify a patient with mumps. This stimulant provokes marked exacerbation of the parotid pain, presumably owing to an acute increase in congestion of the gland resulting from the reflex secretion, together with hyperemia. Current diagnostic strategies involve identification of antibodies to the mumps S (core) and V (outer surface) hemagglutinin antigens or isolation of the virus in cerebrospinal fluid. Treatment for viral mumps is mostly supportive with bed rest, hydration, and avoiding stimulation of the salivary glands with dietary modifications. Fever will resolve before resolution of the parotid swelling. Widespread vaccination programs have significantly reduced the incidence of this viral disease; however, resurgence may be seen with patients not receiving childhood vaccinations. Immunocompromised and pregnant patients should not receive the vaccine for mumps because it is a live attenuated vaccine.

Acute suppurative sialadenitis is a bacterial infection of the salivary glands. Normal salivary flow and the presence of different antimicrobial agents in saliva normally prevents bacterial colonization of the salivary glands. However, reductions in saliva production caused by drug therapy (e.g., antipsychotic drugs with anticholinergic properties) or certain diseases (e.g., diabetes, HIV/AIDS, cystic fibrosis), increase the risk for sialadenitis. The parotid gland appears to have the highest risk for infection. Furthermore, stone development within the ductal system (sialolithiasis) can lead to physical obstruction of salivary flow. These stone are comprised usually of calcium phosphate and carbonate with glycoproteins and mucopolysaccharides. Empiric antibiotics, warm compresses, gland massage, or drugs that increase salivary flow can be used to treat the reduced saliva flow. If left untreated, an abscess may develop in the gland that may require drainage after identification by ultrasound or cross-sectional imaging (computed tomography). Stones may require surgical intervention with gland excision in the case of large or impacted stones, however, combined or endoscopic removal techniques have emerged in the last few years.

Sjogren's syndrome, a chronic autoimmune disease that affects all moisture-producing glands, leads to severe and long-lasting failure of salivation and a dry mouth. Patients suffering from Sjogren's syndrome produce antibodies that elicit immune cell-mediated damage to salivary acini and ducts, and consequently diminish the capacity of the glands to produce saliva. Patients may present with primary Sjogren's or a secondary process due to other connective tissue diseases. Patient can develop symptoms of dry mouth, odynophagia, mastication problems, dental caries, and risk for acquiring oral candidiasis in addition of salivary gland stones. Patients are encouraged to chew gum or use lozenges to moisten the mouth by increasing salivary flow. Water or ice chips should be consumed multiples times per day. Avoidance of sugary foods and acidic foods that promote dental caries should be avoided. Medical treatment with salivary stimulants may be considered. Dental products marketed as toothpaste, mouthwash, or gels are also available that may moisten the mouth.

Deglutition

Given the complex and highly coordinated nature of the events involved in deglutition, it is not surprising that difficulty in swallowing (dysphagia) is a common consequence of diseases of the medulla oblongata, where the swallowing center is located. Oropharyngeal dysphagia is associated with certain neurological disorders such as multiple sclerosis, Parkinson's disease, and muscular dystrophy. Similarly, brain injury resulting from a stroke can impair swallowing. From an anatomical standpoint, a lesion that obstructs the oropharynx, hypopharynx, or upper esophagus must be excluded before considering physiological etiologies for dysphagia. Strictures may develop from neoplasms, prior surgery, injury from ingestions of caustic agents, or congenital deformities. Although swallowing disorders are evident in all age groups, the incidence of dysphagia is much higher in the elderly.

Early in the swallowing sequence the nasopharynx is closed to prevent regurgitation of food into the nasal passages. This is achieved by the action of *levator veli palatini*, a muscle innervated by the vagus nerve. This muscle, along with the tensor veli palatine and palatoglossus muscles, lifts the soft palate upward and backward, closing off the nasal passages. Many of the other small pharyngeal muscles involved in swallowing are also innervated by the vagus. Damage to the motor nucleus or fibers of one or both vagi will cause dysphagia, particularly for liquid foods. A prominent and very distressing feature of this dysfunction, particularly when both vagus nerves are involved, is nasal regurgitation of liquids. Difficulty in swallowing liquid food in such patients is a common cause of choking and aspiration.

Historically, patients with bulbar poliomyelitis, now prevented by the polio vaccine, would develop dysphagia due to weakness in the pharyngeal muscles leading to a high risk of death. In patients with amyotrophic lateral sclerosis (ALS), a progressive neurologic disease with degeneration of the upper and lower motor neurons throughout the nervous system, swallowing difficulties are common, particularly when the trigeminal, facial, hypoglossal, glossopharyngeal, or vagus cranial nerves are involved. The temporal sequence of muscle weakness during the progression of ALS usually follows a pattern of tongue and lips, followed by palatal, jaw, and pharyngeal muscles, and then facial, upper trunk, and laryngeal muscles. ALS patients experience difficulty chewing and swallowing. Initially, solid food is difficult to manage, but over time even pureed foods and saliva become difficult to swallow. Consequently, an alternative feeding mechanism, such a percutaneous gastrostomy tube (direct tube from the skin to the gastric lumen), may be required to maintain nutrition.

REFERENCES

Catalan MA, Amatipudi KS, Melvin JE. Salivary gland secretion. In: Johnson LR, Ghishan FK, Kaunitz JD, Merchant JL, Said HM, Wood JD. Physiology of the Gastrointestinal Tract, 5th Edition, 2012, Chapt. 45, pp 1229–1249.

Jackson NM, Mitchell JL, Walvekar RR. Inflammatory disorders of the salivary glands. In: Flint PW, Haughey BH, Lund V, et al (eds), Cummings Otolaryngology, 6th edition, 2015, Chapter 85. Philadelphia, PA: Elsevier Saunders.

Kahrilas PJ, Pandolfino JE. Esophageal neuromuscular function and motility disorders. In: Sleisenger and Fordtran's Gastrointestinal and Liver Disease, 2010, 9th Edition, Chapt. 42, pp 677–704.

Lee MG, Ohana E, Park HW, Yang D, Muallem S. Molecular mechanisms of pancreatic and salivary gland fluid and bicarbonate secretion. *Physiol Rev* 92: 39–74, 2012. doi: 10.1152 /physrev.00011.2011.

Mirowski GW, Leblanc J, Mark LA. Oral Disease and Oral-Cutaneous Manifestations of Gastrointestinal and Liver Disease, In: Sleisenger and Fordtran's Gastrointestinal and Liver Disease, 2016, 10th Edition, Chapt. 43, pp 377–396.

Pandol SJ, Raybold HE, Yee, H. Integrative responses of the gastrointestinal tract, pancreas, and liver to a meal. In: Podolsky DK, Camilleri M, Fitz JG, Kalloo AN, Shanahan F, Wang TC (editors). Yamada's Textbook of Gastroenterology, 6th edition, 2016, Chapt. 12, pp 185–197. https://doi.org/10.1002/9781118512074.ch12

Proctor GB. The physiology of salivary secretion. *Periodontol* 2000, 2016;70:11–25. doi: 10.1111 /prd.12116.

CHAPTER 3

The Esophagus

INTRODUCTION

The essential function of the esophagus is to transfer solids and liquids from the pharynx to the stomach. Although its main activity is related to eating and drinking, the esophagus occasionally functions to transfer gas (eructation) and gastric contents (vomiting) from stomach to mouth. The esophagus has no storage function and is normally empty and collapsed when not transferring material to the stomach. The esophagus does not have any digestive or absorptive abilities, and its only secretions are mucus, which serves to lubricate the food bolus as it passes to the stomach, and bicarbonate, which protects the esophagus against refluxed gastric acid.

Both secretions originate from submucosal glands that are distributed as clusters in the most proximal and distal ends of the esophagus.

ANATOMIC CONSIDERATIONS

The esophagus is a hollow tube extending from the pharynx to the stomach. It is approximately 20 to 25 cm long in the adult human and its length is not related to an individual's height. At rest, it is empty and collapsed, its transverse diameter being approximately 3 cm and its anteroposterior diameter 2 cm. It is maintained empty by virtue of its two sphincters, the upper esophageal sphincter (UES) and the lower esophageal sphincter (LES) (Figure 3-1). Both sphincters are defined operationally; i.e., they represent zones of high intraluminal pressure. The UES extends for about 3 cm at the junction of the pharynx and esophagus and appears to correspond to the cricopharyngeal muscle and the inferior pharyngeal constrictor. As the esophagus passes through the thorax, it lies in the posterior mediastinum, running in close apposition to the trachea and left main stem bronchus. A unique feature of the esophagus is that it is comprised of both skeletal muscle and smooth muscle (Figure 3-1). The upper part (~4 cm) of the esophagus is entirely striated muscle, whereas the lower part (approximately 11 cm) is entirely smooth muscle. The middle transition zone is comprised of both muscle types. The UES is composed entirely of skeletal muscle, whereas the LES is entirely smooth muscle. Both oxidative (slow twitch) and glycolytic (fast twitch) striated muscle fibers are found in the UES, with the oxidative fibers likely contributing to tonic contraction and glycolytic fibers to the phasic contraction of this sphincter. There is no anatomic equivalent of the lower esophageal sphincter. Like the UES, the lower sphincter is recognized as a zone of high pressure that reflects the intense tone of the muscles that extend over a few centimeters close to the level of the diaphragmatic hiatus, a tear drop-shaped canal about 2 cm long on its vertical axis. A portion of the LES lies below the diaphragm, and this part is exposed to intra-abdominal pressure, whereas the supradiaphragmatic portion is exposed to intrathoracic pressures.

There is a sharp demarcation between squamous epithelium of the esophagus and glandular mucosal epithelium of the stomach, the squamocolumnar junction. In the collapsed esophagus, one can recognize this junction at endoscopy as a serrated line of demarcation, which straightens out as the esophagus is distended. The esophageal mucosa appears as a pink featureless membrane with no obvious blood vessels. The mucosa consists of stratified squamous cells that account for up to 10 per cent of the total thickness of the layer. Extensions of the lamina propria into the epithelium are also seen. Some mucous glands are found in the connective tissue of the lamina propria, particularly at the upper and lower ends of the esophagus. The ducts of these glands extend up

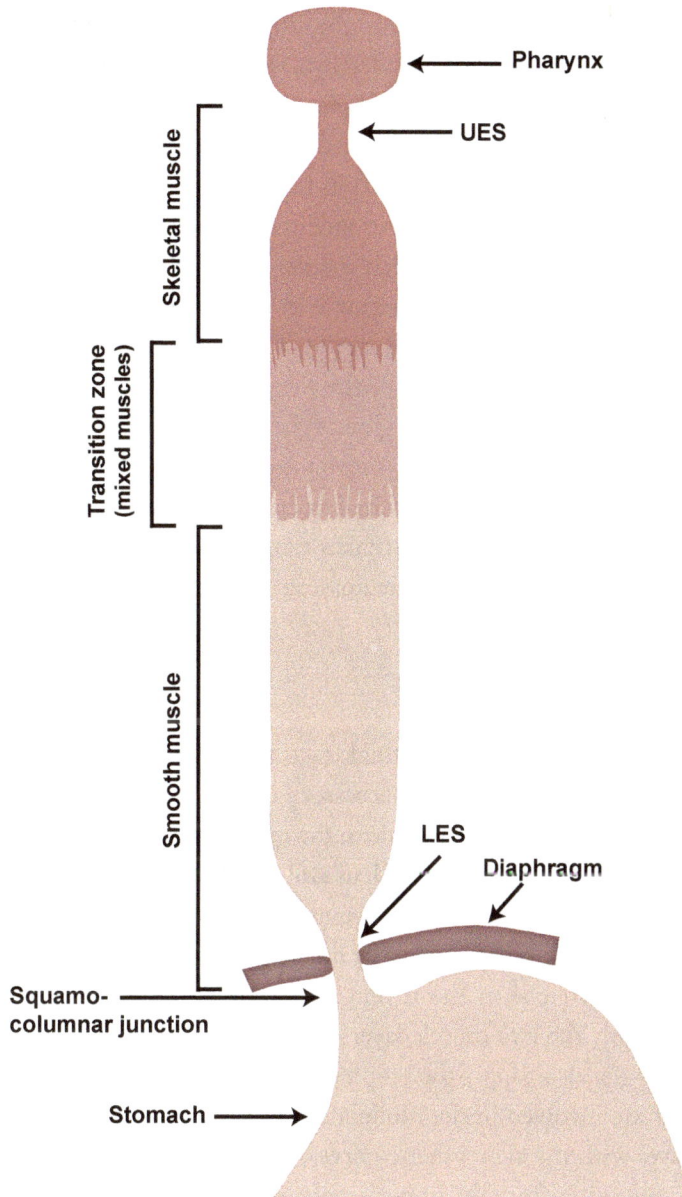

FIGURE 3-1: Anatomical features of the esophagus. The upper (UES) and lower (LES) esophageal sphincters control the flow of material into and out of the esophagus. The upper region of the esophagus is comprised of skeletal muscle and the lower region is comprised of smooth muscle, with a transition zone between the two regions that consists of both muscle types. The LES lies both above and below the diaphragm, which are exposed to intrathoracic and intra-abdominal pressures, respectively. The squamocolumnar junction is the point of transition between the squamous epithelium of the esophagus and columnar epithelium of the stomach.

through the squamous epithelium to empty into the esophageal lumen. The acinus-like structure of the esophageal submucosal gland is surrounded by a ring of myoepithelial cells, which can contract to expel the glandular secretion (mucus, HCO_3^-) into the lumen. Unlike the remainder of the GI tract, ganglion cells of the submucosal plexus are very sparse in the human esophagus.

Beneath the lamina propria is the very thin (2–3 cells thick) muscularis mucosa, which separates the mucosa from the submucosa. An inner circular and an outer longitudinal muscle layer lie beyond the submucosa, each of which is several cells (0.75 mm) thick. The longitudinal muscle layer is thicker than the circular muscle in the proximal part of the esophagus, but the two layers are equally thick in the distal region. The myenteric nerve plexus lies between these two muscle layers. The ganglion cells of the myenteric plexus are more numerous in the smooth muscle region than in the striated region of the esophagus. Since the submucosa is very loose, the mucosa of the esophagus is free to move to some extent in relation to the esophageal muscle. Unlike other areas of the gastrointestinal tract, which have a peritoneal covering, there is no serosal coat on the muscle layer. This makes esophageal surgery somewhat difficult; the peritoneum is very useful in helping to secure anastomoses in the stomach and intestine.

Innervation

Extrinsic

As with much of the rest of the gastrointestinal tract, the most important nerve supplying the esophagus is the vagus, although the spinal accessory nerve supplies the high cervical esophagus. Vagal efferents that supply striated muscle in the upper esophagus are axons of lower motor neurons with cell bodies that reside in the nucleus ambiguous (Figure 3-2). These fibers directly terminate on motor endplates on the striated muscle fibers and release acetylcholine, which stimulates nicotinic cholinergic receptors on the motor endplates. In the distal esophagus, parasympathetic vagal fibers synapse in the myenteric plexus, from which short postganglionic nerve fibers arise to supply the two muscle layers. Sympathetic nerves arising from the cervical and thoracic chain ganglia also supply the esophagus. These fibers do not terminate directly on the muscle fibers but are involved in the modulation of myenteric neurons. Myelinated fibers, which primarily travel with the sympathetic nerves, carry sensory information from the lower esophagus, whereas sensation from the upper esophagus is carried by parasympathetic fibers.

Intrinsic

Enteric neurons that reside between the circular and smooth muscle layers (myenteric plexus) contribute to the control of peristalsis in the esophagus, particularly in the smooth muscle region. These neurons receive autonomic input from both parasympathetic and sympathetic

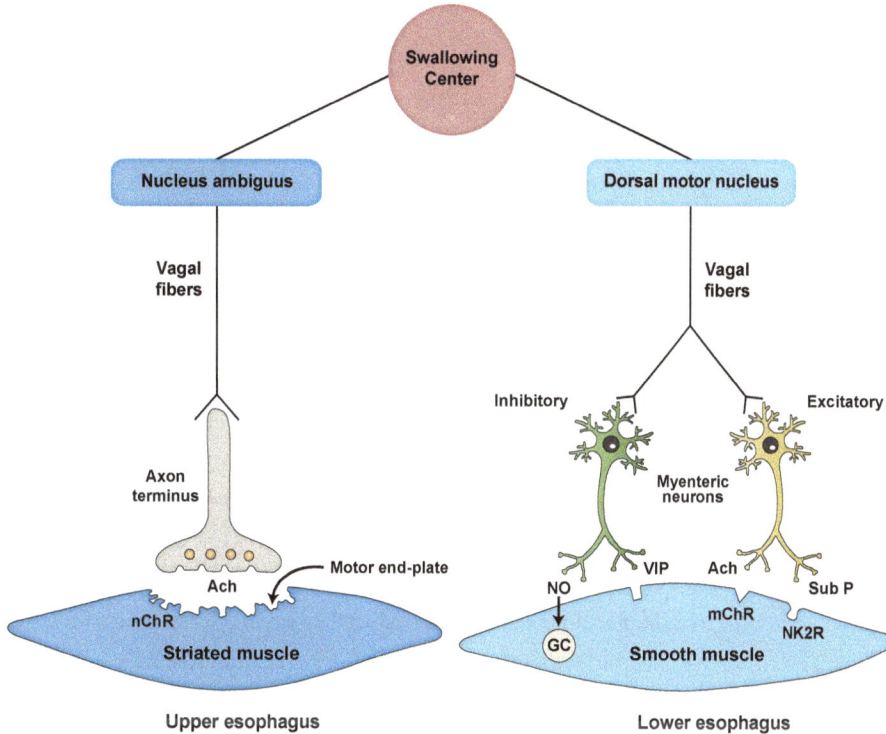

FIGURE 3-2: Differences in the mode of innervation and regulation of the striated and smooth muscle regions of the esophagus. In the striated muscle region, peristaltic contractions are controlled by the swallowing center via the nucleus ambiguus and result from the sequential (oral to gastric) activation of motor units. Cholinergic fibers release acetylcholine (Ach) to stimulate nicotinic cholinergic receptors (nChR) on the motor endplates of the striated muscle cells. In the smooth muscle region, which is activated by the swallowing center via the dorsal motor nucleus, vagal fibers synapse with inhibitory and excitatory myenteric neurons. Here, peristalsis results from a coordination of proximal excitatory neurons that mediate muscle contraction via Ach and substance P (Sub P) and distal inhibitory neurons that mediate muscle relaxation through nitric oxide (NO) and vasoactive intestinal peptide (VIP). GS, guanylyl cyclase; mChR, muscarinic cholinergic receptor; NK2R, neurokinin-2 receptor.

nerves, directly innervate the smooth muscle cells, and communicate with one another. The myenteric plexus in the smooth muscle esophagus exhibits two major types of neurons, excitatory and inhibitory (Figure 3-2). The excitatory neurons release acetylcholine and substance P, whereas nitric oxide and VIP are released by the inhibitory neurons. Enteric neurons also innervate motor end plates on some of the striated muscle fibers in the upper esophagus. These neurons release nitric oxide, CGRP, and other peptides that exert an inhibitory influence on skeletal muscle contraction.

ESOPHAGEAL FUNCTION

Two methods that are routinely used for studying esophageal activity are (1) cineradiography, which follows the transit of a swallowed bolus of barium through the esophagus, and (2) intra-luminal pressure recording by manometry. The latter method uses a tube that is either fluid-filled or solid with small openings or pressure sensors (transducers) that is positioned in the esophagus. As a peristaltic wave moves along the esophagus, the accompanying changes in intraluminal pressure are detected by the intermittently spaced sensors. The pattern of pressure development that is detected along the esophagus reveals both the baseline esophageal muscle tone at specific sites within the organ as well as the speed of propagation of a peristaltic wave (Figure 3-3). With the development of high-resolution manometry (HRM), pressure waves generated within the esophagus can be visualized topographically. This technology uses a computer algorithm that interpolates the pressures detected between closely spaced transducers and presents the data as colorized pressure plots along the entire length of the esophagus (Figure 3-4). These plots reveal how an entire segment (rather than a discrete point) of the esophagus contracts at any given moment once a peristaltic wave is initiated. For example, with HRM, it has been shown that the length of the contracted segment increases as peristalsis progresses distally and may reach over 15 cm.

The Interdigestive Period

At rest, manometric measurement of pressures in the upper esophageal sphincter zone (3 to 4 cm) gives values in a man that range between 40 and 200 mm Hg. Electrical recordings from the UES in the opossum, an animal whose esophagus resembles that of man, shows continuous electrical spike activity at rest. The basal electrical activity is responsible for the sustained muscle contraction and high intraluminal pressures at the UES. The tonically contracted UES normally functions to prevent air from entering the esophagus during breathing and prevents the reflux of esophageal contents into the pharynx to guard against aspiration into the airways. Tonic discharges of the motor neurons of the nucleus ambiguous (via vagal efferent fibers) account for the high resting tone of the striated muscles that comprise the UES. Vagal stimulation causes contraction of UES muscles, whereas vagal section abolishes their activity, indicating that parasympathetic nerve activity contributes significantly to the resting tone of UES muscle.

The body of the esophagus at rest does not exhibit spontaneous contractions and the intraluminal pressure is related to respiration, being −5 to −15 mm Hg during inspiration and −2 to +5 mm Hg during expiration. Because intraesophageal pressure correlates well with intrapleural pressure, pulmonary physiologists use esophageal pressure as an indirect measure of in-

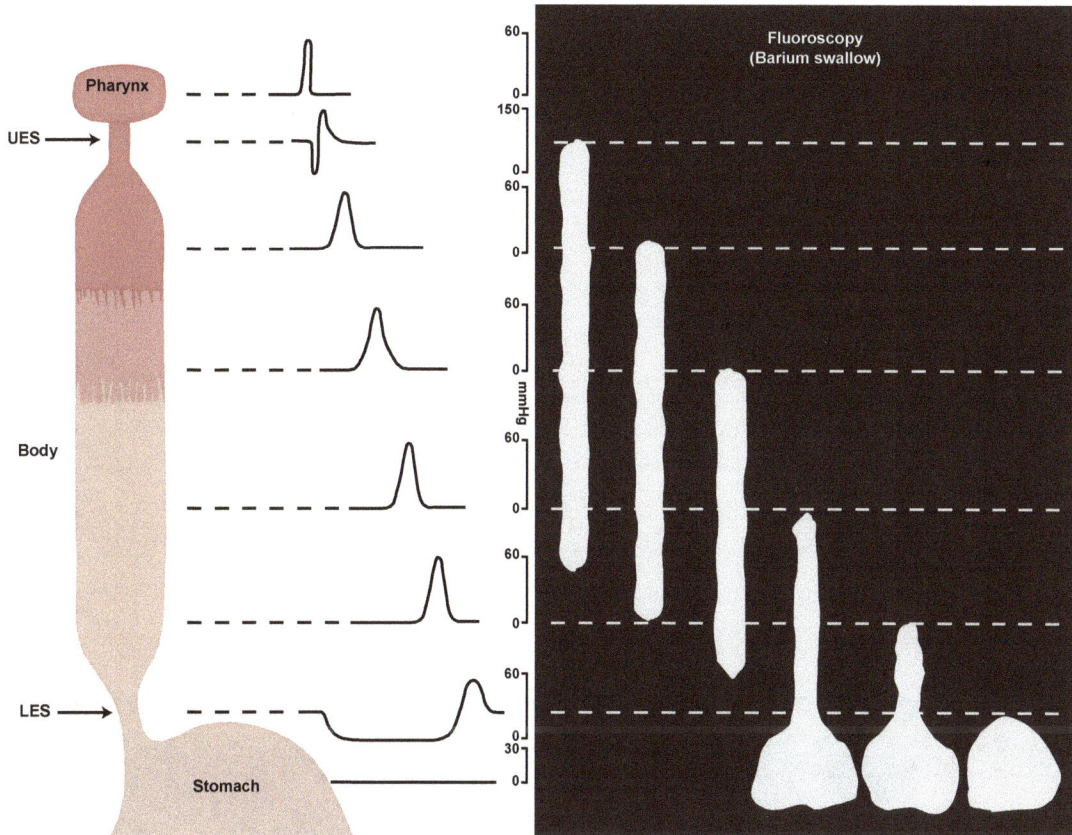

FIGURE 3-3: Manometric recordings (left panel) of the pharynx, esophagus and stomach that depict the movement of a primary peristaltic wave after swallowing. The simultaneous movement of the swallowed barium bolus (detected by fluoroscopy) is shown in the right panel. Note the downstream movement of the pressure wave, the coordinated opening of the LES, and that the passage of the tail of the barium bolus corresponds with the onset of the detected pressure wave at each site.

trapleural pressure. The position and pressures of the lower esophageal sphincter (LES) region are also affected by respiration. If one gradually withdraws a manometric catheter from the stomach (Figure 3-5), the pressure rises from normal intragastric levels of 4 to 10 mm Hg to approximately 15 to 35 mm Hg over a distance of about 4 cm, then falls to the negative values of the body of the esophagus. Pressures in the upper and lower halves of the LES are affected differently by respiration. Inspiration causes an increase in pressure in the lower half and a fall in pressure in the upper half. The effect of respiration on the measured values of LES pressure

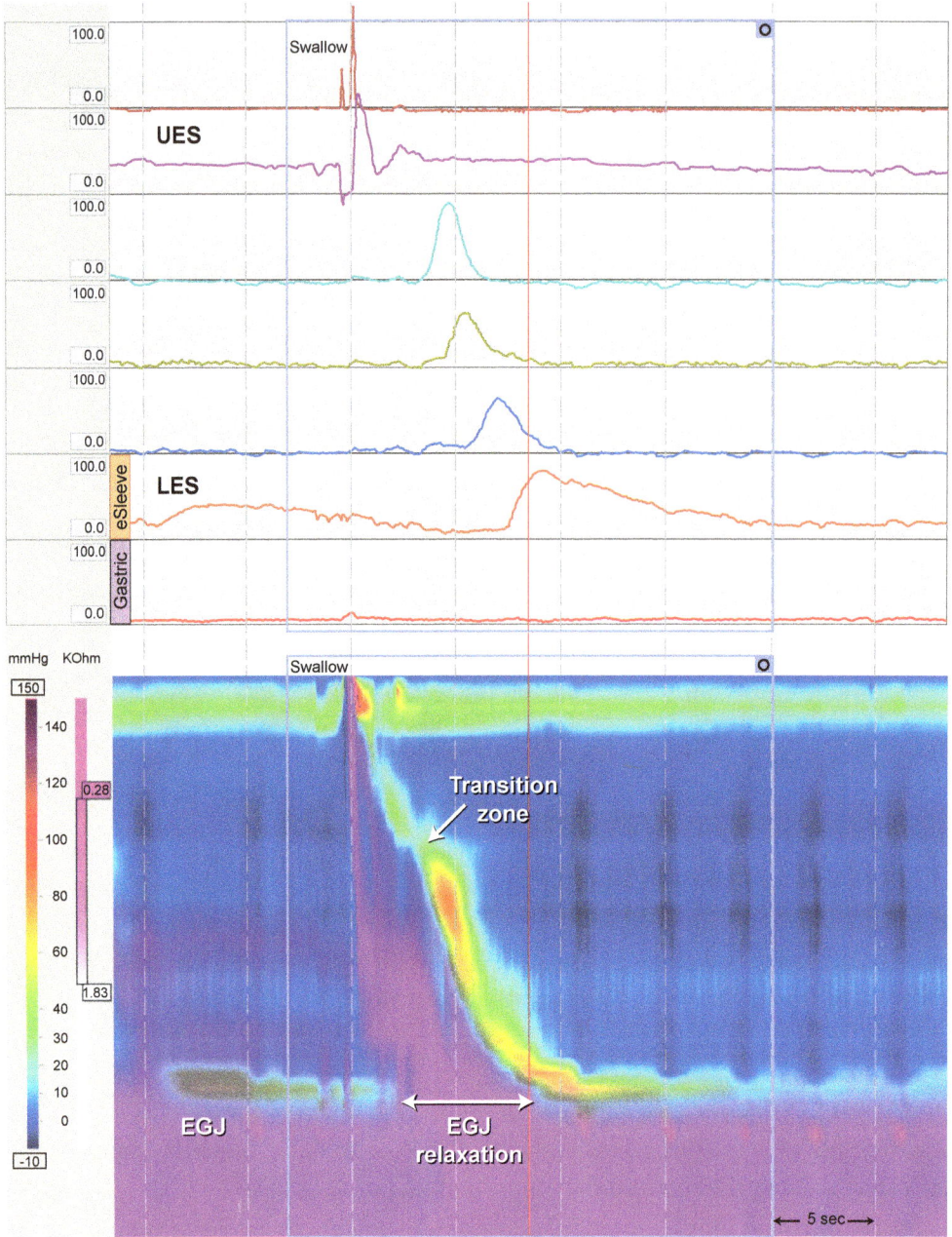

FIGURE 3-4: High-resolution manometric (HRM) analysis of the normal peristaltic sequence in the esophagus after a wet swallow. The image depicts the time-dependent changes in pressure within the upper esophageal sphincter (UES), striated and smooth muscle regions of the esophageal body, the transition zone between the two muscle regions, and lower esophageal sphincter (LES) after a swallow. The pressure topography de-

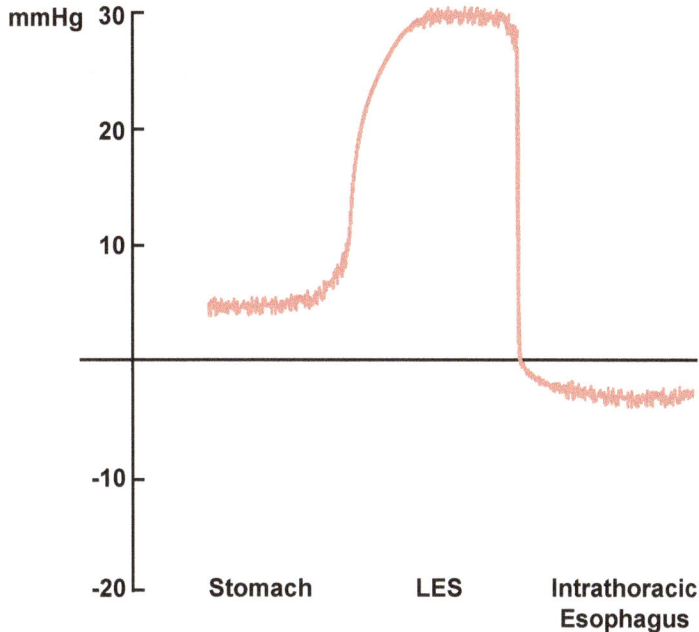

FIGURE 3-5: Pressure measured by a catheter as it is withdrawn from the stomach through the lower esophageal sphincter (LES) and into the body of the esophagus. Note the rise in pressure in the LES and the decline to a negative value in the esophagus body. The high pressure in the LES reflects the strong resting tone of the sphincter, whereas the negative pressure in the esophagus reflects the surrounding pressure in the thoracic cavity.

is understandably complex. The LES moves axially with respiration. The vertical movements of the diaphragm alter, in a cyclical fashion, the external pressures exerted on the esophagus (positive intra-abdominal pressure and negative intrathoracic pressure).

At rest, the tonically contracted LES generates an intraluminal pressure in humans that ranges between 10 and 45 mm Hg. This resting pressure is the result of intrinsic myogenic

picted in the lower panel shows bright red and yellow colors that represent higher pressures that are consistent with contraction of the smooth muscles of the distal esophagus during peristalsis. The tracing generated by the eSleeve (electronic sleeve) is a computer calculation of the highest pressures generated across the endoscopic/anatomic site of the esophagogastric junction (EGJ) and is presented in relation to the gastric pressure. The EGJ represents the point where the distal esophagus meets the diaphragmatic hiatus and the cardia of the stomach. The LES, which is tonically contracted and relaxes with the initiation of a swallow, comprises the last 3 to 4 cm of smooth muscle that is associated with the EGJ.

TABLE 3-1: Agents and Conditions that Can Alter Lower Esophageal Sphincter (LES) Pressure

	INCREASE	DECREASE
Foods	Protein meal Coffee	Chocolate Alcohol Peppermint
Risk factors	Diabetes	Smoking Obesity
GI hormones	Gastrin Motilin Substance P	Secretin Cholecystokinin Vasoactive intestinal peptide
Medications	Metoclopramide Cisapride Domperidone	Nitrates Calcium channel blockers Theophylline
Neural agents	α-adrenergic agonists β-adrenergic antagonists Cholinergic agonists	α-adrenergic antagonists β-adrenergic agonists Cholinergic antagonists

properties (tone originating from the muscle itself) of LES muscle and the influence of autonomic nerves (cholinergic regulation). In the human esophagus, excitatory cholinergic drive dominates over myogenic control of resting LES tone. The major role of cholinergic nerves is evidenced by the observation that atropine treatment (which reduces neural cholinergic drive) reduces LES pressure by as much as 70%. A wide variety of agents (Table 3-1) are known to affect LES pressure. Gastrointestinal hormones have the potential to either increase (eg, gastrin, substance P) or decrease (secretin, CCK) LES pressure. Similarly, some of the risk factors for cardiovascular disease and a variety of medications are known to alter LES tone. It can be seen from Table 3-1 that common after-dinner habits of smoking, drinking alcohol, and eating chocolate or peppermint all lower LES pressure to promote belching and gastroesophageal reflux.

The Ingestive Period

Swallowing is a highly complex set of events controlled by the "swallowing center" in the medulla oblongata and pons. Swallowing begins at 12 weeks intrauterine life. Normal adults swallow about 600 times per day, 350 times while awake, 50 times during sleep, and the remaining 200 times accompany eating and drinking. Swallowing frequency is significantly reduced in the elderly. Inasmuch as swallowing and breathing share the same anatomical space (e.g., pharynx), proper nutrition and prevention of pulmonary aspiration requires fine temporal coordination of swallowing and breathing when consuming a meal. This coordination depends on the interplay between medullary centers controlling swallowing and respiratory activity. Swallowing usually starts during the expiratory phase of breathing, and respiration resumes with continued expiration after swallowing. The pause in breathing during swallowing is not simply due to closure of the airway, but reflects an inhibition of respiration at neural control centers in the brainstem.

With each swallow, the UES relaxes; its pressure may fall to subatmospheric levels, although not as low as esophageal values. The period of relaxation usually lasts 0.5 to 1.0 second and is accompanied by a brief cessation of spike activity in the inferior pharyngeal constrictor and cricopharyngeal muscles (Figure 3-6). The extent of UES opening during a swallow is also determined by the volume of the food bolus and the resulting distension pressure. This brief cessation is due to central inhibition (cessation of motor neuron discharges) rather than the activity of intrinsic inhibitory nerves, thus contrasting with the relaxation of the LES (see below).

In the body of the esophagus, the major motor event associated with swallowing is *primary peristalsis*, which is the continuation of oral and pharyngeal swallowing movements. Once the food bolus passes the UES and the sphincter assumes its resting tone, a ring of contraction is elicited that creates a wave of high pressure (lasting 2 to 4 seconds) that travels the length of the esophagus at a speed of 3 to 6 cm per second, with the greatest velocity achieved in the mid-esophagus (Figure 3-3). The velocity of peristalsis with warm fluids is greater than with cold ones. The pressures generated by the peristaltic wave range between 35 and 70 mm Hg, with higher pressures detected in the upper and lower esophagus, whereas the lowest values are recorded in the region where the striated muscle blends with smooth muscle. The peristaltic wave does not pass through the LES, it reaches approximately 6 seconds after swallowing begins. If a bolus of barium is swallowed, the peristaltic wave follows the tail of the bolus, that is, it propels the bolus forward (see Figure 3-3).

Peristalsis progresses in a seamless fashion, despite separate control mechanisms, from the skeletal to smooth muscle esophagus (Figure 3-2). In the striated muscle region of the esophagus, peristaltic contractions are controlled by the swallowing center via the nucleus ambiguus and result from the sequential (oral to gastric) activation of motor units. Cholinergic

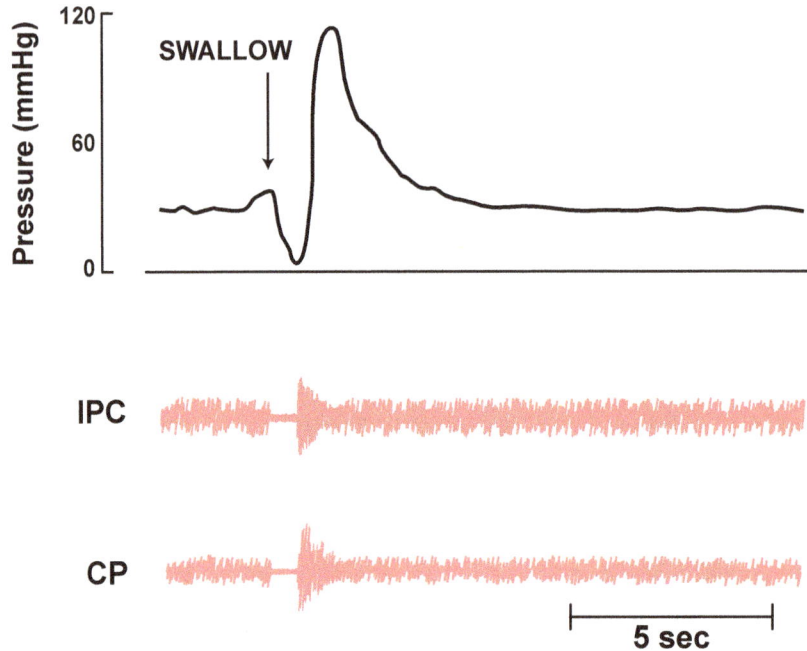

FIGURE 3-6: Simultaneous recordings of pressure in the upper esophageal sphincter (UES) and spike activity of the inferior pharyngeal constrictor (IPC) and cricopharyngeal (CP) muscles of the UES during a swallow. Note that the brief period of UES relaxation is associated with a corresponding cessation of spike activity in IPC and CP muscles. A period of enhanced UES pressure and increased IPC and CP spike activity is seen after the swallow.

fibers release acetylcholine to stimulate nicotinic cholinergic receptors on the motor endplates of the striated muscle cells. The critical role of vagus input in this process is evidenced by the observation that bilateral vagotomy results in paralysis of the striated muscle esophagus. Vagal control of peristalsis in esophageal smooth muscle is more complex and relies on vagal fiber synapses of myenteric neurons, rather than directly on the myocytes, to exert an influence on muscle tone. Here, the peristaltic wave results from a coordination of proximal excitatory neurons that mediate muscle contraction via acetylcholine and substance P and distal inhibitory neurons that mediate muscle relaxation through nitric oxide and VIP (Figure 3-2).

The LES relaxes in conjunction with swallowing (Figure 3-3). The relaxation may begin at the start of deglutition but generally does not occur until about 2 seconds after the initiation of swallowing. A low viscosity bolus swallowed in the upright position may reach the LES with the help of gravity before relaxation has begun, and a transient delay results before the bolus enters the stomach. Relaxation of the LES is a much longer event (5 to 10 seconds) than that of the UES. Like the UES, relaxation is accompanied by a cessation of the continuous spike activity, which occurs at rest. After relaxation, the upper half of the LES develops an after-contraction of 5 to 10 seconds, whereas the lower half returns directly to the resting pressure level.

The LES relaxation elicited by swallowing is a neurally mediated (vagal) reflex. Unlike relaxation of the UES, LES relaxation is not due to the central inhibition of ongoing activity in the central nervous system, but due to the release of an inhibitory neurotransmitter at the neuromuscular junction by myenteric neurons. Nitric oxide is the major inhibitory neurotransmitter that mediates LES relaxation during deglutition. However, NO may not act alone. In addition to acting directly on smooth muscle, NO can stimulate the release of VIP, which is another potent stimulus for LES relaxation, from presynaptic nerve terminals. After the food bolus enters the stomach, the LES regains its resting tone when the excitatory neurotransmitter acetylcholine dominates over nitric oxide.

Secondary peristalsis is the term applied to the peristaltic waves that are initiated in the esophagus by afferent impulses from the esophagus itself. Unlike primary peristalsis, secondary peristaltic waves are not the consequence of the oropharyngeal movements of swallowing. They create no appreciable sensation. Under physiological conditions, secondary peristalsis is observed with either a bolus retained in the esophagus or with distension associated with gastroesophageal reflux. In both conditions, a wave of contraction begins above the bolus and proceeds distally to propel the bolus into the stomach. Regions of the esophagus above the bolus, pharynx, or oral cavity are not involved in secondary peristalsis. Experimentally, one can elicit secondary peristalsis by distending a balloon in the esophagus. This motor response appears to be responsible for clearing the esophagus of retained food material or refluxed gastric content.

Under certain circumstances, the oropharyngeal movements of swallowing can be dissociated from peristaltic activity in the body of the esophagus. During drinking, multiple swallows occur in rapid succession (~1 per second). Esophageal peristalsis is inhibited ("deglutitive inhibition") by the sequence of closely spaced swallows. In addition, the LES remains relaxed during drinking, presumably as a result of inhibitory neuron discharge. The last swallowing movement during drinking, however, results in a peristaltic wave that empties the esophagus; then the LES resumes its resting pressure. During the guzzling of fluid, such as beer in a bar or a sports drink after a marathon, the process of fluid transfer through the esophagus is much more efficient than drinking. With this activity, the head and neck are extended, the larynx and pharynx descend, and both sphincters remain open. A constant column of fluid is maintained, and this pours into the stomach under gravity. Rapid to and fro movements of the tongue and soft palate transfer the fluid from mouth to pharynx. As with drinking, a final peristaltic wave clears the esophagus.

ESOPHAGEAL MOTOR EVENTS ASSOCIATED WITH RETROGRADE TRANSPORT

While the dominant motor patterns (primary and secondary peristalsis) of the esophagus serve to move food, liquids, and air from mouth to stomach, there are other distinct esophageal motor patterns that serve to carry stomach contents toward the mouth. The latter motor events are observed with belching, regurgitation, and vomiting (discussed in Chapter 4). These events begin with spontaneous and transient relaxation of the LES. The transient LES relaxation that accompanies the retrograde motor patterns is significantly longer (>10 seconds) than swallow-induced LES relaxation (<10 seconds), and the esophageal body remains relatively quiescent, whereas the LES is relaxed. Belching (eructation), the venting of intragastric air via the esophagus, is a normal physiologic event that occurs an average rate of 30 times per day in the healthy human adult. Distension of the stomach by intragastric air triggers spontaneous transient relaxation of the LES, which allows gastric air to flow into the esophagus. The increased pressure and distension of the proximal esophagus then results in UES relaxation and the subsequent flow of air from the esophagus into the oral cavity. Some patients, especially those with dyspeptic symptoms (upset stomach), seek medical attention because of excess belching, which may be part of a spectrum of complaints in patients with gastroesophageal reflux disease (GERD).

Regurgitation or "spitting up" is the most common form of retrograde esophageal transport observed in infants. This is considered a normal physiological phenomenon because 70% to 85 % of infants regularly regurgitate, many (up to 30) times per day, within the first 2 months of life. The regurgitation generally occurs postprandially and is commonly caused by overfeeding or the accumulation of air during feeding. As noted with belching, the gastric distension that results from infant feeding triggers transient LES relaxation, the flow of milk into the esophagus, upper esophageal distension, UES relaxation, and ultimately regurgitation. The condition resolves without intervention in about 95 % of infants by 1 year of age, likely reflecting the higher LES pressure (increased sphincter tone) that accompanies maturation.

MECHANISMS PREVENTING GASTROESOPHAGEAL REFLUX

Some degree of gastric acid reflux into the esophagus occurs in all individuals. However, in the absence of effective gastroesophageal anti-reflux mechanisms, abnormal amounts of gastric acid and pepsin are allowed to enter the esophagus, where the acid/pepsin can cause mucosal damage and possibly elicit a vicious cycle of events that can perpetuate the GERD state (Figure 3-7, the pathophysiology of GERD is discussed below). Fortunately, there are some protective mechanisms that exist to reduce the incidence of gastroesophageal reflux. Such anti-reflux mechanisms are needed in light of the fact that the pressure in the body of the esophagus is subatmospheric, whereas the pressure in the stomach is 5 to 10 mm Hg. A unique anatomical feature of the proximal stomach that creates a barrier to acid reflux is a "mucosal flap valve," which reflects mucosal folds at the cardioesophageal junction. Similarly, the acute angle of insertion of the esophagus into the stomach (angle of His) is also considered to function as an anti-reflux mechanism. Of prime importance in the prevention of acid reflux is the sphincter actions exerted by the LES as well as the diaphragm. The diaphragm, through which the

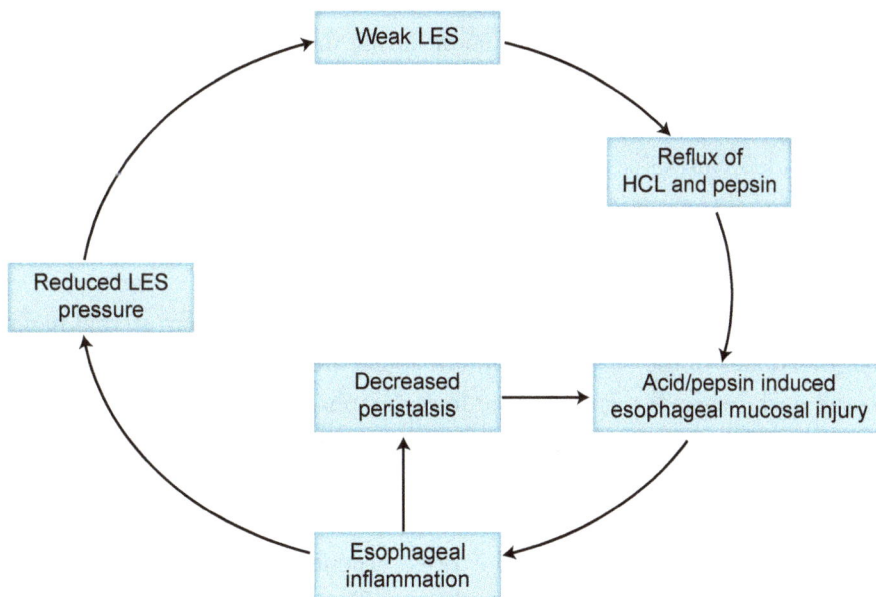

FIGURE 3-7: Vicious cycle of events that initiate and maintain erosive gastroesophageal reflux with esophagitis.

esophagus passes at the hiatus, can function as an "external sphincter" by compressing the gastroesophageal junction and increasing LES pressure by 10 to 100 mm Hg during inspiration. When the two sphincter mechanisms (LES and diaphragm) generate an LES pressure that is less than intragastric pressure, acid reflux occurs.

There are other mechanisms intrinsic to the esophagus or stomach that also serves to minimize either the occurrence of, or esophageal mucosal injury response to, gastroesophageal reflux. Secondary peristalsis can be regarded as a second line of defense that helps to clear the esophagus of refluxed material. The bicarbonate- and mucin-rich secretions of the esophageal submucosal glands, as well as the HCO_3^- in saliva, also protect against refluxed acid. Adding to the resistance of the esophageal mucosa to acid-induced injury are the thick (20–30 cells) squamous epithelial cell layer and the subepithelial blood vessels, the combination of which minimize the entry and accumulation of H^+ ions in the mucosal interstitium.

The phenomenon of receptive relaxation of the stomach, which allows relatively large volumes of food and drink to be accommodated with small increases in intragastric pressure, serves to minimize the development of a gastroesophageal pressure gradient that favors acid reflux. Also, as intragastric pressure rises, the resting LES pressure rises, probably as a result of a neural reflex. When intra-abdominal pressure increases, unlike intragastric pressure, the force is applied both to the stomach and to that segment of the LES, which lies below the diaphragmatic hiatus. For example, in deep inspiration or during the Valsalva maneuver (forced expiration against a closed glottis), the pressure gradient between the stomach and the sphincter is unchanged, thus preventing reflux.

ESOPHAGEAL PAIN

Pain from the body of the esophagus is characteristically felt in the retrosternal region and, if severe enough, radiates to the back in the interscapular area. From the upper esophagus, pain radiates to the neck, whereas pain from the lower esophagus is experienced in the region of the xiphisternum. Heartburn is a retrosternal burning sensation that is usually attributed to esophageal reflux of gastric acid. It is the most common symptom associated with esophageal dysfunction, with over 40% of the US population experiencing heartburn at some point. It is particularly frequent in late pregnancy and is often experienced after large meals, particularly in the recumbent or bending position. In heartburn sufferers, the discomfort can be reproduced by perfusing the esophagus with 0.1 N HCl (the Bernstein test), which is used to answer the question: are the patient's symptoms due to reflux esophagitis? The sensation of heartburn is the result of stimulating specific acid-sensitive nerve endings located in the mucosa and lamina propria that carry the sensation of heartburn to the central nervous system. Vanilloid and anion-sensing ion channel receptors, which are present on the acid-sensitive nerve endings, are likely involved in transducing the acid-induced pain sensation.

PATHOPHYSIOLOGY AND CLINICAL CORRELATIONS

Gastroesophageal Reflux Disease (GERD)

Gastroesophageal reflux disease (GERD) is attributed to esophageal injury resulting from repetitive reflux of gastric acid and is very common in the Western world and in affluent societies. Complications of GERD can include esophagitis (small breaks in the esophageal mucosa), stricture (narrowing of the esophagus by scar tissue) formation, bleeding, Barrett's esophagus (a premalignant condition—called metaplasia—in which the squamous mucosa is replaced by specialized columnar epithelium), and esophageal cancer. The detection of esophagitis and other complications can be achieved by endoscopic evaluation. Primary empiric treatment of GERD symptoms can include weight reduction, smoking cessation, avoiding foods that may trigger symptoms, and elevating the head of the bed. However, these strategies may only provide modest benefit for symptom relief. Medications that inhibit gastric acid secretion, such as histamine (H_2 receptor) blockers and proton pump inhibitors (PPIs), help to relieve GERD symptoms. PPIs have been shown to be superior to H_2 blockers in GERD resolution and healing esophagitis. PPIs are relatively safe drugs, although some evidence suggests that they increase the risk of *Clostridium difficile* infection, pneumonia, hypomagnesemia, and hip fractures.

After many years of repeated acid exposure, GERD patients may develop strictures that narrow the lumen of the esophagus and offers resistance to the passage of solid materials. Common foods that are problematic are dry breads, chicken, steak, or other solid cuts of meat. An esophageal lumen diameter less than 13 mm is a clinical predictor of patients who report dysphagia. A stricture may also be a harbinger of an underlying malignant process. Commonly, GERD patients will point at or below the level of a stricture when attempting to describe the point where food sticks in the esophagus.

There are a number of factors that are important for the development of GERD. The most fundamental among this list are gastric acid production, the function and competence of the LES, and the integrity of the esophageal mechanisms for acid clearance and mucosal defense. Likely initiating events in the development of GERD (Figure 3-7) is dysfunction (weakening) of the anti-reflux barriers (LES, diaphragmatic sphincter). The weakened anti-reflux barriers, coupled to an increased frequency of transient LES (non-swallow associated) relaxations, allow small increments in intragastric pressure to drive gastric contents into the esophagus. Along with hydrochloric acid, gastric contents include pepsin and possibly bile as well as pancreatic enzymes from the duodenum. The combination of acid plus pepsin and/or bile is potentially more injurious to the esophageal mucosa than acid alone. With repeated and prolonged exposure to gastric contents, esophageal mucosal injury and erosion can occur, leading to inflammation (esophagitis). The presence of inflammatory cells in the mucosa can then lead to more tissue damage and result in a hypomotile state (decreased peristalsis) of the esophageal body as well as a further reduction in LES pressure (tone). Alterations in both cholinergic excitatory and nitrergic inhibitory neurotransmission pathways appear to account for the inflammation-induced motor abnormalities. The resulting impairment of esophageal peristalsis diminishes the clearance of refluxed acid and further prolongs the contact of acid with the mucosa, whereas the heightened incompetence of the LES favors more acid reflux. Hence, in the presence of mucosal injury (erosion) and inflammation, a vicious cycle is established, which can maintain and extend the GERD state (Figure 3-7).

Less than a third of GERD patients exhibit erosive esophagitis and most (60–70%) of the remaining population suffer from a non-erosive form of the disease. Although the pathophysiological distinction between erosive esophagitis and non-erosive GERD remains poorly defined, there is evidence that the frequency of acid reflux events and duration of esophageal acid exposure are lower in the non-erosive form of the disease, which may account for the minimal level of erosive mucosal injury and inflammation seen in these patients. Despite the absence of mucosal injury and inflammation, patients with the non-erosive form exhibit the classic symptoms of GERD. These patients typically present with a mildly reduced resting LES pressure and a slightly higher rate of abnormal peristaltic contractions, suggesting the existence of mildly impaired anti-reflux acid clearance mechanisms. Some patients with non-erosive disease exhibit a hypersensitivity (heartburn symptoms) to physiologic amounts of acid, suggesting a lower perception threshold for esophageal pain, compared with normal individuals. Therapy for both erosive esophagitis and non-erosive GERD is usually targeted toward controlling gastric acid secretion with PPIs, which minimize acid-induced injury and allows time for the esophageal mucosa to heal. Although PPI treatment is highly effective in reducing heartburn symptoms in patients with erosive esophagitis, patients with non-erosive disease generally show less or no symptom improvement, suggesting that the heartburn symptoms are unrelated to acid reflux.

Motility Disorders

Achalasia (a Greek word for "failure of relaxation") is an uncommon swallowing disorder that is characterized by a gradual and insidious loss of peristalsis in the lower esophagus and incomplete LES relaxation. Pressure in the LES is usually elevated, and the sphincter fails to relax adequately in response to swallowing. Organized peristaltic waves in the smooth muscle portion of the esophagus are absent. Thus, the fluid or food bolus is not propelled forward, and the swallowed material is held up by the LES. Food only enters the stomach when the column of esophageal content is sufficiently high to overcome, by gravity, the resistance offered by the LES. During the later stage of advanced cases of achalasia, the distal esophagus appears to be obstructed (bird's beak appearance on fluoroscopy) with a dilation of the more proximal esophagus. Patients develop dysphagia to solids and liquids, possible regurgitation of undigested food, and there is a risk that esophageal contents will be aspirated into the trachea. The esophageal mucosa usually shows inflammatory changes due to chronic stasis of food material. Diagnosis is made by esophageal manometry and by demonstration of the absence of both peristalsis and LES relaxation with swallowing. There are different treatment options for achalasia, including dilation of the distal esophagus with a special balloon, surgical correction with an esophageal myotomy (cutting the muscle fibers of the lower esophageal sphincter), peroral endoscopic myotomy (POEM) using a submucosal tunneling method with endoscopic submucosal cutting of the muscle fibers, or injection of Botulinum toxin directly into the sphincter to cause relaxation.

The underlying pathophysiologic basis for the development of achalasia remains poorly understood. However, there is evidence suggesting that the disorder results from an autoimmune response or viral infection that ultimately leads to the degeneration and loss of inhibitory (NO-producing) ganglia in the myenteric plexus. The consequences of this neurodegenerative process are a clear reflection of the critical role that NO plays in the propagation of a peristaltic wave and LES relaxation after swallowing. Support for the importance of NO in this disease process comes from manometric studies in human subjects and experimental animals that demonstrate an esophageal motility pattern that closely mimics achalasia after inhibition of the NO-producing enzyme, nitric oxide synthase. In the initial stages of achalasia, the neurodegeneration is limited to the inhibitory neurons, which results in an unopposed action of excitatory neurotransmitters, such as acetylcholine, leading to high-amplitude non-peristaltic contractions (vigorous achalasia). This is followed by a progressive loss of excitatory cholinergic neurons that accounts for the dilation and low-amplitude simultaneous contractions in the esophageal body that is seen in "classic" achalasia.

Another esophageal motility disorder that can cause difficulty in swallowing is *diffuse (or distal) esophageal spasm* (DES). The esophageal motor pattern that is characteristic of DES includes high-amplitude, repetitive, simultaneous, and/or prolonged contractions of the body of the esophagus. This disorder is not life-threatening and can be treated pharmacologically with calcium channel blockers or antidepressants. DES, sometimes triggered by very hot or cold fluids, is characterized by periodic attacks of severe retrosternal pain due to strong uncoordinated contractions of the esophagus (sometimes called tertiary contractions). The patient may become pale and sweaty, and the pain may radiate to the neck, face, and arms, all characteristics of myocardial ischemia. The fact that DES and myocardial ischemia may coexist in an older patient, and that nitroglycerine may also relieve the pain of DES, can create a common diagnostic problem in view of the differences in management and prognosis of these two diseases. Esophageal manometry is a useful diagnostic tool in this situation.

REFERENCES

Conklin J, Pimentel M, Soffer E. Color Atlas of High Resolution Manometry. 2009. Springer.

Diamant NE. Pathophysiology of gastroesophageal reflux disease. Part 1: Oral cavity, pharynx and esophagus. *GI Motility* online (2006) doi:10.1038/gimo21

Fass R, Dickman R. Nonerosive reflux disease. Part 1: Oral cavity, pharynx and esophagus. *GI Motility* online (2006) doi:10.1038/gimo42

Katz P, Gerson L, Vela M. Diagnosis and management of gastroesophageal reflux disease. *Am J Gastroenterol* 2013; 108:309–328.

Mittal R. Motor function of the pharynx, esophagus, and its sphincters. In: Colloquium Series on Integrated Systems Physiology (DN Granger & JP Granger, eds.) Morgan & Claypool Publishers (2011). doi: 10.4199/C00027ED1V01Y201103ISP016.

Vaezi M, Pandolfino J, Vela M. Diagnosis and Management of Achalasia. *Am J Gastroenterol* 2013; 108:1238–1249. doi: 10.1038/ajg.2013.196

CHAPTER 4

The Stomach

INTRODUCTION

Although the esophagus serves simply as a conduit for food on its rapid passage from oropharynx to stomach, the principal roles of the stomach are to sterilize, store, and process food prior to its entry into the small intestine. During the hours after a meal, the small intestine receives a flow of gastric chyme, a semifluid mixture of solutes, emulsion particles, and suspended material, which are the products of digestive activity of gastric juice and mechanical action of the stomach. By regulating the delivery of food to the duodenum, the stomach protects the small intestine from being suddenly overwhelmed by the arrival of large amounts of poorly prepared material. A patient with a total gastrectomy (removal of the stomach) must substantially modify his eating habits to compensate for the loss of storage and digestive functions of the stomach. More importantly, loss of the antimicrobial action of gastric acid renders the patient more susceptible to enteric infection by pathogens.

GENERAL ANATOMICAL FEATURES OF THE STOMACH

As illustrated in Figure 4-1, the human stomach can be divided into a number of anatomically distinct regions: the fundus, body, and antrum. The cardia, a short transition zone ($\approx$10 mm) between the esophagus and the fundus, is characterized by the presence of both stratified squamous epithelium of the esophagus and mucous cells similar to those of the fundus. However, only approximately 50% of individuals have an anatomically identifiable cardia; prompting the proposal that the cardia may be an abnormal feature attributed to local damage by reflux of gas-

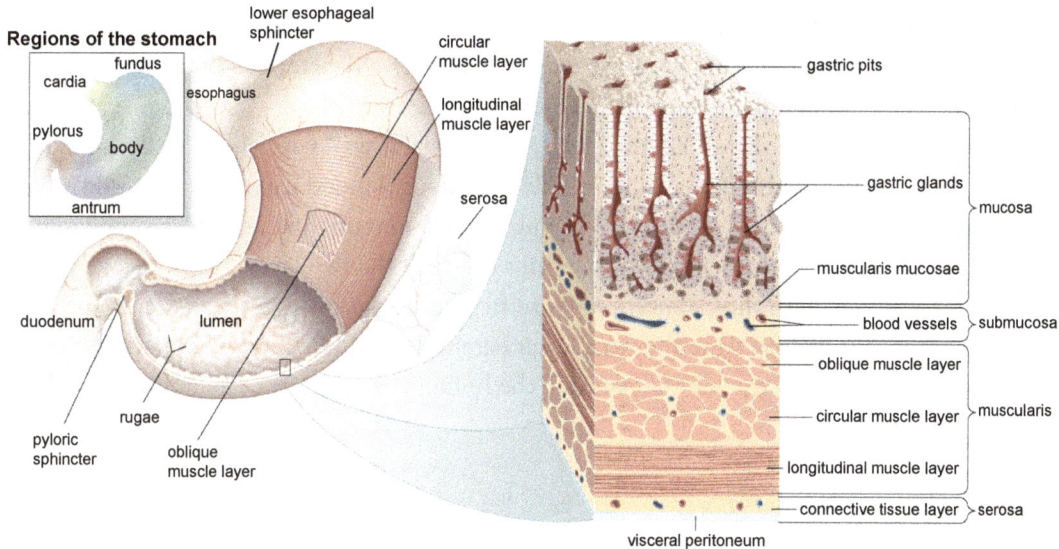

FIGURE 4-1: Gross and microanatomy of the stomach. The inset depicts the three major layers of the gastric wall; the mucosa and associated glands, the submucosa containing blood vessels, and the muscularis with its oblique, circular and longitudinal arrangement of the muscle layers. Reprinted with permission from the Encyclopædia Britannica, © 2003 by Encyclopædia Britannica, Inc.

tric contents. At the gastroesophageal junction, the lower esophageal sphincter (LES) separates the esophagus from the stomach. The LES is a functional, rather than an anatomical, sphincter; represented by a zone of high pressure that creates a barrier to gastroesophageal reflux. At the lower end of the stomach is the pyloric sphincter, which separates the stomach from the duodenum. The pyloric sphincter is a well-defined ring of smooth muscle that regulates the passage of chyme into the duodenum.

The mucosal surface of the empty stomach appears to be thrown up in folds of the mucosal and submucosal layers (rugae), which are most conspicuous in the fundus and body of the stomach (Figure 4-1). When the stomach is distended by accommodation of ingested food, the rugae unfold or spread. Upon emptying, the stomach returns to its original fasting (empty) size and the rugae reform. The stomach is also the only organ of the digestive tract with three muscle layers, rather than two. In addition to a circular and longitudinal muscle layer, the stomach also has an inner oblique layer that is most prominent in the fundus and body (Figure 4-1). The functional advantage of containing three muscle layers oriented in different directions is to facilitate gastric emptying. Collectively, these features of the gastric wall are analogous to that of the urinary bladder. The urinary bladder contains three layers of crisscrossing bundles of smooth muscle and the inner lining of the empty bladder is characterized by rugae. The rugae flatten as it accumulates urine and re-form when it empties.

Like other parts of the alimentary tract, the stomach wall consists of well-defined tissue layers: the mucosal, submucosal, and muscle layers, surrounded externally by a thin serosal (or peritoneal) layer (Figure 4-1; inset). The submucosa contains the arterioles supplying the mucosa and muscularis regions as well as the veins draining the two layers. The mucosal layer is principally concerned with the elaboration of gastric juice to sterilize gastric contents and initiate digestion of ingested proteins and, to a lesser extent, lipids. The muscle layers serve to mix ingested food with the gastric juice and empty the homogenate into the small intestine.

GASTRIC SECRETION

The epithelium of the gastric mucosa (Table 4-1) is chiefly concerned with secretory processes, that is, the elaboration of gastric juice, a digestive mixture of hydrochloric acid and the enzymes, pepsin and lipase. The parietal cells secrete acid, whereas the chief cells produce pepsinogen and lipase. The acid and pepsin elaborated by the gastric mucosa can reach concentrations that can be corrosive to the epithelium. The elaboration of an alkaline mucus by the gastric epithelium provides a protective barrier against gastric mucosal injury.

Mucosal Glands

The mucosa of the stomach can be functionally divided into oxyntic and pyloric glandular regions (Figure 4-2). The oxyntic glandular mucosa encompasses the fundus and body and represents approximately 80% of the total surface area of the stomach; the remaining 20% being the pyloric glandular mucosa of the antrum. The entire mucosal surface of the stomach is lined by a layer of simple columnar epithelium, the surface mucous cells, which secrete an alkaline mucus. This mucus layer protects the gastric mucosa from chemical and physical injury. The mucus-secreting cells dip down from the surface to form the gastric pits, into which drain one or more gastric glands. Mucous cells are also found in the neck of the glands. Located in the isthmus of the glands are undifferentiated, granule-free stem cells, which are the precursors for the various mature cells of the glandular epithelium. In addition, a subpopulation of mucus neck cells in the isthmus region can also serve as stem cells, generating precursors for parietal and chief cells, as well as mucous cells. The progenitor zone of the oxyntic gland is located close to the gastric pit, whereas the stem cell niche of the pyloric gland resides in the basal region of the gland.

	TABLE 4-1: Contribution of Gastric Mucosal Epithelial Cells to Gastric Juice	
CELL TYPE	**PRODUCT**	**FUNCTION IN STOMACH**
Parietal	*HCl*	Bactericidal
		Protein denaturation
		Pepsin(ogen) → pepsin
	Intrinsic factor	Vitamin B_{12} absorption
Chief	*Pepsin(ogen)*	Bactericidal
		Endopeptidase
	Gastric lipase	Triacylglycerol hydrolysis
Mucous cells*	*Mucus*	Diffusional barrier
		Lubricant
	HCO_3-	Neutralization of HCl
	Trefoil factors	↑ Mucus viscosity
		↑ Epithelial cell migration

* Both surface mucous cells and mucous neck cells. However, the mucin composition is slightly different; surface mucous cells (MUC5AC) versus mucous neck cells (MUC 6).

Stem cells migrate upward to become surface mucus cells or downward to become parietal, chief, or enteroendocrine cells. The stem cells are constantly replacing the gastric epithelium with a turnover rate of the surface mucus cells of 3 to 5 days; chief and parietal cells turnover at a slower rate (on the order of months).

In the oxyntic gland, the parietal cells have a conical shape and are found predominantly in the neck region of the gland (Figure 4-2). A vital product secreted by parietal cells is intrinsic factor (IF) (Table 4.1). Intrinsic factor binds vitamin B_{12} (cobalamin) and is critical for vitamin B_{12} absorption in the ileum; it can only be absorbed as an IF/cobalamin complex. Vitamin B_{12} deficiencies can lead to pernicious anemia and severe neurological disorders. Thus, a patient who has undergone a total gastrectomy has to have routine injections of vitamin B_{12}.

Parietal cells are also the source of hydrochloric acid. The parietal cell can secrete acid against a huge concentration gradient (up to 10^6-fold). The high-energy demand of acid secretion is supplied by numerous mitochondria; an estimated one third of the cytoplasm being occupied by mitochondria (Figure 4-3). In the resting state, the bulk of the parietal cell cytoplasm is occupied by a membrane network consisting of flattened tubules and associated cisternae and

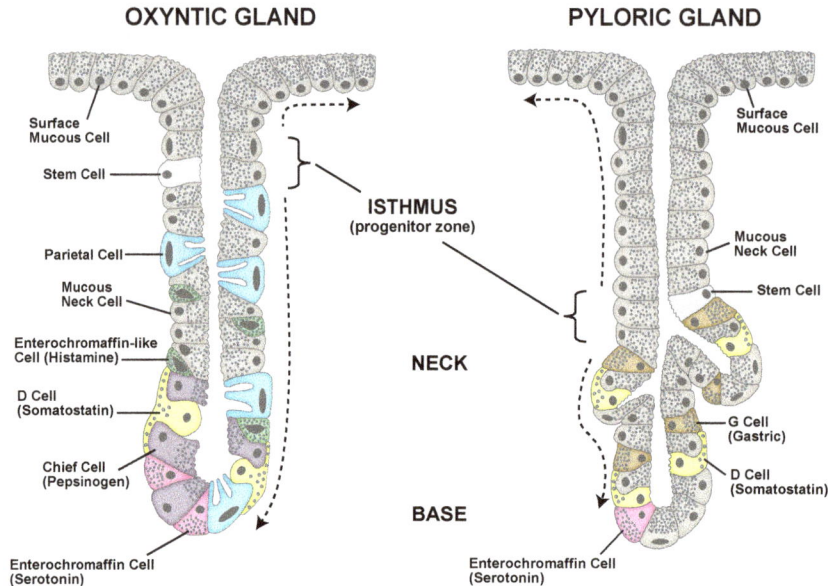

FIGURE 4-2: Mucosal glands of the functional regions of the stomach: the oxyntic region (fundus and corpus) and the pyloric region (antrum). Cell populations that are unique to oxyntic glands include parietal, chief, and enterochromaffin-like cells, whereas G cells are unique to the pyloric glands [modified from Schubert ML, In: *Physiology of the Gastrointestinal Tract* (LR Johnson, ed.), 2012, Elsevier Inc., used with permission].

vesicles; referred to as tubulovesicles (Figure 4-3A). Embedded in the tubulovesicles are numerous H^+, K^+ ATPases or "proton pumps." Invaginating from the apical membrane are small, interconnected canaliculi lined with short microvilli. When stimulated to maximally secrete HCl, the parietal cell morphology undergoes a dramatic transformation (Figure 4-3B). The tubulovesicles translocate and fuse with the intracellular canaliculi, distending them, elongating their microvilli, and greatly expanding the apical surface area, thereby aligning the proton pumps with the gastric lumen. When acid secretion returns to basal levels, the intracellular morphology also reverts to the presecretion configuration; the apical canalicular membrane collapses to reform the tubulovesicle membrane system.

The chief cell is found mostly in the deeper regions of the gland and is the source of pepsinogen and gastric lipase (Figure 4-2). The morphology of the chief cell is characterized by an extensive rough endoplasmic reticulum and Golgi apparatus with zymogen granules clustered in the apical region of the cell. The synthesis, intracellular transport, and storage of the enzymes in zymogen granules follow a pattern typical of exocrine cells. With appropriate stimulation, the chief cell discharges its products into the gastric lumen by exocytosis of the zymogen granules.

A variety of enteroendocrine cells populate the neck and base regions of the oxyntic gland. Of particular importance are the cells whose products regulate acid secretion by the parietal cell; specifically, the enterochromaffin-like (ECL) cells that secrete histamine to increase acid secretion and D cells, which produce somatostatin (SST) to inhibit acid secretion (Figure 4-2). The D cells contain cytoplasmic extensions that are in direct contact with ECL and parietal cells and exert a tonic inhibitory (paracrine) influence on both of these cells. D cells and ECL cells in the oxyntic region are of the "closed" type, that is, they do not have extensions communicating with the lumen and thus are not directly affected by luminal contents.

The pyloric gland contains substantial numbers of surface and neck mucous cells, but is relatively devoid of parietal and chief cells (Figure 4-2). The predominant enteroendocrine cells of the pylori gland are also involved in regulation of acid secretion. The G, or gastrin, cell has a narrow apical membrane facing the gland lumen ("open" type of EEC) and thus can sense and respond to luminal stimuli. The G cell has numerous gastrin-containing secretory granules stored in the basal region of the cell. Upon stimulation, gastrin is released into the interstitium, where it gains access to the microcirculation to act as a hormone that enhances acid secretion, primarily by stimulating the histamine-producing ECL cells in the oxyntic mucosa. As in the oxyntic gland, the pyloric gland contains D cells, which secrete SST to exert a tonic inhibitory (paracrine) influence on the G cells, thereby restraining acid secretion. In contrast to the D cells of the oxyntic region, the D cells in the pyloric region are of the open type and capable of responding to luminal stimuli.

Acid Secretion

The H^+, K^+ ATPase on the apical membrane of the parietal cell actively extrudes H^+ into the gastric lumen in exchange for K^+. Neural, paracrine, and endocrine regulatory factors are involved in eliciting HCl secretion by parietal cells. To this end, the basolateral parietal cell membrane contains (1) muscarinic (M_3) receptors for acetylcholine released from enteric cholinergic neurons, (2) CCK_2 receptors for the hormone, gastrin, released from antral G cells, and (3) H_2 receptors for the paracrine, histamine, released by adjacent ECL cells (Figure 4-3). Ligation of the CCK_2 or M_3 receptors results in increases in intracellular Ca^{++}, whereas activation of H_2 receptors elevates cAMP. These two intracellular signaling components work in concert to induce acid secretion (Figure 4-3B). The relative impact of these regulatory factors on parietal acid secretion is as follows: histamine (most important) > acetylcholine > gastrin (least important). Nonetheless, during periods of enhanced acid secretion, more than one of the aforementioned secretagogues is involved in stimulating parietal cells and their combined effect is more than additive.

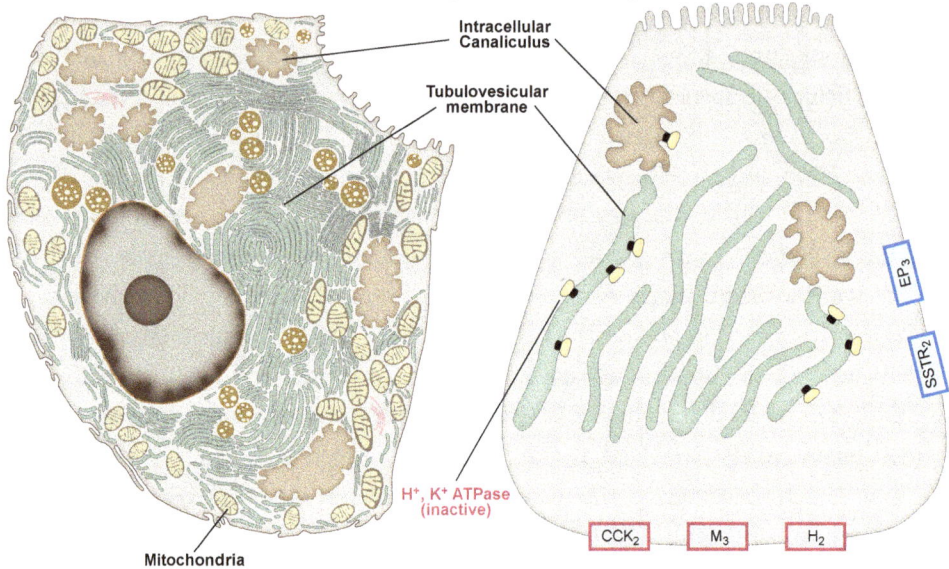

A. Resting, non-secreting

FIGURE 4-3A: The resting, nonsecreting parietal cell. Within the cytoplasm is an extensive membrane network of flattened tubules (tubulovesicles). Imbedded in the tubulovesicles are numerous inactive H^+, K^+ ATPases or "proton pumps." In addition, there are small, interconnected canaliculi (IV) lined with short microvilli. The basolateral membrane of the parietal cell contains receptors for factors that stimulate (red) or inhibit (blue) parietal function.

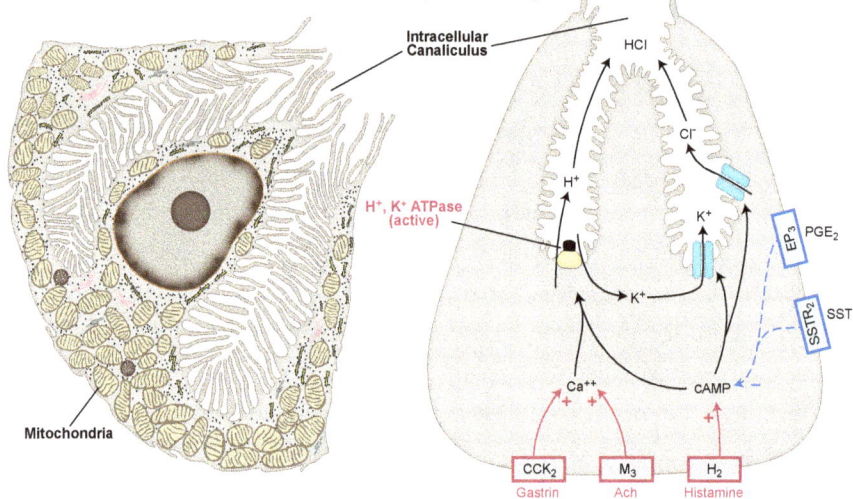

B. Actively secreting

FIGURE 4-3B: The actively secreting parietal cell. Engagement of the CCK_2, M_3, and/or H_2 receptors by gastrin, acetylcholine (ACh), and/or histamine activate the parietal cell (red pathways). Engagement of the somatostatin (SST) or prostaglandin (PGE_2) receptors, $SSTR_2$ or EP_3, respectively inhibit the parietal cell

The parietal cell is also under the influence of inhibitory regulating factors, such as somatostatin and prostaglandins of the E series (e.g., PGE_2) (Figure 4-3B). Somatostatin (SST) produced by D cells, is the major inhibitor of acid secretion. SST exerts tonic restraint on parietal cell HCl secretion by decreasing intracellular cAMP levels. SST also inhibits antral G cells and oxyntic ECL cells, resulting in tonic inhibition of gastrin and histamine production, respectively. Thus, to attain optimal acid secretion rates in response to secretagogues (ACh, histamine, and gastrin), the effects of SST must be nullified.

Receptor activation of cAMP and Ca^{++} signaling pathways results in tubulovesicular membrane fusion with the intracellular canaliculi, which enlarge to form microvilli at the apical membrane; an event required for optimum localization and activation of H^+, K^+ ATPase (Figure 4-3B). The H^+, K^+ ATPase consists of two subunits: a catalytic α-subunit and a regulatory β-subunit that mobilizes the enzyme to the appropriate membrane. Activation of H^+, K^+ ATPase requires the localization of both K^+ and Cl^- channels to the apical membrane. K^+ channels are required for the recycling of K^+, thereby continuously replenishing the extracellular K^+ needed for exchange with H^+ by the proton pump. The Cl^- channels are required so that Cl^- can leave the cell and enter the lumen to combine with H^+ to form HCl. Both the K^+ and Cl^- channels are activated by cAMP–mediated mechanisms.

The H^+ ions required for the proton pump are derived from the intracellular hydration of carbon dioxide by the enzyme, carbonic anhydrase (Figure 4-4). The carbonic acid formed as a result of this reaction disassociates into H^+ and HCO_3^-. The H^+ is exchanged for K^+ by the H^+, K^+ ATPase; if K^+ is unavailable for exchange, proton pumping ceases. An adequate amount of K^+ is maintained in the lumen by transfer of K^+ from the interstitial space. Two basolateral transporters move K^+ from the interstitium into the cytoplasm: the Na^+, K^+ ATPase, which exchanges $3Na^+$ for $2K^+$ and the Na^+, K^+, $2Cl^-$ transporter (NKCC). These transporters keep intracellular K^+ above its electrochemical equilibrium. The presence of K^+ channels in the apical

---

(blue pathways). Gastrin and ACh activate cell-signaling pathways that elevate intracellular Ca^{++}, whereas histamine increases cAMP levels. Both of these events result in a dramatic morphological transformation of the parietal cell. The tubulovesicles translocate and fuse with the intracellular canaliculi, distending them, elongating their microvilli, greatly expanding the apical surface area and aligning the H^+, K^+ ATPases with the gastric lumen. The H^+, K^+ ATPases become active due the cAMP-mediated localization of a K^+ and Cl^- channel in the close proximity to the H^+, K^+ ATPases in the apical membrane. The inhibitory effects of PGE_2 and SST are mediated via inhibition of the cAMP pathway.

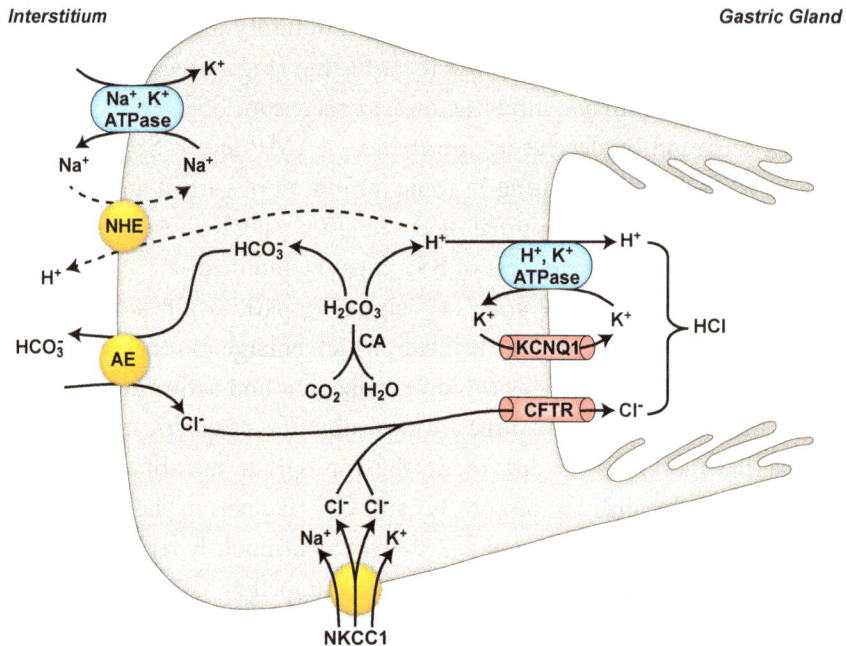

FIGURE 4-4: The H^+ ions required for the proton pump (H^+, K^+ ATPase) are derived from the intracellular hydration of carbon dioxide by the enzyme, carbonic anhydrase (CA). The carbonic acid formed as a result of this reaction disassociates into H^+ and HCO_3^-. The H^+ is exchanged for K^+ by the H^+, K^+ ATPase. An adequate amount of K^+ is maintained in the lumen by recycling the K^+ using the K^+ channel (KCNQ1). The excess HCO_3^- is removed from the cell by the Cl^-/HCO_3^- exchanger. The Na^+, K^+, $2Cl^-$ transporter (NKCC) and Cl^-/HCO_3^- exchanger in the basolateral membrane transport Cl^- into the cell. The accumulating Cl^- leaves the cell via the Cl^- channel (CFTR) and interacts with H^+ to form HCl. The Na^+/H^+ exchanger (NHE) appears to serve a "housekeeping" function to prevent excessive intracellular accumulation of H^+ ions.

membrane allows K^+ to move into the lumen, creating a recycling system that maintains K^+ at luminal concentrations sufficient to meet the needs of the H^+, K^+ ATPase. Chloride channels are also present in the H^+, K^+ ATPase-rich apical membrane. These channels allow Cl^- to enter the lumen and associate with the secreted H^+ to form HCl. Chloride enters the parietal cell from the interstitium via the NKCC transporters and Cl^-/HCO_3^- exchangers imbedded in the basolateral membrane. The exact identity of the K^+ and Cl^- channels responsible for efflux of these ions has not been established with certainty. However, the available evidence suggests that a likely candidate for the K^+ channel is the KCNQ1 channel; a member of a family of voltage-gated K^+ channels. When KCNQ1 forms a complex with a regulatory subunit (KCNE2) in the apical parietal cell membrane, it is converted from a voltage-gated channel to a constitutively open channel. The most likely candidate for the Cl^- channel is the cystic fibrosis transmembrane conductance regulator (CFTR); an anion channel that is defective in the disease, cystic fibrosis.

The secretion of H^+ by the parietal cell leaves behind HCO_3^-, which, if not disposed of, would lead to intracellular alkalinization, metabolic dysfunction, and ultimately cell death. The Cl^-/HCO_3^- exchanger, which transports Cl^- into the cell, extrudes HCO_3^- into the interstitium, thereby maintaining intracellular pH to within normal levels. The HCO_3^- that enters the interstitium diffuses into the blood vessels supplying the gastric glands. The flow of HCO_3^- into the mucosal circulation raises the pH of gastric venous blood and is termed the "alkaline tide."

Pepsinogen and Lipase Secretion

Pepsin, an endopeptidase, is synthesized and stored as a zymogen (pepsinogen) in chief cells. Pepsin is not a single molecule, but represents a family of at least 5 isoenzymes; pepsin I and II are the best characterized. Both isoenzymes are found in chief cells; pepsin II represents approximately 80% and pepsin I accounts for less than 5% of the total pepsin in gastric juice. Secreted pepsinogen is rapidly converted to pepsin at a pH of less than 4. This activation involves the autocatalytic cleavage of a peptide fragment from pepsinogen, thereby uncovering the catalytic site. Once activated, pepsin catalyzes its own conversion from pepsinogen initiating a cascade of pepsin activation. As an endopeptidase, pepsin cleaves only internal peptide bonds, thereby reduces dietary proteins to a mixture of peptides. The pH optimum of pepsin activity lies between 2 and 5. As pH progressively increases, pepsin activity declines and at pH of 6.5 to 7, the protease is inactive. This is a reversible inactivation process, because pepsin remains stable and its activity can be restored by re-acidification (pH 3). However, the enzyme is irreversibly destroyed in more alkaline solutions (pH > 7). Stimulation of chief cells by vagal stimulation (acetylcholine) or food ingestion, leads to degranulation of the chief cells; thereafter, continued output of the enzyme is maintained by an increase in synthesis. Pepsinogen exocytosis is mediated by activation of both the cyclic AMP and Ca^{++} signaling pathways (see Figure 1-4); cyclic AMP is more important for continued pepsinogen synthesis. As was the case for parietal acid secretion, both positive and negative regulatory factors influence chief cell pepsinogen secretion. The major stimulants of pepsinogen secretion are ACh and gastrin, whereas SST is inhibitory. The effects of histamine are equivocal. In general, pepsinogen secretion is increased in conjunction with acid secretion.

Gastric lipase is co-localized with pepsinogen in chief cells and is released in an active form into the gastric juice. As with acid and pepsinogen secretion, gastric lipase secretion is stimulated by gastrin and cholinergic nerves. It hydrolyzes 10% to 20% of ingested triacylglycerides (TAG), generating primarily diacylglycerides and free fatty acids (released from sn-3 position). Gastric lipase is admirably suited for lipolytic activity in the stomach. It is pepsin resistant, stable and active at pH values from 2 to 7 (pH optimum is 5.0–5.5), and does not require bile salts or cofactors (e.g., co-lipase) for activity. However, its activity is limited to the stomach, because it is rapidly degraded in the intestine by trypsin. In the intestine, pancreatic lipase is responsible for the digestion of most (80%–90%) of the dietary TAG. The contribution of gastric lipase to overall TAG hydrolysis becomes more important when there is a deficiency in pancreatic lipase. For example, in newborns, the immature pancreas fails to synthesize sufficient lipase or in adults with pancreatic exocrine insufficiency caused by disease. Interestingly, in adults, with negligible amounts of pancreatic lipase due to chronic pancreatitis, there is a compensatory increase in gastric lipase secretion, allowing for digestion of as much as 30% of ingested fat.

Functions of Gastric Acid and Pepsin

A major digestive function of gastric juice is hydrolysis of ingested protein. As mentioned previously, the secretion of pepsinogen occurs in conjunction with HCl secretion. The role of HCl acid is two-fold. It is required for the autocatalytic conversion of pepsinogen to pepsin. In addition, HCl facilitates peptide bond hydrolysis by denaturing ingested proteins and maximally exposing peptide bonds to the action of pepsin. Denaturation is particularly important for hydrolysis of collagen, because the triple helix structure shelters the internal peptide bonds from the endopeptidase. However, the importance of HCl acid and pepsin to protein digestion is questionable, because achlorhydria or total gastrectomy does not substantially compromise an individual's nutritional status. The intestinal capacity for digestion/assimilation of protein appears to be more than sufficient to maintain normal nitrogen balance.

Gastric juice is also bactericidal. *In vitro* studies indicate that HCl alone (pH < 2–3) can kill various pathogenic bacteria in a concentration-dependent manner; the bactericidal effect is enhanced if pepsin is also present. *In vivo* studies indicate that the colonization of pathogenic bacteria in the gastrointestinal tract is significantly enhanced in mice rendered hypochlorhydric by genetic blockade of H^+, K^+ ATPase activity. On the other hand, the susceptibility to infection is substantially reduced in wild type (control) mice that are rendered hyperchlorhydric by histamine administration. Similar observations have been made in patients. For example, chronic use of proton pump inhibitors renders patients more susceptible to enteric bacterial infections.

It is generally considered that the bactericidal role of gastric juice is far more important than its role in protein digestion. However, if gastric juice is bactericidal, how do microbes survive the gauntlet of gastric juice to establish a niche in the more distal segments of the gastrointestinal tract? The most common mode of microbial entrance into the GI tract is in conjunction with food ingestion. When food enters the stomach, its buffering capacity can increase the pH of gastric contents to 4.0 or higher. This less acidic environment would diminish the bactericidal effect of gastric juice for a length of time that is likely sufficient to allow for any surviving microbes to colonize the gastrointestinal tract.

Mucus Secretion

The epithelial lining of the gastric mucosa also contains mucus-secreting cells (Figure 4-2 and Table 4-1). The main organic constituent of gastric mucus is a group of high molecular weight glycoproteins, the mucins (MUC). There are several different secreted, gel-forming mucins, of which MUC5AC and MUC6 are present in the stomach. MUC5AC is secreted by the surface mucous cells, whereas MUC6 is secreted by the mucous neck cells. The surface mucus protects the gastric surface epithelium proper, whereas the mucus secreted within the glands by the mucous neck cells protects the glandular epithelium. Immunofluorescence of fixed gastric mucosal samples indicates the mucus adherent to the epithelial lining consists of alternately layered MUC5AC and MUC6; with MUC5AC being the dominant mucin. The major secretagogues (acetylcholine, gastrin, and histamine) that induce parietal cells to secrete acid also increase mucus output by isolated gastric mucous cells.

Secretion of mucus is accompanied by HCO_3^- secretion. This latter point is important for the efficient dispersal of a mucus gel to cover the gastric epithelium. The mucin constituents of mucus tend to repel each other due to the negative charges of the carbohydrate moieties. Thus, the condensed packing of mucin in the secretory granules requires H^+ and Ca^{++}, which interact with the carbohydrate moieties of mucin and neutralize the negative charges. Upon secretion, the H^+ and Ca^{++} are removed by interacting with the co-secreted HCO_3^- and electrostatic repulsion of the mucin molecules is reinstated. This allows for the rapid expansion (> 1,000-fold) and spreading of the secreted mucins.

The mucus covering the gastric epithelium can be divided into a firmly adherent inner layer abutting the epithelium and a rather loosely adherent outer layer extending into the lumen (see Chapter 1). The mucus layer of the gastric antrum is almost twice as thick as that of the gastric body; primarily due to an increase in the thickness of the firmly adherent mucus layer. It is generally believed that the inner layer gives rise to the more displaceable outer layer via proteolytic dissolution of the mucin gel network. The outer layer is readily removed by shear stress (e.g., motility), whereas the more firmly adherent (to epithelial cells) inner layer is difficult to displace. The mucus layer is a dynamic entity, that is, as the outer layer is removed, mucus secretion is increased to replenish the mucus layer.

The mucous cells of the gastric mucosa also secrete phospholipids, which represent a minor organic constituent of gastric mucus (less than 3%); the major phospholipid is phosphatidylcholine. Phospholipids are amphiphilic; in the case of phosphatidylcholine, the choline group is positively charged, whereas the fatty acid chains are nonpolar. This allows the charged group to electrostatically bind to the negatively charged mucin glycoproteins, whereas the nonpolar hydrocarbon tails extend toward the lumen. This creates a hydrophobic, phospholipid monolayer covering the gastric mucus, limiting the interaction of hydrophilic substance (e.g., acid) with mucus.

The peptides of the trefoil factor family (TFF) are also important constituents of gastric mucus. They are so named because of the three-leaved, clover-like loops in their globular structure. The expression of TFFs in the gastric mucosa is specific and complimentary. TFF1 is expressed in the surface mucous cells and colocalized with MUC5AC, whereas TFF2 is expressed in the mucous neck cells and colocalized with MUC6. TFFs can interact with their mucin partners to increase mucus viscosity. Although specific receptors for TFFs have not been identified, they are important inducers of epithelial cell migration (motogens) and proliferation (mitogens).

Functions of Mucus

The mucus layer covering the gastric epithelium is continuous and its major purpose is to protect the epithelial lining from mechanical abrasion or chemical irritation from ingested substances. The protective role of mucus is underlined by the fact that mechanical stimulation (gentle rubbing) or exposure to potentially injurious agents provokes copious mucus secretion. The viscous property of mucus provides lubrication such that the mixing of ingested food particles by gastric motility does not damage the epithelium. In addition, the gel-like network of mucus can impede the movement of noxious macromolecules or microbes to such an extent that, before these agents can reach the epithelium, they are swept downstream into the duodenum by gastric contractions (gastric emptying).

Mucus is believed to be a significant "barrier" preventing the corrosive effects of acid and pepsin from impacting the epithelium. It is generally held that the HCO_3^- secretion into the mucus gel establishes a protective "mucosal HCO_3^- (or pH) gradient" (Figure 4-5). The HCO_3^- is taken up from the interstitium by various transporters located at the basal aspect of the surface cells and subsequently transported into the immediately adjacent mucus layer by Cl^-/HCO_3^- exchangers. If the bicarbonate secreted by the surface epithelial cells were to

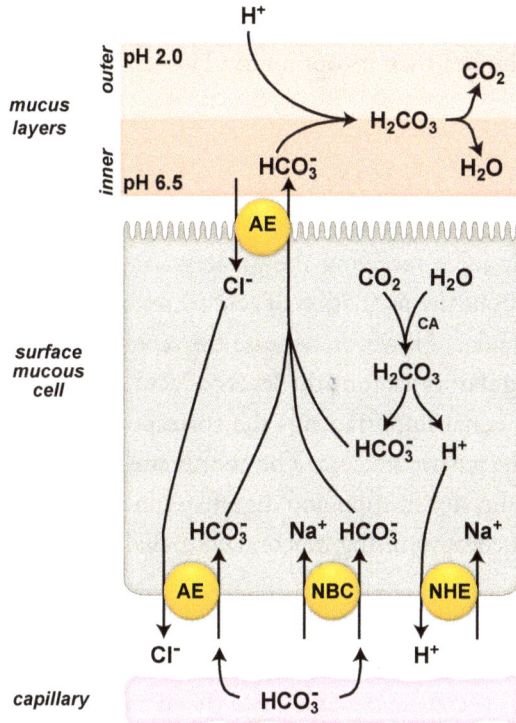

FIGURE 4-5: The gastric mucus pH gradient is generated by the transport of HCO_3^- from the surface epithelium into the adjacent firmly adherent mucus layer. Epithelial cell HCO_3^- is derived from hydration of CO_2 by carbonic anhydrase (CA) and via uptake by exchangers and transporters on the basal aspect of the cell membrane. Subsequently, the HCO_3^- is exported into the adjacent mucus in exchange for Cl^- by an anion exchanger (AE) on the apical membrane. The HCO_3^- entering the mucus from the epithelial cells reacts with the H^+ diffusing into the mucus from the lumen to form H_2O and CO_2. This neutralization reaction within the inner adherent mucus layer creates a gradient of several pH units between the lumen (~pH 2) and the surface of the mucosal cells (~pH 6.5).

enter directly into the gastric lumen, the H^+ secreted by the parietal cells would quickly neutralize and overwhelm the alkalinizing function of HCO_3^-. However, the relatively dense and unstirred inner mucus layer delays the immediate neutralization reaction between H^+ entering the mucus from the lumen and the HCO_3^- entering the mucus from the epithelial lining. This creates a gradient of several pH units between the lumen (~pH 2) and the surface of the mucosal cells (~ pH 6.5). The mucus HCO_3^- also protects the gastric epithelium from the actions of pepsin. Although pepsin can also slowly diffuse through the mucus layer to reach the epithelium, its enzymatic activity will diminish as it nears the epithelium proper. Even if a small fraction of pepsin manages to penetrate the dense inner mucus layer, the pH near the epithelium will be significantly higher than its optimum pH.

INTEGRATED REGULATION OF GASTRIC SECRETION

The component of gastric juice receiving the greatest attention has been hydrochloric acid. Because much is known about the regulation of acid secretion by parietal cells, we will focus on this component of gastric juice. However, because the secretion of pepsinogen by chief cells is influenced by the same regulatory factors, the levels of acid and pepsin generally rise concomitantly in gastric juice. Of equal importance is the corresponding liberation of alkaline mucus that coats the surface of the gastric mucosa. The coordinated secretion of acid, pepsinogen and an alkaline mucus allows the sterilization and digestive functions to proceed simultaneously in the gastric lumen without compromising mucosal epithelial integrity.

Interprandial (Basal) Acid Secretion

The stomach continuously secretes an acidic juice (basal secretion), but superimposed on this are phases of enhanced secretion that are associated with feeding. Basal acid secretion appears to result from the influence of resting vagal tone. Vagotomy or atropine can almost completely abolish basal acid secretion. Let us consider the neural, endocrine, and paracrine pathways that enable the vagus to regulate acid secretion (Figure 4-6), keeping in mind that pepsinogen and mucus secretin are regulated by similar factors and occur concurrently with acid secretion.

As shown in Figure 4-6, efferent vagal fibers emanating from the CNS synapse and activate gastric intramural cholinergic neurons (acetylcholine; ACh) and peptidergic neurons (gastrin-releasing peptide; GRP). In the antrum, ACh and GRP stimulate gastrin release from G cells, whereas ACh concurrently inhibits D cell secretion of SST, nullifying its inhibitory influence on G cells. Gastrin enters the circulation and is delivered to the fundus where it stimulates both parietal cells and ECL cells; the latter effect (gastrin-histamine axis) is the most important. In the fundus, cholinergic neurons directly stimulate parietal cells to secrete acid. Simultaneously, ACh inhibits D cell secretion of somatostatin (SST) which removes the restraining paracrine influence of SST on (1) parietal cells and (2) histamine-secreting ECL cells, thereby facilitating parietal cell secretion. Histamine stimulates parietal cells (H_2 receptors) to enhance acid secretion, and simultaneously inhibits SST secretion from D cells (H_3 receptors). Collectively, the efferent vagal input to the gastric intramural neurons results in simultaneous increase in stimulatory factors and decrease in inhibitory factors at the level of the parietal cell.

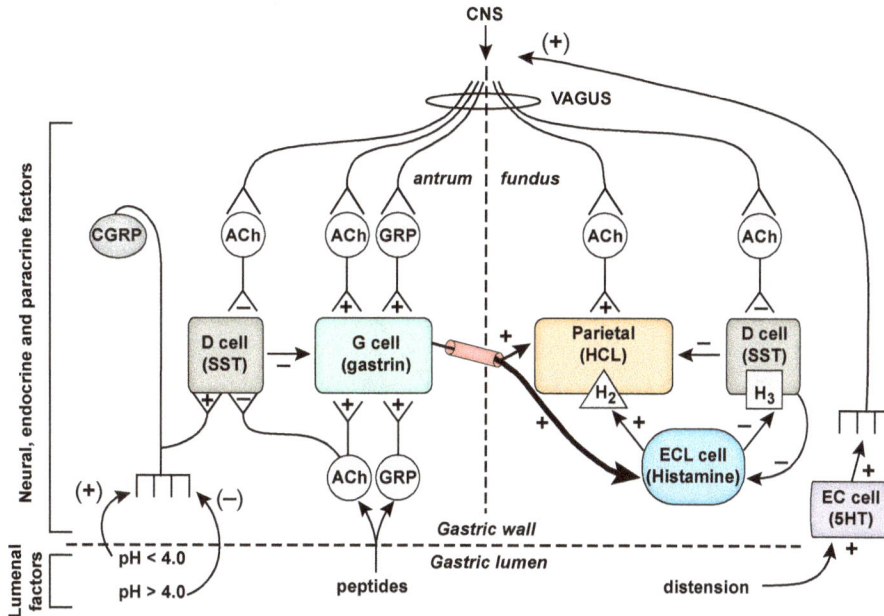

FIGURE 4-6: Interaction of neural, endocrine, and paracrine regulation of acid secretion by parietal cells. Basal acid secretion is primarily driven by the vagus which activates gastric intramural cholinergic neurons (acetylcholine; ACh) and peptidergic neurons (gastrin-releasing peptide; GRP). In the antrum, ACh and GRP stimulate gastrin release from G cells, whereas ACh simultaneously inhibits D cell secretion of somatostatin (SST), a paracrine inhibitor of D cells. In the fundus, ACh-producing neurons and the gastrin-histamine axis (H_2 receptor) stimulate parietal cells to secrete acid. Cholinergic (ACh) nerves and histamine (via H_3 receptors) inhibit D cells, thereby removing the restraining influence of SST. A low luminal pH activates afferent sensory neurons that release calcitonin gene-related peptide (CGRP), which stimulates SST-releasing D cells. In response to sensory (e.g., taste, olfactory) and conditioned cues (cephalic phase), the CNS enhances vagal efferent drive. During the gastric phase, luminal cues stemming from the entrance of food into the stomach come into play. Distension of the gastric wall elicits a vago-vagal reflex, which further augments efferent vagal drive. Peptides stimulate gastric intramural neurons to activate G cells (ACh and GRP), further enhancing parietal acid secretion via the gastrin-histamine axis. Finally, the buffering effect of food raises intragastric pH and nullifies the effect of CGRP neurons on D cells [modified from Schubert ML, In: Physiology of the Gastrointestinal Tract (LR Johnson, ed.), 2012. Elsevier Inc., used with permission].

Acid secretion is kept well below maximum levels due to a negative feedback mechanism elicited by low pH in the antrum (Figure 4-6). D cells in the antrum are stimulated by low intragastric pH (pH < 4); SST secretion increases in proportion to further decreases in pH. The activation of D cells (and SST secretion) in response to luminal acid is mediated by interstitial sensory neurons that release CGRP. Because afferent sensory neurons do not project terminals

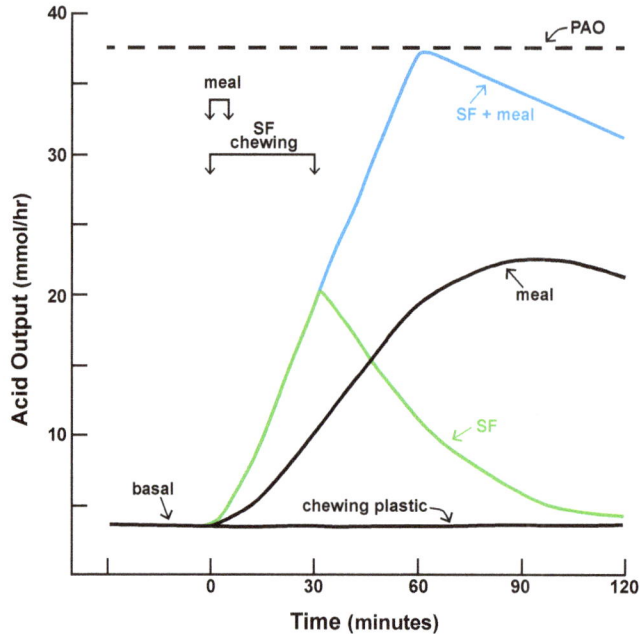

FIGURE 4-7: The cephalic and gastric phases of acid secretion. Healthy volunteers were sham fed (SF) a steak and potatoes meal (chewed but spit; not swallowed) for 30 minutes, or a 600 ml homogenate of the meal (meal) was instilled intragastrically over 5 minutes. SF induced a secretory response that began to wane immediately after cessation of SF. Intragastric placement of the meal produced a comparable secretory response, albeit delayed. The combination of SF and a meal produced a near maximal response (PAO; peak acid output in response to pentagastrin). Basal secretion is approximately 10% of maximum [data from Richardson CT et al., *Journal of Clinical Investigation* 60: 435–441, 1977, Elsevier Inc. used with permission].

into the lumen, it has been proposed that enterochromaffin cells respond to protons in the lumen and release serotonin to activate the afferent sensory neurons. Nonetheless, during the interdigestive period, the balance of stimulatory and inhibitory influences is dominated by an SST-induced restraint on parietal cell HCl secretion, resulting in a gastric juice of low volume and low pH. The dominating influence of SST on basal gastric acid secretion is exemplified by the 5- to 10-fold greater basal acid secretion in 1) mice genetically deficient in the SST receptor ($SSTR_2$) or 2) normal mice treated with an antibody directed against SST.

The amount of hydrochloric acid secreted under basal (fasting) conditions is approximately 10% of the maximum secretory response induced by secretagogues (Figure 4-7). It is higher in men than in women, attributed to the higher mass of parietal cells in men. Basal secretion follows a circadian rhythm, it is highest in the evening and lowest in the morning. Regulation of this rhythm is unclear. Despite the low rate of acid secretion, the pH of the gastric juice is 1.5 to 2.5; presumably because buffering is minimal in an empty stomach. The sustained and metabolically costly acid secretion that occurs during fasting likely serves to maintain a relatively sterile intragastric milieu.

Postprandial Secretion

The secretory response of the stomach to meals has traditionally been divided into three phases: *cephalic*, *gastric*, and *intestinal*. This rather arbitrary classification is based on the source of the signals impacting on gastric acid secretion. However, these are not discrete phases, but a continuum of secretory responses dictated by the relative influence of stimulatory and inhibitory factors that act on parietal cells during the course of a meal. For example, both cephalic and gastric stimuli are generated concurrently during ingestion of a meal to ensure adequate gastric secretion. This further complicate matters, because up to 40% of a meal can empty into the intestine before the meal is finished, signals from the intestine serve to restrain the acid secretion induced by the cephalic and gastric phases. Despite these limitations, division of the postprandial secretory responses into phases is useful for discussion purposes.

The *cephalic* phase is initiated by sensory cues that impact on vagal centers in the CNS, for example, the sight, smell, and taste of food. The method of choice for demonstrating and studying the cephalic phase is sham feeding (SF). With SF, the subject is allowed to ingest and chew the meal, but food is not allowed to enter the stomach. The cephalic phase accounts for 30 to 50% of the maximum gastric secretory response to a meal (Figure 4-7). When the components of the cephalic phase are individually assessed, the sight and smell of food alone elicits approximately 30% of the secretory response noted with sham feeding (chewing and spitting). Surprisingly, simply the thought (or discussion) of appetizing food is a more potent stimulus for gastric secretion than visual or olfactory cues. This "Pavlovian" response can elicit up to 70% of that noted with sham feeding. Chewing per se is not believed to be a good stimulus, because chewing plastic does not elicit a secretory response. In general, an appetizing meal elicits a greater secretory response than a bland meal.

In response to sensory (e.g., taste, olfactory) and conditioned cues, the CNS responds by enhancing vagal efferent drive to initiate the cephalic phase of acid secretion (Figure 4-6). Thus, even before food enters the stomach, the enhanced efferent vagal activity increases HCl secretion above basal levels. In the antrum, the activity of intramural neurons is enhanced, further stimulating G cells (via ACh and GRP) and inhibiting D cells (via ACh). Thus, the impact of the gastrin-histamine axis on parietal cell acid secretion is increased. An analogous effect of enhanced efferent vagal activity occurs in the fundus. Parietal cells are further stimulated to secrete acid; an effect facilitated by removal of the restraining influence of D cells (decreased SST secretion). The cephalic phase is best characterized as tipping the balance of stimulatory and inhibitory influences on parietal cells in favor of the stimulatory factors, resulting in enhanced acid secretion.

The *gastric* phase of acid secretion accounts for 50% to 70% of the maximum gastric secretory response to a meal. It is initiated by the entrance of food into the stomach and it amplifies the secretory response initiated by the cephalic phase (Figure 4-7). Both mechanical (distension) and chemical (nutrients/pH) stimuli linked to the presence of food in the stomach trigger the augmented acid secretion of the gastric phase (Figure 4-6).

When ingested food begins to accumulate in the stomach, distention of the gastric wall activates tension receptors or mechanoreceptors and elicits a long vago-vagal reflex (both afferent and efferent components of the reflex arc reside in the vagus nerve). Enterochromaffin cells of the mucosa are believed to be the mechanoreceptors that transduce the tension signal by releasing serotonin. Serotonin impacts on afferent sensory neurons eliciting a vago-vagal reflex. The vago-vagal reflex of the gastric phase augments the events initiated by the vagus during the cephalic phase (Figure 4-6), that is, the stimulatory factors (ACh, gastrin, and histamine) are enhanced and there is a further reduction in the major inhibitory factor, SST.

Pepsin initiates the hydrolysis of proteins in food, releasing peptides of various sizes; little or no amino acids are generated. The presence of protein hydrolysates in the antrum is an important stimulus for gastrin release from G cells (Figure 4-6). The peptide-induced activation of G cells has been attributed to short intramural reflexes releasing GRP and ACh. The cholinergic neural component increases gastrin release, in part, by inhibiting SST secretion from D cells. Because the enteric sensory nerve terminals are located in the interstitium and do not penetrate into the lumen, they cannot directly sense luminal contents. Consequently, a yet un-identified EEC cell may act as the sensor for peptides and activate interstitial cholinergic and peptidergic neurons.

As food accumulates in the gastric lumen, intragastric pH begins to increase despite the augmented HCl output by parietal cells. This occurs because food components (especially proteins) tend to buffer the acid produced. The buffering capacity of food is such that gastric pH can increase from the basal levels of 1 to 2 to as high as 5 during the ingestion of a meal. This increase in pH tends to nullify the acid-activated reflex that stimulates SST-secreting D cells via the release of CGRP (Figure 4-6).

The *intestinal* phase of gastric secretion commences as the meal begins to leave the stomach and enter the duodenum. The emptying of food from the stomach gradually removes the luminal stimuli (distension, nutrients, pH) that drive the gastric phase of secretion. In addition, the entrance of chyme into the duodenum elicits the release of factors that mediate feedback inhibition of gastric acid secretion.

As a result of the transfer of chyme from the stomach to duodenum, the gastric secretory response wanes and eventually returns to basal levels (usually within 4–5 hours). Several neurocrine and paracrine pathways activated during the cephalic and gastric phases of secretion are de-activated to restore the SST-mediated restraint on gastric acid secretion. There is also a diminishing influence of the vagus due to reductions in the (1) sensory cues of the cephalic phase of vagal activation and (2) distension-induced vago-vagal reflex of the gastric phase. The progressive decline in intragastric peptide levels (due to gastric emptying) removes the luminal stimulus for G cell activation and gastrin secretion. The diminishing influence of stimulatory regulators of gastric acid secretion is accompanied by an increase in the impact of inhibitory regulators; specifically, SST. The waning influence of the vagus allows for the restoration of SST-mediated inhibition of G cells, ECL cells, and parietal cells. The buffering capacity of food is also lost as the stomach empties and the intragastric pH begins to decrease. The decreasing pH re-activates the CGRP-containing sensory neurons that further activate D cells to secrete SST.

The entrance of chyme into the duodenum can also inhibit gastric acid secretion, particularly if it is acidified and/or lipid-rich. Acidification of the duodenum results in the release of secretin from S cells, whereas long chain fatty acids elicit the release of CCK from I cells (Table 1-1). Both secretin and CCK can reduce acid secretion by the stomach, either via humoral or neural mechanisms. The inhibitory effects of secretin and CCK on acid secretion appear to be largely mediated by an afferent vagal pathway that results in SST release from D cells.

GASTRIC MOTILITY

There are three main physiologic aspects of gastric motility: (1) accommodation of a meal with a relatively small rise in intraluminal pressure, (2) contractile activity, which is responsible for mixing of food with gastric juice and reducing particle size, and (3) controlled gastric emptying of chyme (a slurry of food and gastric secretions) into the duodenum. In general, the proximal half of the stomach has little contractile activity and is chiefly involved in food storage, whereas the distal stomach is the site of vigorous contractions, especially in the postprandial state.

Microscopic Anatomy of the Muscularis
The muscle coat of the stomach has three discrete layers of smooth muscle: (1) an outer longitudinal layer, (2) a middle circular layer, and (3) an inner oblique layer (Figure 4-1). The outer

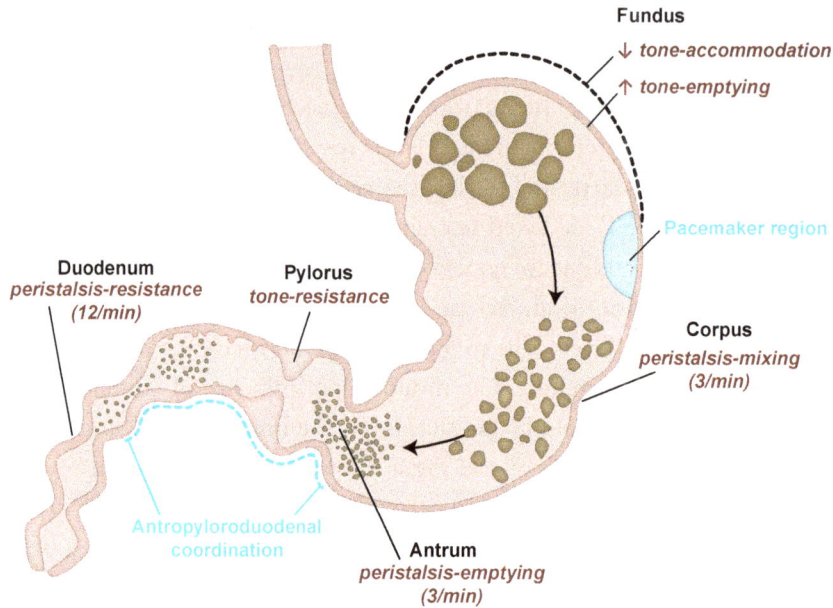

FIGURE 4-8: Regional changes in gastric motor function during the processing of a meal. During ingestion of a meal, the fundus relaxes to accommodate the food. Subsequently, the fundus slowly contracts, moving the accumulated food toward the corpus and antrum for mixing with gastric secretions. Finally, corpus→antrum peristaltic contractions propel the least viscous/smaller material through the pylorus and into the duodenum. Efficient gastric emptying requires coordination of the antropyloroduodenal contractile activity [modified from Parkman HP and Jones MP, *Gastroenterology* 136: 1526–1543, 2009, Elsevier Inc., used with permission].

longitudinal and inner oblique layers are rather incomplete, but the circular muscle layer is well-developed. There are features of the circular and oblique muscle layers that are of particular functional relevance to gastric motility. The thickness of the circular muscle layer progressively increases as it approaches the duodenum. The relative thickness of the circular muscle layer in the corpus and antrum allow for strong contractions to mix and break up ingested food. At the gastroduodenal juncture, the circular muscle forms a thick, ring-like sphincter that occludes the lumen, the pylorus. Relaxation (opening) of the pylorus allows the emptying of gastric contents into the duodenum. The inner oblique muscle layer appears to originate from the circular muscle layer just below the LES, runs along the fundus and fuses again with the circular muscle at the corpus/antrum interface. Because it encloses the saccular portion of the fundus, it contributes substantially to the regulation of fundic volume. When it relaxes the fundus can accommodate large volumes of ingested material, whereas when it contracts it forces fundal contents toward the antrum (Figure 4-8).

Intrinsic nerve plexuses are found in the wall of the stomach, the most prominent of which is the myenteric plexus that lies between the longitudinal and circular muscles. Also located between the longitudinal and circular muscles are the interstitial cells of Cajal (ICC). The ICCs are considered to be the generators of gastric slow wave activity, which is conducted to the adjacent smooth muscle cells via gap junctions. Parasympathetic and sympathetic nerves make synaptic connections with both excitatory and inhibitory neurons of the myenteric plexus as well as the ICC. Electron microscopy has revealed that intramural neurons either lie in close apposition to ICC or make contact with them. Thus, in addition to acting as pacemakers, the ICC may play a role in neurotransmission; relaying signals from extrinsic nerves to the myenteric plexus.

Electrical and Contractile Activity of Smooth Muscle

As shown in Figure 4-9, gastric slow waves originate in the corpus region and spread toward the pylorus; no slow waves can be detected in the fundus. There is a transition zone in the corpus between the absence and presence of slow waves; the "pacemaker" region (Figure 4-8). Although slow waves can travel both proximally and distally, they soon die out in the proximal direction. The ICC cells located in the myenteric region between the circular and longitudinal muscle are likely responsible for generating this pacemaker activity (slow waves). The density of the ICCs increases from the corpus to the antrum, and they fire at the highest frequency at the level of the mid-corpus. The pacemaker region is the basis for the functional division of proximal and distal gastric contractile activity (Figures 4-8 and 4-9). The motility of the proximal fundus is characterized by either tonic relaxation or contraction, allowing for increases and decreases in gastric volume, respectively. The contractile activity of the more distal corpus/antral region is propulsive; the peristaltic activity follows the basal electrical rhythm (BER) set by the pacemaker (3 cycles/min). The contractions in this region spread from the pacemaker toward the pylorus, thereby emptying gastric contents into the duodenum.

Recordings of slow waves from equi-distant points from the pacemaker region to distal antrum exhibit a shorter lag time between successive recordings of a given wave (red line in Figure 4-9A). Thus, the speed of conduction of the slow waves originating from the pacemaker region increases as they travel distally, that is, approach the terminal antrum. Because smooth muscle contraction is dependent on slow waves, the velocity of the corresponding peristaltic wave also increases as it approaches the pylorus (red line in Figure 4-9B). Furthermore, the number and amplitude of the action potentials on the slow waves increases as the slow waves approach the pylorus (Figure 4-9A), resulting in a progressive increase in the strength of the contractions as they move distally (Figure 4-9B). To summarize, both the speed and strength of the peristaltic contractions increase as they approach the pylorus.

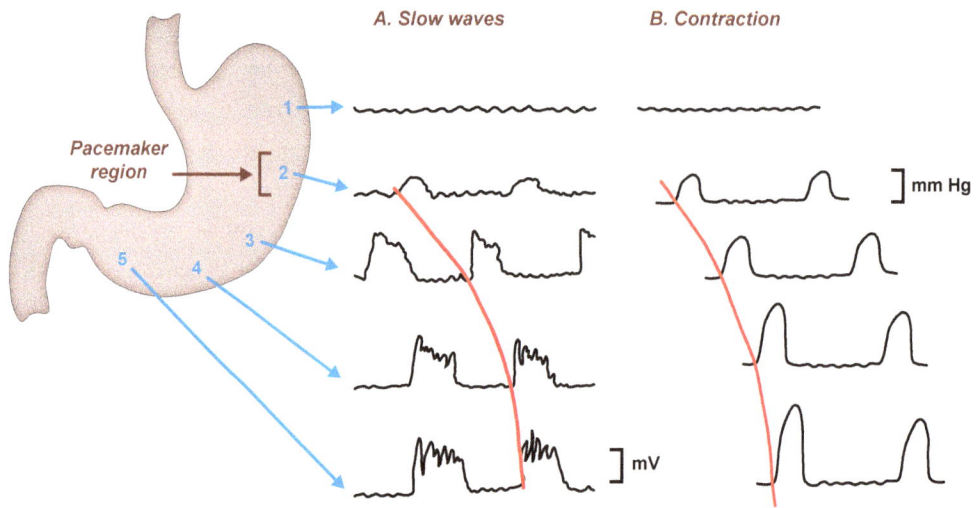

FIGURE 4-9: Electrical activity (A) and intraluminal pressure (B) recorded at 5 equi-distant points along the stomach. Proximal to the "pacemaker" region, no slow waves or peristaltic contractions are evident. Distal to the "pacemaker" region, both electrical slow waves and peristaltic contractile activity are evident. Both the velocity of the slow waves and the number and amplitude of their action potentials increase as they progress toward the pylorus (red line in A). The corresponding peristaltic contractions also increase in velocity and force as they progress toward the pylorus (red line in B).

Gastric Accommodation

The ability of the stomach to accommodate meals is generally attributed to activity of the proximal portion of the stomach. As opposed to the more distal regions, the fundus has a relatively thin muscular layer and is highly compliant; up to 1.5 liters of a meal can be accommodated with little, if any, increase in pressure. Using single-photon emission computed tomography (SPECT), trans-axial images can be reconstructed that yield sufficient detail to quantitate the volumes of proximal and distal regions of the stomach. As shown in Figure 4-10A, both proximal and distal regions of the stomach increase in volume after ingestion of a liquid meal. However, the proximal portion of the stomach (fundus) accommodates 80% of the total volume ingested. To maintain a low intragastric pressure while accommodating a meal, proximal gastric tone decreases. Gastric accommodation can be quantitated using an intragastric balloon of almost infinite compliance connected to a barostatic apparatus. The barostat maintains a constant pressure within the bag by either removing (increases in tone) or adding (decreases in tone) fluid to the bag. As shown in Figure 4-10B, almost immediately after ingestion of a liquid meal, the volume of a bag positioned in the proximal stomach increases; representing a decrease in fundic tone.

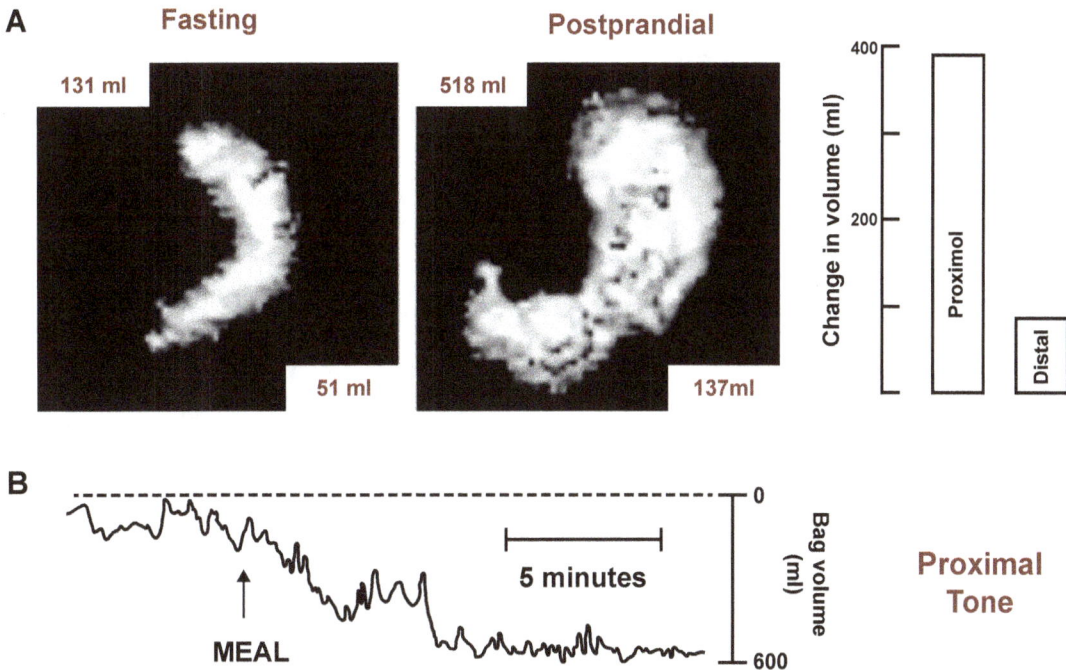

FIGURE 4-10: Accommodation of a meal by the stomach. A. Reconstructed gastric transaxial images acquired by single-photon emission computed tomography (SPECT) in the fasted and postprandial state. Volumes of the proximal and distal stomach are depicted in the upper left and lower right of each image, respectively. The histograms to the right indicate that the postprandial increase in volume is greater in the proximal than the distal stomach; most of the meal is accommodated in the proximal stomach (fundus). B. Gastric accommodation as measured by a balloon positioned in the proximal stomach (fundus). The intra-balloon pressure is maintained constant by a barostat regulating volume added or removed from the balloon. An increase in balloon volume is indicative of a decrease in tone [modified from Camillieri M, *Gastroenterology* 131: 640–658, 2006, Elsevier Inc., used with permission].

Transpyloric Movement of Chyme

The churning and emptying of an ingested meal is attributed, in large part, to the peristaltic contractions of the more distal regions of the stomach: mid-corpus, antrum, and pylorus (Figure 4-11). The initial contraction in the pacemaker region of the corpus begins to move chyme toward the pylorus. The liquid portion of the gastric chyme is expelled through the narrow pyloric opening ($\approx$1–2 mm). As the contractile wave moves to the antrum, the pylorus begins to close and the transpyloric movement of chyme becomes bidirectional. The more liquefied component

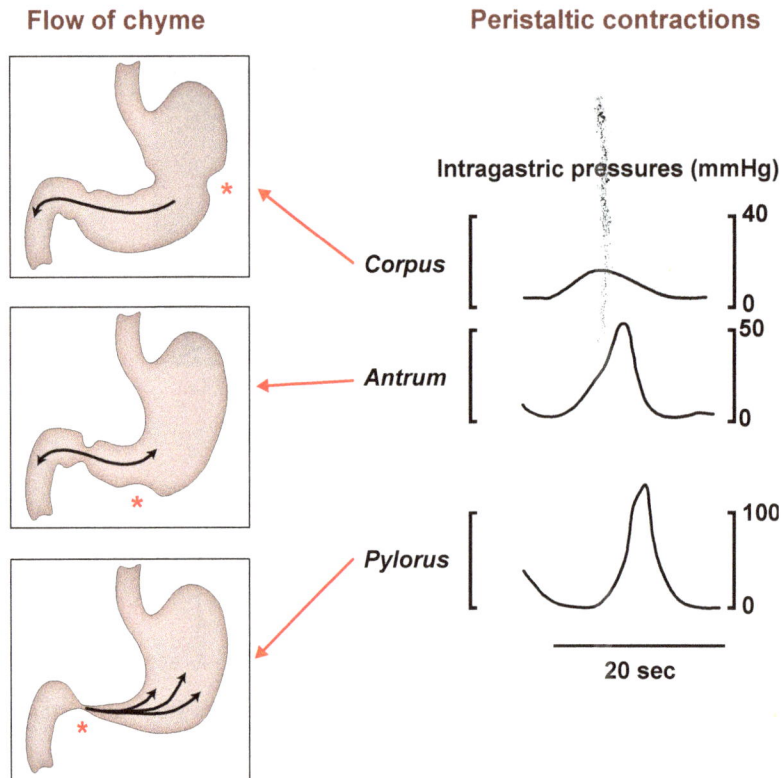

FIGURE 4-11: Transpyloric flow and retropulsion of chyme. The flow of chyme through the pylorus (left side) is dependent on the pressures generated by the peristaltic contractions of the distal stomach (right side). The red asterisk indicates the location of the contraction corresponding to the intragastric pressure generated. The peristaltic wave originating in the mid-corpus and progressing to the antrum results in transpyloric movement of chyme, until the contractile wave occludes the pylorus and propels the chyme back toward the corpus (retropulsion) for further processing. This cycle of contractions is repeated (3 cycles/min) until the chyme is emptied from the stomach.

of chyme continues to empty into the duodenum, but the more viscous portion, containing larger solids, is forced back toward the fundus. When the contractile wave reaches the distal antrum and pylorus, the increased strength of the contractions occlude the pylorus, and all of the chyme is propelled back to the fundus (retropulsion). This retropulsion is believed to be important for the continued mixing and digestion of the viscous components of chyme, thereby, reducing it to a consistency more readily emptied by a subsequent contractile wave.

Gastric Emptying

Gastric emptying of an ingested meal involves the coordinated motor activities of both the proximal and distal stomach as well as the duodenum (Figure 4-12). During the ingestion of a meal the fundus relaxes to accommodate the meal, but subsequently begins to contract (increase in tone) pushing gastric contents toward the antrum. The peristaltic contractions of the antrum force the less viscous component of chyme through an open pylorus. The opening of the pylorus is transient; followed by closure and retropulsion of residual chyme. The pulsatile

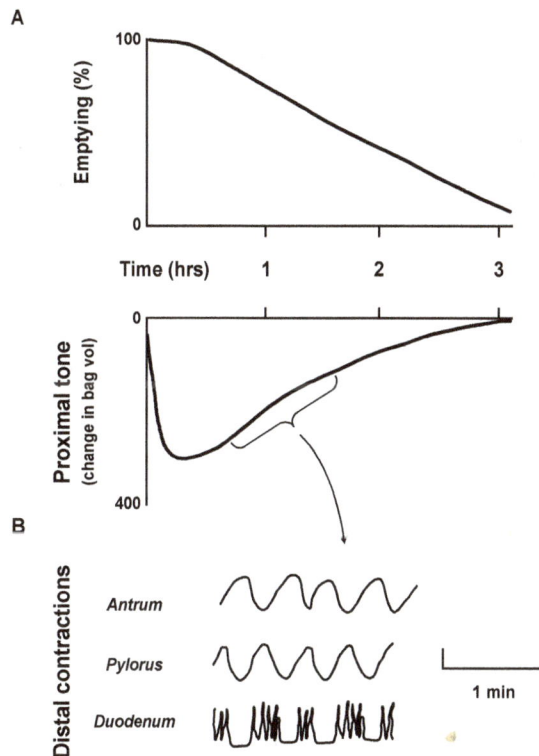

FIGURE 4-12: Gastric emptying is dependent on fundic tone and coordinated antropyloroduodenal contractions. A. Relationship between gastric emptying (upper panel) and fundic tone measured by barostat apparatus attached to an intragastric bag (lower panel). After the initial accommodation of the meal in the proximal stomach (increase in bag volume), fundic tone begins to increase (decrease in bag volume) and parallels gastric emptying. B. Antropyloroduodenal coordination refers to the coordination of the antral peristaltic wave, pyloric opening and inhibition of duodenal contractions to facilitate emptying of gastric chyme into the duodenum [data from Moragas G et al., *Am J Physiol* 264: G1112–1117, 1993 and H. Ehrlein and M. Schemann, *Gastrointestinal Motility*. Technische Universität München, Munich, 2005, used with permission].

nature of transpyloric flow results in the delivery of chyme to the duodenum at a maximum rate of 3 times a minute (gastric BER). Duodenal contractions can occur as often as 10 to 12 times a minute (duodenal BER) and would provide a significant resistance to the entrance of chyme. However, during the antral contractions that propel chyme through the pylorus, duodenal contractions are inhibited (antropyloroduodenal coordination) (Figures 4-8 and 4-12).

INTEGRATED REGULATION OF GASTRIC MOTILITY

Interprandial (Basal) Motility

During the interdigestive period (between meals), the stomach is relatively quiescent; there is little rhythmic propagating or nonpropagating contractile activity. However, the stomach is not flaccid, but has tone. Inhibition of vagal activity (or atropine) relaxes the stomach; an effect prevented by atropine. Thus, the resting tone of the stomach is the result of a vagal cholinergic pathway.

The relative quiescence of the gastric musculature is interrupted periodically by a pattern of motor activity that begins in the pacemaker region of the stomach and traverses the entire small intestine (Figure 4-13). This migrating motility complex (MMC) occurs approximately every 90 to 120 minutes. The MMC is generally divided into three phases based on the contractile pattern. During phase I, the stomach is relatively devoid of contractile activity (aside from tone). Phase I persists for half of the MMC cycle and is followed by a slightly shorter phase II (40% of the MMC cycle) that is characterized by erratic, nonpropulsive, rhythmic contractions. Phase II eventually leads to a very short period (5–10 minutes) of intense peristaltic activity, referred to as phase III. During phase III, the frequency of contractions can reach maximum levels as dictated by the local BER (3/min in the stomach, 11–12/min in duodenum). As a result of the propulsive contractions of phase III, it is regarded as the "activity front" of the MMC. In healthy volunteers, 70% of recorded that phase III contractions begin in the stomach, whereas only 30% originate from the small intestine.

Because of both intraspecies and interspecies variability of the MMC patterns, the mechanisms by which it is initiated and propagated remain elusive. Nonetheless, there is a consensus that both endocrine and neural mechanisms play a role. Motilin and ghrelin are the major enteroendocrine factors implicated in the MMC. Although they are structurally related (share 21% sequence identity), they do not cross-react with each other's receptors. Motilin expression is highest in the duodenum, whereas ghrelin is localized to the stomach (Table 1-1); both are expressed in the local endocrine cells as well as the myenteric plexus. Administration of

Phase   **II**   **III**   **I**

Antrum

75 mm Hg

Proximal duodenum

1 min

Distal duodenum

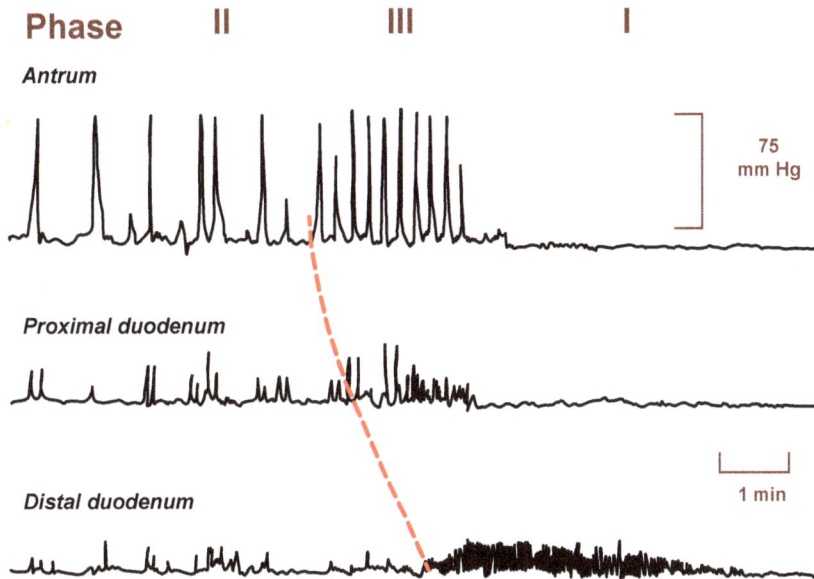

FIGURE 4-13: The migrating motility complex (MMC). Every 90 to 120 minutes during the interdigestive period, a migrating motility pattern begins in the pacemaker region of the stomach, which moves undigested residue through an open pylorus into the duodenum. The MMC is divided into phases: phase I is a period of quiescence, phase II is a period of sporadic activity that builds up to phase III, a period of intense peristaltic contractions. Phase III is the activity front (dashed red line) responsible for the movement of lumenal contents downstream [modified with permission from SpringerNature: Rees WD, Malagelada J-R, Miller LJ, and Go VL. *Dig Dis Sci* 27:4, 321–329, 1982].

either motilin or ghrelin to human volunteers can induce phase III contractions in the stomach reminiscent of MMCs of gastric origin. Circulating levels of both motilin and ghrelin increase during the interdigestive period and decrease postprandially. However, although plasma levels of motilin fluctuate with different phases of the MMC, plasma levels of ghrelin do not. Motilin levels peak just prior to phase III with gastric, but not duodenal, onset MMCs. The precise mechanisms underlying motilin release from the duodenum and how it influences the more proximal and remote antral musculature are not clear. A cholinergic vagal pathway has been implicated because vagotomy (and atropine) can abolish gastric phase III contractions induced by motilin or its agonists (e.g., erythromycin). It has been proposed that duodenal release of motilin activates afferent sensory fibers to initiate a vago-vagal reflex that impacts the gastric musculature.

A major function of the MMC is to evacuate the stomach (and small intestine) of any residual material during the interdigestive period ("housekeeper" function). During phase III antral contractions, the pylorus relaxes markedly. The pyloric relaxation is prevented by antral transection, indicating that a descending intramural inhibitory reflex is involved. Thus, phase III antral contractions serve to clear any undigested roughage not emptied during the postprandial period. The pylorus can open to such an extent that even fairly large items inadvertently ingested (e.g., chewing gum, small coins, etc.), pass into the duodenum. It has also been proposed that the MMC can contribute to the sensation of hunger ("hunger contractions"). Phase III activity of MMCs of gastric origin, but not those of intestinal origin, have been correlated with sensations of hunger.

Postprandial Motility

The MMC is abolished upon ingestion of a meal and gastric contractile activity assumes a pattern that is more conducive to processing the meal, that is, accommodation, mixing/digestion, and emptying. As with gastric secretion, gastric motility after a meal has been rather arbitrarily divided into phases: cephalic, gastric and intestinal. However, these phases can overlap significantly. Cephalic stimulation is ongoing as the stomach is filling with food and the gastric phase will be modulated by the intestinal phase due to concurrent gastric emptying. Although such extensive overlaps preclude strict partitioning of gastric motility into cephalic, gastric and intestinal phases, such a partitioning is useful for discussion purposes.

The *cephalic* phase of gastric motility is characterized by relaxation (reduced tone) of the proximal stomach (fundic region) in anticipation of the need to accommodate incoming food; sometimes referred to as "receptive relaxation." In humans, sham feeding decreases the tone of the proximal stomach. This response is consistent with the decrease in fundic pressure noted by manometry during the initiation of swallowing; the fundus relaxes even before the food traverses the esophagus and reaches the stomach. Sensory cues from the oropharynx appear to be more important than simply the smell and sight of food. In contrast to relaxation of the fundus, sham feeding elicits an increase in antral and duodenal contractions. These responses are mediated by vago-vagal reflexes that involve regional enteric nerves; VIP and NO releasing enteric neurons relax the fundus, whereas a cholinergic mechanism underlies the antral contractions.

During the *gastric* phase, the continuous filling of the stomach further decreases fundic tone; a response referred to as "gastric accommodation." The gastric phase is primarily driven by mechanoreceptors activated by meal-induced distension; the chemical composition of the meal plays a rather minor role. In humans, balloon distension of the fundus to pressures that do not produce sensations of fullness, increase both antral and duodenal contractions. In a similar vein, canine antral and pyloric contractions are induced by ingestion of a meat meal. Simultaneous contractions of the antral/pyloric region would force gastric contents back toward the fundus (retropulsion) for further mixing and digestion (Figure 4-11). However, during the gastric emptying phase, peristaltic contractions of the antrum are associated with relaxation of the pyloric sphincter (Figure 4-12). The pyloric relaxations occur 1 to 2 seconds after the antral contractions and appear to be mediated by NO and VIP released from enteric neurons. Gastric emptying is also facilitated by transient cessation of duodenal contractions that are synchronized with antral contractions and pyloric relaxations, that is, antropyloroduodenal coordination (Figures 4-8 and 4-12).

Collectively, gastric motility is coordinated in such a manner that an ingested meal is accommodated, mixed with gastric secretions, ground to reduce the size of particulate matter and eventually expelled into the duodenum (Figure 4-8). As shown in Figure 4-14, there is a delay in the onset of gastric emptying of solid meals as compared to homogenized solids or liquids. The delay is presumably due to the necessity of "gastric homogenization" of the solid meal; retropulsion of chyme by antro/pyloric contractions serves this purpose. Subsequently, antropyloroduodenal coordination of contractile activity assures the emptying of the less viscous homogenate. Antropyloroduodenal coordination is not important for efficient emptying of a liquid meal; the small pressure gradient developed by an increase in fundic tone is sufficient to drive emptying of liquids into the duodenum.

The *intestinal* phase begins when the stomach empties chyme into the duodenum and reflexes elicited from the intestine restrain gastric emptying. This inhibitory reflex is attributed to the chemical/physical composition of chyme entering the duodenum, rather than distension, per se. The non-nutrient factors that inhibit gastric emptying when present in the duodenum are acid and osmotic stress (hypertonic or hypotonic). Of the macronutrients contained in food (lipids, proteins, carbohydrates), lipids are the most potent inhibitors of gastric emptying. Hydrolytic products of long chain triglycerides (e.g., oleic acid) are more effective than those of medium chain triglycerides and the inhibitory effects are concentration-dependent. Although peptides and disaccharides can also inhibit gastric emptying, their effects are generally attributed to an increase in osmolality resulting from their hydrolysis in the duodenum. Nonetheless, lipids have a much more dramatic inhibitory effect on gastric emptying than acidic or hyperosmolar solutions. This is because lipid hydrolysis continuously generates fatty acids, which prolongs their inhibitory effects, whereas acid is rapidly neutralized by pancreatic and duodenal bicarbonate, and osmolality is promptly returned to isotonicity by transepithelial movement of water.

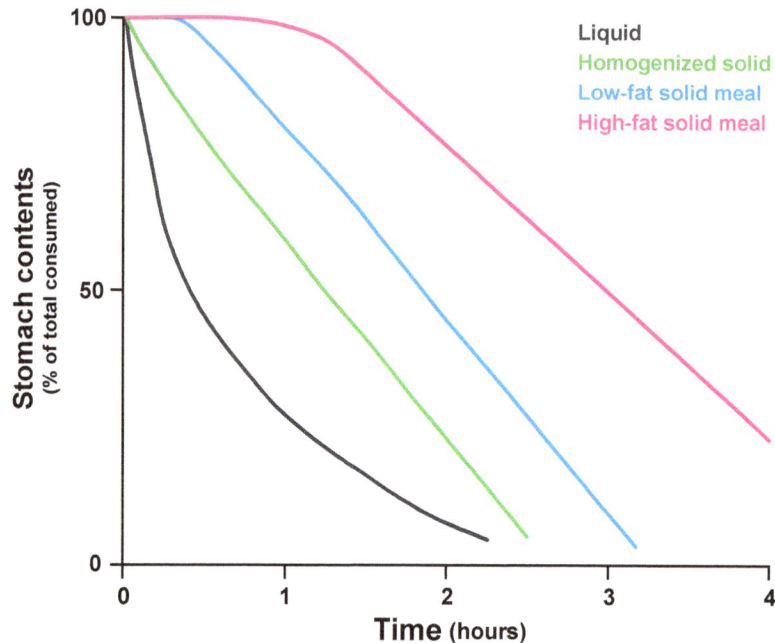

FIGURE 4-14: Gastric emptying of liquids and solids. In general, liquids and homogenized solids empty faster than solid meals. A portion of the delay in emptying of solid meals is the time required for mixing and grinding of the ingested material to produce chyme of appropriate consistency to pass through the pylorus. The lipid content of the meal appears to be an important determinant of gastric emptying; the higher the fat content, the slower the gastric emptying.

Gastric emptying can be inhibited by either a reduction in fundic tone, disruption of antro/pyloric/duodenal coordination, or a combination of both. Perfusion of the duodenum with lipid-rich emulsions decreases fundic tone. In addition, lipid emulsions disrupt antropyloroduodenal coordination by decreasing antral contractions while increasing pyloric and duodenal contractions. Collectively, these effects on gastric motility during the intestinal phase likely account for the prolonged gastric emptying observed with lipid-rich meals (Figure 4-14). The known non-nutrient factors that inhibit gastric emptying when perfused into the duodenum are acidic and hypertonic solutions. The effects of these solutions on proximal and distal gastric motility are consistent with those noted with lipids.

The lipid-induced inhibition of gastric emptying involves activation of afferent vagal sensory fibers, because functional ablation of these nerves (capsaicin) prevents the response. Capsaicin-sensitive afferents are also implicated in the inhibitory effects of acid and hypertonicity in the duodenum. However, as mentioned previously, the terminals of afferent sensory fibers do not reach into the intestinal lumen. The enteroendocrine cells (EECs) lining the intestinal epithelium are believed to be responsible for sensing various constituents of chyme, releasing their secretory products into the interstitium, and activating the appropriate afferent fibers that reside therein. Lipids can release CCK from duodenal I cells and exogenous CCK can relax the proximal stomach and increase contractions of the pylorus. Consistent with this scenario, the inhibition of gastric emptying induced by intestinal lipid emulsions can be blunted by a CCK_A receptor antagonist. It is now generally held that the CCK released in response to lipids interacts with CCK_A receptors on adjacent enteric neurons to elicit an inhibitory vago-vagal reflex that targets the stomach. The most prominent feature of lipid-induced gastric emptying is relaxation of the fundus that is mediated by NO and VIP release from the efferent arm of the vago-vagal reflex. Similar EEC-neural reflexes appear to play a role in gastric inhibition by the presence of hypertonic glucose or acid in the duodenum. Secretin released by acid and serotonin released by hypertonic glucose can activate afferent sensory neurons to elicit the inhibitory vago-vagal reflex.

OVERVIEW OF THE GASTRIC RESPONSE TO A MEAL

In the fasting state, the human stomach contains only a small amount of swallowed air, 20 to 30 ml of gastric secretion, saliva, and debris. The pH of the gastric content is low, generally between 1 and 2 (Figure 4-15). The motor activity of the stomach oscillates between quiescence and the periodic activity associated with the MMC, which serves to remove any undigested debris.

From the start of a meal, gastric acid secretion increases and the fundus of the stomach relaxes. These events may even precede eating if there is a pronounced psychic component (anticipation) in the cephalic phase. Secretion accelerates and the fundus continues to relax during gastric filling. As the stomach is filling, acid secretion begins to accelerate (Figure 4-15). Ingested foodstuffs undergo limited enzymatic hydrolysis. Starch digestion by salivary amylase can continue to some extent, but as acidity increases, this activity declines. Lipolysis, however, can continue to a moderate extent, because oral lipase can function at acidic pH. In a similar vein, the acidic peptidase, pepsin, hydrolyzes proteins to peptides.

Gastric acid secretion peaks by one hour after ingestion of food (Figure 4-15). One can infer that during this accelerating phase stimulatory regulators are dominant over inhibitory regulators (Figure 4-6). Specifically, the acid secretion of the cephalic phase is reinforced by (1) gastric distention as food arrives in the stomach and (2) release of gastrin by peptides generated

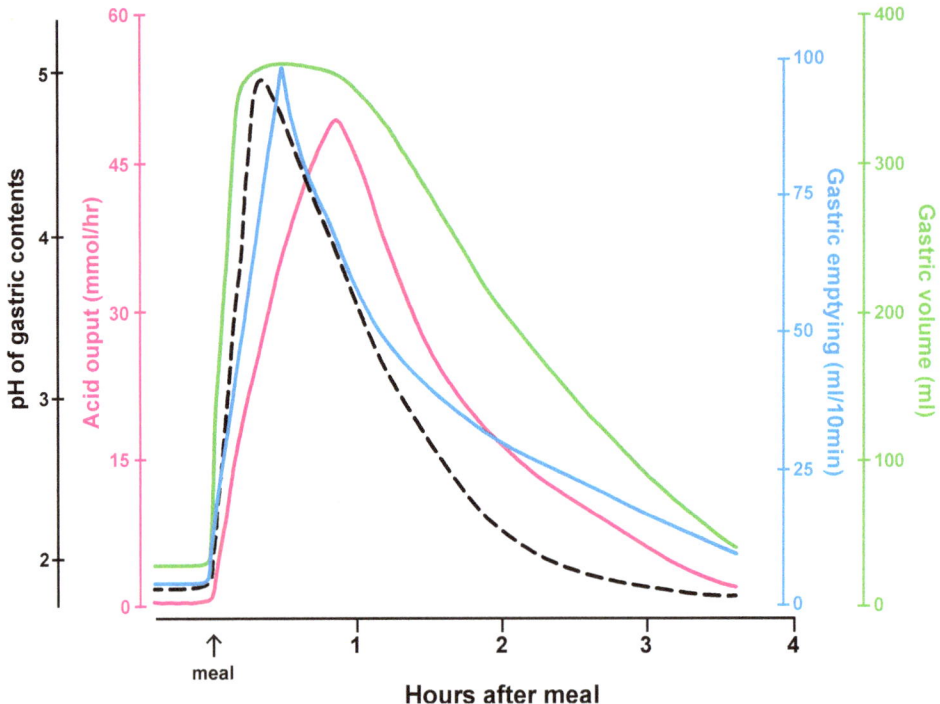

FIGURE 4-15: Summary of gastric secretory and motor activities after ingestion of a meal (40% carbohydrate, 40% fat, and 20% protein). Acid secretion peaks within an hour of ingestion of the meal, but pH of the gastric contents increases due to the buffering capacity of food. The rate of gastric emptying also increases sharply, peaking within 30 minutes of food intake and gradually declining thereafter. Despite the rapid increase in the rate of gastric emptying, there is a lag time before the volume of the stomach begins to decrease; an effect attributed to enhanced volume secretion by the stomach and time required to homogenize the food [data from Malagelada J-R, *Gastroenterology* 70: 203–210, 1976, Elsevier Inc., used with permission].

by the actions of pepsin. Despite the increased acid secretion, the arrival of food in the stomach quickly raises the pH owing to dilution and buffering of HCl.

Shortly after food enters the stomach, vigorous antral contractions occur with intraluminal pressures rising as high as 100 mm Hg. These antral contractions force the more liquid chyme through the pylorus, whereas the more viscous material is forced back toward the corpus (retropulsion) for further processing. The emptying of chyme from the stomach is dependent on coordination of antral, pyloric and duodenal contractions. Nondigestible solids are retained until very late in the postprandial period. If solids cannot be greatly reduced in size, they tend to leave the stomach only when the interdigestive pattern of motor activity (MMC) resumes.

The exit of chyme from the stomach removes the physical (distension) and chemical (peptides) stimuli for acid secretion. Despite the decrease in acid secretion, the pH of gastric contents decreases; an effect attributed to loss of the buffering capacity of food (Figure 4-15). As chyme enters the duodenum, pancreatic and biliary secretions are delivered locally to complete the bulk of the hydrolysis of proteins, carbohydrates, and lipids. The presence of acidic and/or lipid-rich chyme in the duodenum can further decrease gastric acid secretion and delay gastric emptying. Acidic chyme can release secretin, whereas long chain fatty acids can release CCK, both of these enterogastrones can feedback to the gastric mucosa and muscularis to inhibit acid secretion and emptying, respectively. All things considered, the time frame of emptying of a mixed meal and return of gastric acid secretion (and pH) to basal levels is approximately four hours (Figure 4-14 and 4-15).

GASTRIC BLOOD FLOW

In general, gastric blood flow increases after a meal; presumably due to the high oxygen demand of acid secretion. Although administration of secretagogues increases gastric oxygen consumption in proportion to the increment in acid secretion, a hyperemia is not always elicited. For example, histamine-induced acid secretion is associated with a hyperemia, whereas pentagastrin-induced secretion is not. Intrinsic vasoregulatory mechanisms can meet increases in oxygen demand by either increasing oxygen extraction or blood flow or both (Chapter 1). Thus, the oxygen demand of acid secretion induced by histamine (vasodilator) would be met by increasing blood flow, whereas that induced by pentagastrin (not vasoactive) would require an increase in oxygen extraction and perfused capillary density.

The microvascular organization of the gastric mucosa and muscularis is shown in Figure 4-16A. The arteries supplying the stomach pierce the serosa, pass through the muscularis and form an arterial submucosal plexus. Arterioles arising from the submucosal plexus supply either the mucosa or the muscularis. This parallel arrangement allows for independent control of mucosal and muscle blood flows. Under basal conditions (between meals), approximately 75% of total gastric blood flow is distributed to the mucosa; the remaining 25% supplies the muscularis. Ingestion of a meal is associated with an increase in total gastric blood flow, which is preferentially distributed to the mucosa, presumably to meet the oxygen demand of acid secretion.

A unique feature of the gastric mucosal circulation is that it facilitates the matching of mucosal bicarbonate production to acid secretion (Figure 4-16B). As mentioned previously, parietal cells secrete acid into the lumen of the gland while simultaneously secreting bicarbonate

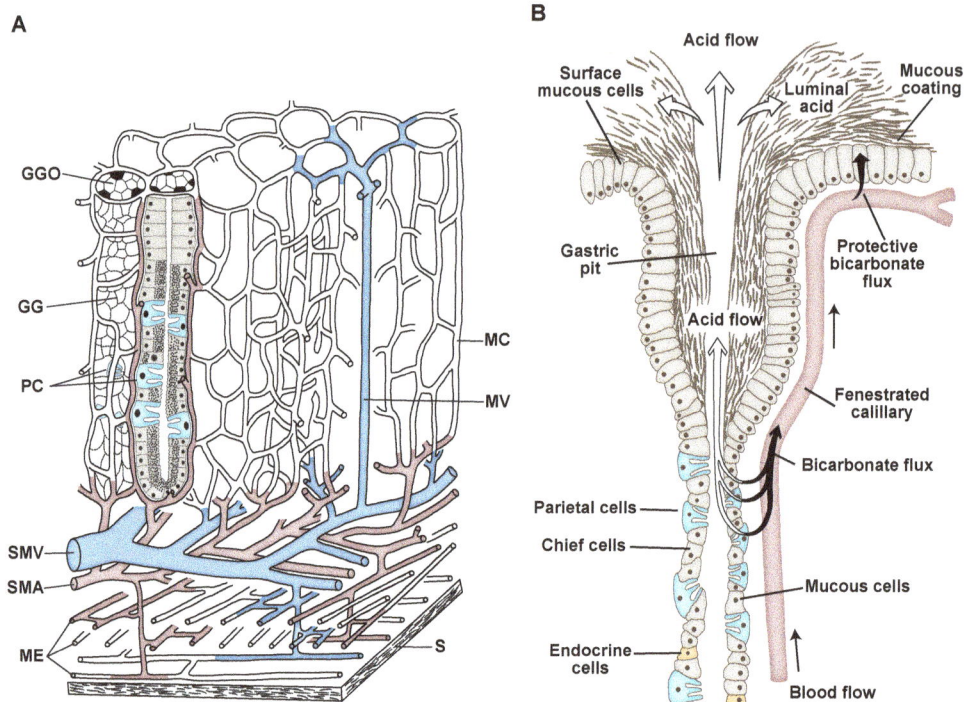

FIGURE 4-16: The gastric microcirculation. A. The submucosal arterioles (SMA) supply the mucosa and muscularis, whereas these regions are drained by the submucosal veins (SMV). The arteries and veins of the muscularis externa run parallel to the smooth muscles of the inner circular and outer longitudinal muscle layers (ME), which is surrounded by a thin serosal layer (S). The arterioles entering the mucosa give rise to mucosal capillaries (MC) which form a network around the gastric glands (GG) containing the parietal cells (PC). When the capillaries reach the surface they surround the gastric gland openings (GGO) giving a "honeycomb" appearance to the capillary network at the surface. The mucosal venules (MV) drain into the submucosal veins. B. The microvascular transport of HCO_3- from the acid-secreting portion of the gastric gland to the surface epithelium. The surface epithelial cells can transport HCO_3- into the overlying mucus layer, creating a protective barrier to acid [modified from Ohtani O et al *Arch Histol Jpn* 46: 1–42, 1983 and Gannon B et al., *Gastroenterology* 86: 866–875, 1984, Elsevier Inc., used with permission].

into the adjacent interstitial space. The discharged bicarbonate subsequently enters the capillaries surrounding the gastric pits. Because of the directional flow of blood in the glandular microvessels, the bicarbonate is transported to the basal aspect of the surface epithelium. Thus, the alkaline tide ensures the availability of bicarbonate for uptake by the surface epithelium and its subsequent secretion into the overlying mucus layer. As gastric acid secretion increases, more bicarbonate is made available to maintain the mucus bicarbonate gradient.

GASTRIC MUCOSAL GROWTH

Chronic increases in the functional activity of the gastric mucosa are associated with increases in its mass. Fasting is associated with a decrease in cell proliferation in the oxyntic mucosa, which is reversed by refeeding. The mucosal proliferative response to feeding, which occurs in the stem cell niche (isthmus), has been attributed to gastrin. Exogenous administration of gastrin induces oxyntic mucosal hyperplasia, but has no effect on the pyloric mucosa. Further, gastrin increases parietal and ECL cell densities, principal cellular targets of gastrin. Chronic pharmacologic inhibition of gastric acid secretion results in hypergastrinemia and an increase in oxyntic mucosal mass, an effect prevented by blockade of the gastrin receptor (CCK$_2$). Antrectomy (removal of gastrin-producing tissue) results in atrophy of oxyntic mucosa; again, reversed by exogenous administration of gastrin. Finally, genetic inhibition of gastrin production results in a decrease in parietal cell number, whereas genetic overexpression of the hormone increases parietal cell density.

GASTRIC MUCOSAL DEFENSE

The primary functions of the stomach are to prevent bacterial colonization and initiate the hydrolysis of ingested food, particularly proteins. To accomplish these tasks the gastric mucosa secretes acid and pepsin. The question arises: how does the gastric mucosa protect itself from its own destructive secretions? Fortunately, a "mucosal defense" system is in place that serves to prevent or, at least, minimize mucosal injury. The mucosal defense system is generally stratified into preepithelial (alkaline mucus) and epithelial (restitution) components. The outer loose layer of mucus tends to trap bacteria in the gastric lumen where they are susceptible to the antibacterial action of acid and pepsin. The bicarbonate gradient of the more viscus, inner mucus layer serves to maintain surface epithelial pH near neutral thereby minimizing the corrosive effects of acid and pepsin. If some epithelial cells are injured and shed from the epithelial lining, the defect is rapidly repaired by a process referred to as "restitution." During restitution, noninjured epithelial cells adjacent to the breach adopt a migratory phenotype, extend lammelipodia, and spread over the denuded basement membrane.

When the first two lines of mucosal defense fail and a breach of the epithelial lining occurs, a neural reflex comes into play to curb the further progression of mucosal injury (development of frank ulceration). The interstitium contains afferent sensory fibers with specific receptors/channels that can be activated by acid (pH < 6.0). The efferent arm of this neural reflex increases mucus and HCO_3^- secretion and it elicits an increased mucosal blood flow (hyperemia); the

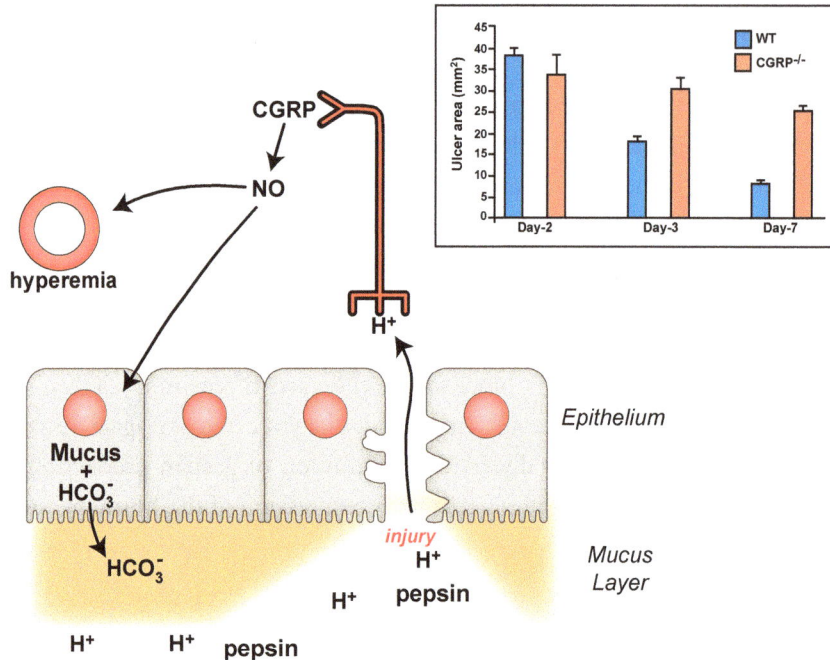

FIGURE 4-17: Injury to the epithelium results in a local neural reflex that facilitates mucosal repair. Afferent sensory neurons respond to H^+ in the interstitium and elicit a local reflex that releases CGRP/NO to (1) dilate submucosal arterioles supplying the mucosa and (2) stimulate mucus/HCO_3^- secretion by surface epithelial cells. This neural reflex facilitates repair of the mucosal injury as illustrated by the delay in ulcer healing in CGRP deficient ($CGRP^{-/-}$) mice (inset) [modified from Granger et al., *Compr Physiol* 5: 1541–1583, 2015 and Ohno T et al., *Gastroenterology* 134: 215–225, 2008, Elsevier Inc., used with permission].

major neurotransmitters involved are CGRP and NO (Figure 4-17). The enhanced secretion of an alkaline mucus coats the injured epithelium, thereby limiting further damage and facilitating restitution. The mucosal hyperemia serves to dilute (via capillary filtration) and neutralize (via enhanced delivery of bicarbonate) any acid and pepsin that accesses the interstitium. Both of these responses serve to enhance healing of the injured mucosal epithelium.

VOMITING (EMESIS)

Although visual, olfactory, and gustatory cues limit ingestion of potentially noxious material, this screening does not completely prevent toxic substances from entering the stomach. In addition, some pharmaceuticals can also irritate the gastrointestinal tract (e.g., alcohol, NSAIDs).

The vomiting reflex is a protective mechanism that responds to gastric mucosal irritation by forcefully expelling gastric contents via the mouth. Mucosal chemosensors detect and respond to various irritants (acids, alkali, toxins). The major enteroendocrine cell implicated in detecting irritants is the EC cell, which releases serotonin into the interstitium, where it activates the afferent extrinsic vagal nerves. The vagal afferents stimulate the area postrema ("chemoreceptor trigger zone"), which relays information to the brain stem "vomiting center" to elicit a programmed vomiting response.

The act of vomiting is generally preceded by retching. This involves contraction of the diaphragmatic, abdominal and thoracic muscles against a closed glottis (Valsalva maneuver). Because the fundus and LES are relaxed, the increase in abdominal and thoracic pressures force the gastric contents up into the esophagus. No material is expelled however due to the closed UES. The contracted muscles then relax, and the esophageal contents return to the stomach. Vomiting is an extension of the retching process. With vomiting, the abdominal and thoracic contractions are sustained and so strong that the pressures generated overcome resistance from the UES, and the gastric contents are expelled. Simultaneously, respiration is inhibited and appropriate laryngeal and pharyngeal responses (e.g., lowering of the epiglottis and raising of soft palate) prevent aspiration. After ejection, esophageal peristalsis resumes and clears any remaining material back into the stomach. Additional cycles may be necessary to evacuate the stomach completely, after which abdominal muscle contractions cease, and normal respiration ensues.

Vomiting is associated with a variety of medical conditions, including infections, brain tumors or brain injuries, myocardial infarctions, noxious or toxic ingestions, bulimia, and many other disease processes. The excessive and prolonged vomiting that accompany these conditions, as well as prolonged nasogastric suction, can lead to significant disturbances in acid-base balance. Considering of the composition of gastric juice, it can be seen that the loss of gastric secretions will include water, H^+, Na^+, K^+, and Cl^-, with the major metabolic disturbances being dehydration and hypokalemic alkalosis. This alkalosis is perpetuated by contraction of the extracellular fluid volume due to water and sodium chloride loss in gastric juice, and the development of potassium deficiency due to secondary aldosteronism and the loss of K^+ in the gastric juice. Because the renal compensatory mechanisms do not correct this complex disturbance, careful fluid and electrolyte replacement is necessary in such patients.

PAIN SENSATIONS FROM THE STOMACH

Most people suffer periodically from pain or discomfort in the area of the upper abdomen, often referred to as "indigestion" or "dyspepsia." The symptoms range from early satiety, bloating, nausea, vomiting, to severe pain. The major causes of dyspepsia are chronic peptic ulcer disease (PUD) and medications, such as NSAIDs. Prompt neutralization of gastric acid relieves the pain, indicating that acid is activating interstitial nerve endings. Gastric distension can also be detected by mechanoreceptors in stomach muscle, and if severe enough, cause pain. Both branches of the autonomic nervous system are involved in transmitting the pain sensations. Gastric spinal afferents transmit mechano-nociception (gastric distention), whereas gastric vagal afferents convey chemo-nociception (acid).

PATHOPHYSIOLOGY AND CLINICAL CORRELATIONS

Peptic Ulcer Diseases

In the experimental setting, gastric mucosal injury can be demonstrated only when either (1) physiologically irrelevant aggressive substances are applied to the mucosa (e.g., HCl at pH < 1.0) or (2) one or more of the components of the mucosal defense system are compromised (e.g., decreased mucus production). Thus, in physiologically relevant scenarios, the gastric mucosa has an adequate defense system in place to protect itself from the acidic/proteolytic juice that it secretes. However, when the gastric mucosal defense system is impaired by exogenous factors, peptic ulcer disease (PUD) ensues. The two major causes of PUD are *H. pylori* infection and chronic use of nonsteroidal anti-inflammatory drugs (NSAIDs).

Helicobacter Pylori (H. pylori)

Ingestion of food can introduce exogenous pathogens to the gastrointestinal tract. However, most bacteria entering via the oral route are prevented from colonizing the stomach due to the bactericidal action of gastric acid/pepsin, entrapment in mucus, and eventual expulsion during gastric emptying. Nonetheless, the stomach harbors up to 200 bacterial species, and *H. pylori* is one such species. Humans have harbored *H. pylori* for approximately 60,000 years, and currently, it is estimated that 50% of the global population is infected by the bacteria. Although it is not known why *H. pylori* selectively colonize the stomach, it is clear that *H. pylori* induces gastritis and PUD in roughly 20% of infected individuals. Because of the prevalence of this clinical condition, considerable effort has been made to define the mechanisms by which *H. pylori* can overcome the gastric mucosal defense system to induce inflammation and ulcer formation (Figure 4-18).

In vitro studies indicate that, like other species of bacteria, *H. pylori* rarely survive longer than 30 to 60 minutes in an acidic environment containing pepsin (pH < 3.5). Thus, to success-

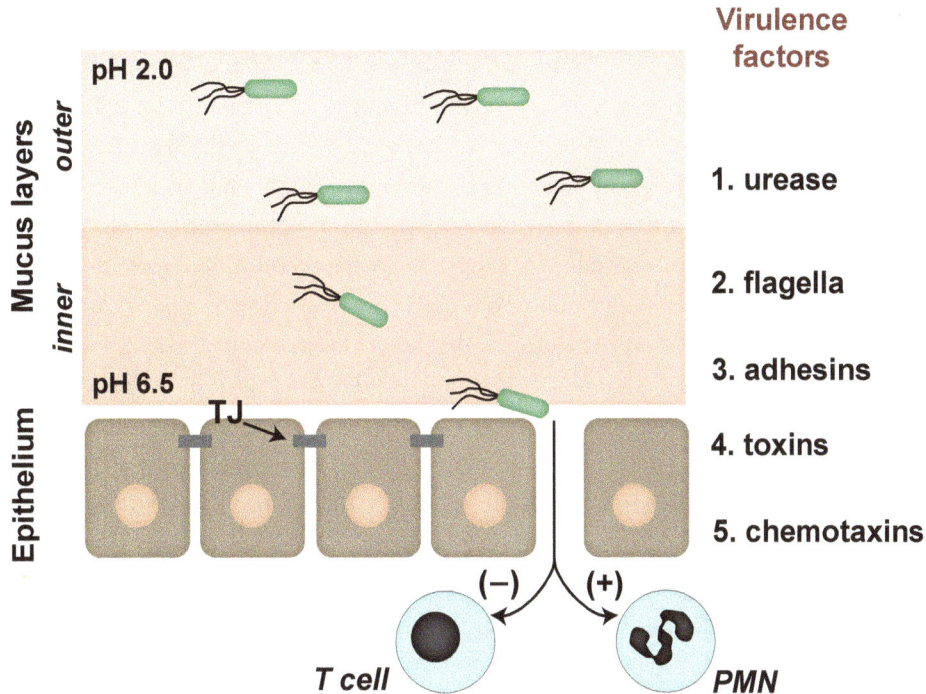

FIGURE 4-18: The five virulence factors involved in *H. pylori* colonization of the stomach and pathogenesis of mucosal injury: (1) the local production of NH_3 by urease facilitates survival in the acidic environment of the stomach, (2) flagella allow migration toward the alkaline milieu of the inner mucus layer, (3) adhesins allow adherence to epithelium near tight junctions, (4) release of toxins disrupt tight junctions (TJ) and damage epithelium, and (5) release of chemotactic factors induce inflammation (i.e., recruit neutrophils, PMN), whereas toxins entering the interstitium suppress T cell function [modified from Kao C-Y et al., *Biomed J* 39: 14–23, 2016 and Dunne C et al., *World J Gastroenterol* 20: 5610–5624, 2014, Elsevier Inc., used with permission].

fully colonize the stomach, *H. pylori* must "escape" the acidic/peptic milieu of the lumen and enter the more alkaline mucus layer within a very limited time-frame. The successful egress of *H. pylori* from the gastric lumen into the mucus layer is facilitated by two factors operating in concert: bacterial urease and flagellar motility (Figure 4-18). The ability of *H. pylori* to survive in an acidic setting is dependent on intracellular urease and its substrate, urea. Urea present in gastric juice enters the cytoplasm of *H. pylori* via an H^+-activated channel; extracellular acidification increases urea entry. Urease hydrolyzes urea to NH_3 and CO_2; the local increase in NH_3 allows the bacteria to maintain an intracellular pH near normal in the face of an acidic external environment. The migration of *H. pylori* into and through the mucus layer is accomplished by the rotational movement of its flagella. The bacteria have a preferred niche within the mucus, being confined primarily in the inner mucus layer adjacent to the epithelium. This spatial orientation is attributed to the ability of *H. pylori* to detect and respond to the mucus pH gradient by directional movement to a region of higher pH. Despite these advantages, as little as 0.5% to 1.0 % of a gastric inoculum of *H. pylori* successfully colonizes the juxtamucosal (inner) mucus layer.

Most of the H. pylori are noted as "swimming" in the inner mucus layer, with very few actually attached to the gastric epithelium. The pathogen uses multiple adhesion molecules to attach to cognate ligands on the epithelium, preferably near junctional regions. Specific "toxins" derived from the bacteria can disrupt interepithelial junctions to compromise the integrity of the epithelial lining. The epithelial injury and chemotactic factors released by *H. pylori* result in the recruitment of inflammatory cells to the affected site (Figure 4-18). The local inflammatory response (gastritis) further exacerbates mucosal injury. The *H. pylori*-induced gastritis is a chronic condition and, when severe enough, can result in gastric ulceration. The chronic nature of the infection and inflammation is due to the ability of *H. pylori* to evade clearance by the immune system. *H. pylori* uses molecular mimicry to avoid detection by recruited PMN; both LPS and flagella contain humanized components and are recognized by PMN as "self." In addition, the recruited T cells are rendered hyporesponsive by H. pylori toxins entering the interstitium.

Nonsteroidal Anti-Inflammatory Drugs (NSAIDs)

NSAIDs are a diverse class of pharmaceuticals with analgesic, anti-inflammatory, and antipyretic properties. They are the most widely used medications for acute conditions, ranging from intermittent headaches and bouts of the common cold or flu, to more chronic illnesses such as rheumatoid arthritis or gout. The widespread use of NSAIDs has uncovered adverse gastrointestinal side effects, specifically gastrointestinal mucosal injury and ulceration. The mucosal injury induced by NSAIDs is attributed to both (1) inhibition of cyclooxygenase (COX) resulting in reduced levels of prostaglandins and (2) a direct toxic effect on the epithelium.

Prostaglandins (PGs) are lipid autocoids derived from membrane phospholipids. In brief, arachidonic acid is released from plasma membranes by phospholipase A_2 and subsequently converted to various prostaglandins by sequential actions of cyclooxygenases (COX-1 and COX-2) and specific synthases/isomerases. Once synthesized, they are released into the extracellular space and act on specific receptors in an autocrine or paracrine manner. Because NSAIDs inhibit COX, it is generally held that the resultant decrease in prostaglandins (esp. prostaglandin E_2; PGE_2) plays a critical role in the gastropathy associated with the use of these drugs.

Prostaglandins augment many of the components of gastric mucosal defense and diminish the impact of the major aggressive factor, acid/pepsin. Prostaglandins increase both mucus and bicarbonate secretion by the epithelium, thereby contributing to the maintenance of the mucus pH gradient. Prostaglandins are also vasodilators and their continuous local generation ensures the maintenance of mucosal blood flow. If the epithelial lining is abraded, PGs accelerate epithelial restitution. Finally, PGs inhibit acid secretion by a direct action on parietal cells (Figure 4-3B). Thus, the NSAID-induced reduction in local levels of prostaglandins results in an unfavorable shift in the balance between aggressive and defensive factors, and consequently promotes gastric mucosal injury.

NSAIDs are also able to cause gastric mucosal injury independently of COX inhibition and lowering of local prostaglandin levels. NSAIDs can interact with phospholipids in mucus to reduce mucin hydrophobicity and render it more permeable to acid. NSAIDs can also penetrate cell membranes and cause intracellular injury; with mitochondria a critical target. Intracellular NSAIDs uncouple oxidative phosphorylation and decrease mitochondrial respiration, ultimately resulting in cell death and disruption of the epithelial lining. The increase in epithelial permeability allows acid and pepsin to enter the interstitium and aggravate mucosal injury.

H. pylori and NSAIDs represent the underlying cause of PUD in over 90% of patients suffering from this condition. Other, less common, conditions that associate with PUD are Crohn's disease, cocaine usage, radiation injury, and viral infections. Clinically, mucosal breaks in the stomach or duodenum are classified as *erosions* if less than 5 mm wide and *ulcers* if larger than 5 mm. Diagnosis is made by endoscopic evaluation with a gastroscope placed into the lumen of the esophagus and advanced into the stomach and duodenum. Although radiological investigation with barium was used to identify ulcers prior to the introduction of the flexible endoscope in the 1970s, this approach is not routinely used in most patients. Pain patterns associated with the development of an ulcer are fairly inconsistent and unreliable. Treatment of an ulcer generally requires an antisecretory medication (PPI). Ulcers secondary to *H. pylori* or patients found to have *H. pylori* identified by pathologic examination, immunohistochemical staining, or stool antigen identification should receive an antimicrobial regimen. The World Health Organization has classified *H. pylori* as a carcinogen and thus a target for eradication. Treatment regimens are chosen based on local resistance patterns, penicillin allergy history, and prior macrolide antibiotic exposure. Multiple regimens have been proposed, which should be combined with a PPI and/or bismuth product. Resistance is becoming an increasing problem resulting in treatment failures (especially with macrolides). The long-term risk of *H. pylori* infection can include gastric carcinoma or MALT (mucosal associated lymphoid tissue) lymphomas that are the result of two different pathways in the development of malignancies. Early MALT lymphoma can be effectively treated with antibiotics.

NSAID-induced PUD can be treated successfully with proton pump inhibitors (PPIs) and discontinuation of the inciting NSAID, attesting to the importance of acid penetration across a compromised mucosal defense in the gastropathy. An alternative approach to gain the therapeutic benefits of NSAIDs and avoid the acid-induced gastropathy includes the use of slow-release or enteric-coated NSAID preparations. However, NSAIDs can also induce intestinal injury if they are allowed to reach this region. The intestinal cytotoxicity of NSAIDs is independent of acid and PPIs do not alleviate, or can even worsen, the damaging effects of NSAIDs in the small bowel. The interactions of NSAIDs with bile salts and bacteria in the intestine as well as their enterohepatic circulation are important aggravating factors in NSAID-induced injury at this site. As little as 10 mg of aspirin is sufficient to injure the intestinal mucosa.

Ulcers may bleed with increasing size or with progression into a vessel beneath the surface of the mucosa. On endoscopy, the presence of a visible vessel portends a high risk for recurrent bleeding. An ulcer may also extend through all the gastric or duodenal wall layers and result in a *perforation*. A perforation results in the release of gastric luminal contents into the peritoneum; surgical repair is often required for treatment when this condition develops. Ulcer development may also lead to an obstruction, particularly if located in the pyloric channel. The obstructive process is a result of edema and inflammation. The narrow channel created by this process leads to vomiting, weight loss, and problems associated with early satiety.

Gastroparesis (Impaired Gastric Emptying)

Impaired gastric emptying (gastroparesis), an uncommon (prevalence per 100,000 = 9.6 in men and 37.8 in women) and generally non life-threatening condition, can significantly impair quality of life and well-being. Although approximately one third of gastroparesis cases are considered idiopathic, this condition is known to be associated with prior gastric surgery (including bariatric surgery), certain medications (opioids, antidepressants), connective tissue disease (e.g., scleroderma), and infections. Gastroparesis (Gp) is also a complication in patients with a long history of diabetes mellitus and uncontrolled glucose levels for prolonged periods. Although the underlying mechanisms that elicit Gp remain poorly understood, diminished activity of neuronal nitric oxide synthase (nNOS) and the loss and/or dysregulation of interstitial cells of Cajal (ICC) are considered likely causes of this condition. A gastroparesis-like state can be induced in normal mice by pharmacologic inhibition of nNOS and is also observed in mice that are genetically deficient in nNOS. Vagus nerve damage has also been invoked as an underlying cause of Gp.

The primary symptoms of Gp are nausea, vomiting, feeling full early after starting a meal, and abdominal pain. In diabetic patients, a history of retinopathy, peripheral neuropathy, or nephropathy is a risk factor for the development of diabetic gastroparesis. Initial testing for Gp involves endoscopic evaluation with identification of food in the gastric body, even after a prolonged fasting period prior to the endoscopy. A testing process using a meal with low-fat, egg-white labeled with technetium-99m sulfur colloid with subsequent radio-labeled imaging can be used to detect prolonged gastric emptying. Treatment of Gp involves maintaining normal glucose levels (diabetic gastroparesis), avoiding fat-laden meals (which prolong gastric emptying) and consuming liquid, rather than solid, calories (which would also facilitate emptying) (see Figure 4-14). The patient may also benefit from prokinetic (peristalsis-enhancing) medications, although significant side effects are noted with these agents. Implantable gastric electrical stimulation (GES) devices have also been shown to improve gastric emptying and symptom severity in cases of refractory diabetic gastroparesis. In more severe forms of Gp, weight loss occurs, and the patient may require nutritional support to maintain body weight.

REFERENCES

Camilleri M, Parkman H, Shafi M, Abell T, Gerson L. Management of gastroparesis. *Am J Gastroenterol* 2013;108:18–37. doi: 10.1038/ajg.2012.373.

Chey W, Wong B. Management of Helicobacter pylori Infection. *Am J Gastroenterol* 2007; 102:1808–1825. doi: 10.1111/j.1572-0241.2007.01393.x.

Cryer, B, Feldman, M. Effects of very low dose daily, long-term aspirin theraphy on gastric, duodenal, and rectal prostaglandins levels and on mucosal injury in health and humans. *Gastroenterology* 1999; 117:17–25. doi: 10.1111/j.1572-0241.2007.01393.x

Ehrlein H and Schemann M. Gastrointestinal motility, Technische Universität München, Munich, 2005.

Hunt RH, Camilleri M, Crowe SE, El-Omar EM, Fox JG, Kuipers EJ, Malfertheiner P, McColl KEL, Pritchard DM, Rugge M, Sonnenberg A, Sugano K, and Tack J. The stomach in health and disease. *Gut* 64: 1650–1668, 2015.

Kvietys PR, Yaqinuddin A, and Al-Kattan W. Gastrointestinal Mucosal Defense System. In: Colloquium Series on Integrated Systems Physiology (Granger DN and Granger JP, eds.). Morgan and Claypool Publishers, 2015. https://doi.org/10.4199/C00119ED1V 01Y201409ISP058

Lanza F, Chan F, Quigley E. Prevention of NSAID-related ulcer complications. *Am J Gastroenterol* 2009; 104:728–738.

Rayner CK, Hebbard GS, and Horowitz M. Physiology of the antral pump and gastric emptying. In: Physiology of the Gastrointestinal Tract (LR Johnson, ed.), 5th edition, 959–996, Academic Press, 2012.

Schubert M. Regulation of gastric acid secretion. In: Physiology of the Gastrointestinal Tract (LR Johnson, ed.), 5th edition, 1281–1310, Academic Press, 2012.

CHAPTER 5

The Pancreas

INTRODUCTION

The pancreas is both an exocrine and an endocrine organ. The endocrine activity of the pancreas is localized in clusters of specialized cells (islets of Langerhans) interspersed throughout the gland. The islet cells, which comprise only 2% to 3% of the parenchyma, secrete several important hormones of which insulin and glucagon are notable for their roles in the regulation of blood glucose. Pancreatic polypeptide and somatostatin, which modulate the function of insulin- and glucagon-producing cells, are also produced by islet cells. The exocrine pancreas elaborates a secretion that is delivered to the duodenum, where it plays an important role in the digestive process. This exocrine secretion, termed pancreatic juice, contains two functionally important constituents: bicarbonate ions and digestive enzymes. The bicarbonate ions neutralize the acidic chyme entering the duodenum from the stomach. Raising the pH of chyme not only protects the mucosal epithelia from injury but also provides an optimal pH for the activity of the pancreatic enzymes. The digestive enzymes of the juice are capable of degrading all of the major nutrients of ingested food, that is, carbohydrates, proteins, and lipids. Although the salivary glands and stomach also elaborate enzymes that participate in the digestion of a meal, pancreatic enzymes play an essential role in this process and with the loss of pancreatic exocrine function nutrient absorption is significantly impaired, resulting in malnutrition.

ANATOMY

The pancreas, which weighs approximately 80 grams in the adult human, is situated in close apposition to the duodenum. The four anatomical regions of the pancreas are the head, tail, body, and uncinate process (Figure 5-1). The pancreas is a compound acinous gland whose lobules (account for >80% of pancreas) are bound together by loose connective tissue. The lobules are composed of grape-like clusters of acini, which are structurally similar to those of the serous salivary glands. Each acinus is a sphere that consists of 20 to 50 pyramidal cells (with their apices facing the lumen) and a few irregularly shaped centroacinar cells (Figure 5-1, inset). The acinar cells are specialized for protein secretion and contain an elaborate network of rough endoplasmic reticulum in the basal portion of the cell, a Golgi complex in the midportion, and numerous zymogen granules in the apical region. Each acinus is drained by a ductule. The centroacinar cells, which are the most proximal cells lining the ductules, are characterized by a sparse cytoplasm devoid of zymogen granules but containing many large mitochondria. The centroacinar cells and the cuboidal cells of the interlobular (intercalated) duct, which drains the acinus, appear to be the primary sites of electrolyte and water secretion.

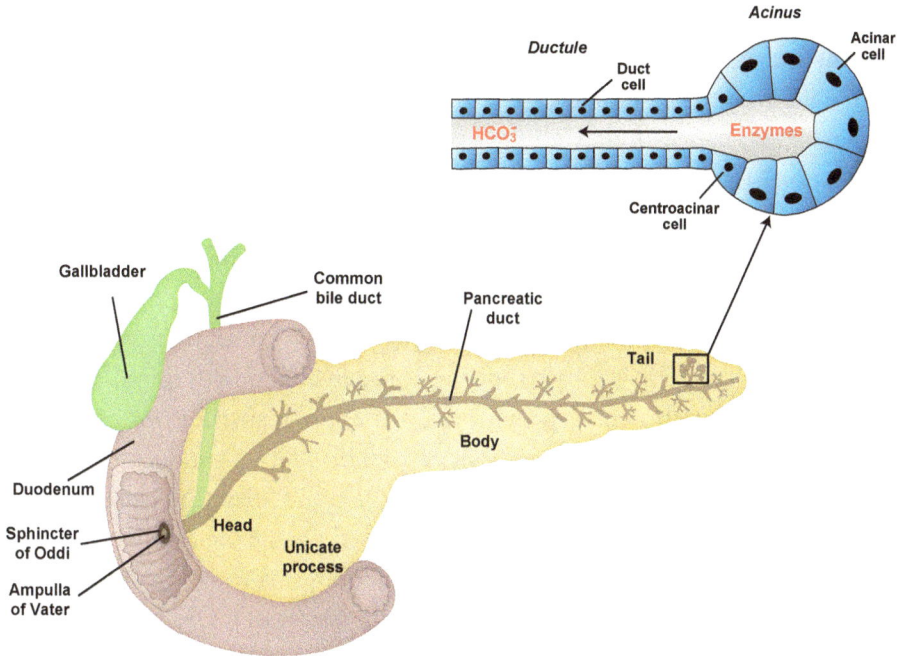

FIGURE 5-1: Gross and microscopic anatomies of the pancreas. The main pancreatic duct and common bile duct are illustrated. The inset depicts the functional secretory unit of the exocrine pancreas, which is comprised of an acinus and an intercalated duct, the sources of enzymes and fluid/bicarbonate respectively.

The intralobular ducts converge to form the interlobular ducts, which empty into the main pancreatic ductal system. There are two main ducts that deliver the secretion to the duodenum. The major pancreatic duct (duct of Wirsung) arises from the tail of the pancreas and enters the duodenum at the ampulla of Vater, alongside the common bile duct; both are surrounded by a ring of smooth muscle, called the sphincter of Oddi. In some individuals (< 40%), the major pancreatic duct and the common bile duct join before entering the duodenum. The minor duct (duct of Santorini) enters the duodenum a few centimeters cephalad to the sphincter of Oddi. In 70% to 90% of individuals, the two pancreatic ducts are in structural continuity within the gland.

Regulation of Pancreatic Mass

Pancreatic size is regulated through genetic programing and environmental influences such as food intake, and the mass of exocrine tissue is regulated independently of islet mass. From birth to adulthood, pancreatic growth occurs roughly in proportion to body growth. During this period, the increase in pancreatic mass primarily reflects cell division, with little change in cell size. An adaptive growth response is elicited in the pancreas when the need for digestive enzymes is increased, such as following the consumption of a high protein diet. Similarly, the removal (diversion of pancreatic juice) or inhibition of trypsin in the duodenal lumen elicits a trophic response in the pancreas. Trypsin mediates this influence on pancreatic growth through a feedback system that involves CCK and is geared toward maintaining a constant activity of trypsin in the intestinal lumen. The adaptive growth responses to a high-protein diet and trypsin inactivation/removal are blunted by treatment with a cholecystokinin (CCK) antagonist. The possibility that CCK is a key mediator of pancreatic growth is also supported by the observation that administration of exogenous CCK is a strong stimulus for pancreatic growth. Administration of cholinergic analogues (e.g., carbachol) will also induce modest pancreatic growth, suggesting that parasympathetic nerves may also modulate the pancreatic growth response. Although secretin per se has little or no effect on pancreatic growth, it appears to significantly potentiate the trophic actions of CCK. Furthermore, pancreatic growth can result from feeding soybean trypsin inhibitor, which stimulates the release of CCK.

COMPOSITION AND FORMATION OF PANCREATIC JUICE

The human pancreas can secrete roughly 1 l of juice, which is equivalent to ~10 times its weight per day. The secreted juice contains a variety of electrolytes and proteins, the concentrations of which vary with the mode of stimulation. The principal anions are Cl^- and HCO_3^-, the concentrations of which depend on the flow rate. The principal cations are Na^+ and K^+, which are secreted at concentrations similar to their plasma concentrations over a wide range of secretion rates. Pancreatic enzymes account for more than 90% of the proteins in the juice. The electrolytes and enzymes are secreted primarily by the ductular and acinar cells, respectively. Therefore, the final composition of the juice is dependent on the relative secretory activities of the ductules and acini.

Electrolytes

In contrast to saliva, pancreatic juice is isotonic with plasma at all rates of secretion (Figure 5-2). The most likely explanation for this phenomenon is that water moves freely across pancreatic epithelia when an osmotic gradient is generated by active transport of solutes, primarily

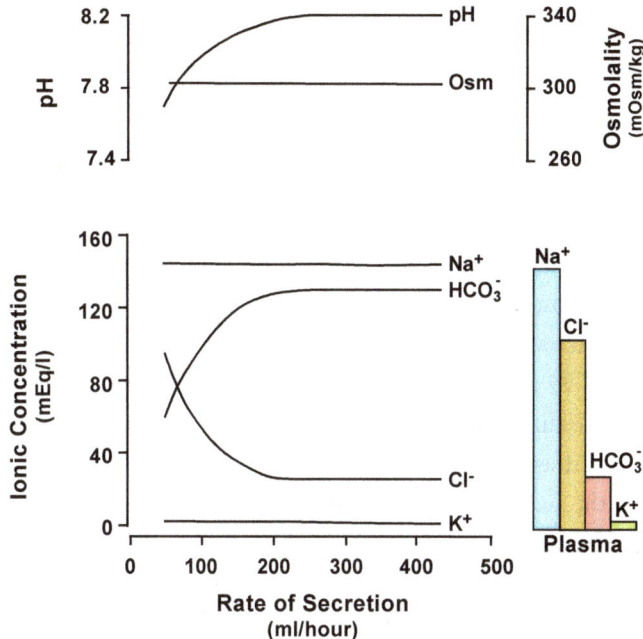

FIGURE 5-2: The pH, osmolality, and electrolyte composition of pancreatic juice at different rates of secretion. Plasma levels for electrolytes are provided for reference. [Modified from Pandol SJ. *The Exocrine Pancreas*. In: Colloquium Series on Integrated Systems Physiology: From Molecule to Function to Disease. (Granger DN, Granger JP, eds.), Morgan & Claypool Publishers, 2011. https://doi.org/10.4199/C00026ED1V01Y201102ISP014]

electrolytes (see Fig. 1-11 for discussion of standing osmotic gradient theory). The sum of the concentrations of cations equals that of the anions in the secreted juice and this balance is maintained at all rates of secretion. Although the Na^+ and K^+ concentrations of pancreatic juice are not affected by secretion rate, the concentrations of the major anions, HCO_3^- and CI^-, exhibit a secretion rate-dependent reciprocal relationship. The flow-related reciprocal relationship between HCO_3^- and Cl^- is likely due to changes in the relative contributions of acinar and ductal cell secretions. In the unstimulated (resting) state, most of the juice is derived from acinar cells with anion concentrations similar to plasma (Figure 5-2). However, in response to a meal, ductal cells elaborate a high volume HCO_3^- rich secretion. The meal-induced rise in HCO_3^- concentration is asymptotic, reaching a maximum value characteristic of a given species at high secretion rates. In man, the maximum bicarbonate concentration achieved in pancreatic juice is approximately 140 mEq per liter, which corresponds to a pH of 8.2. The relationship between Cl^- concentration in the juice and the rate of fluid secretion is the mirror image of that observed for bicarbonate. This reciprocal relationship between the concentrations of Cl^- and HCO_3^- ions is such that the sum of the two anions remains the same (~150 mEq per liter) regardless of the rate of secretion.

The most striking feature of the electrolyte composition of pancreatic juice is the high concentration of bicarbonate at high flow rates. The pancreatic ductal cell is equipped with a number of membrane transporters and channels to achieve this task. The delivery of large amounts of HCO_3^- into the duct lumen involves two mechanisms on the apical cell membrane: 1) Cl^--HCO_3^- exchange, and 2) activation of the cystic fibrosis transmembrane conductance regulator (CFTR) anion channels (Figure 5-3). Bicarbonate enters the ductal cell as a result of carbon dioxide diffusion and the subsequent generation of H^+ and HCO_3^- within the cell by carbonic anhydrase. The H^+ thus generated is removed from the cell via a Na^+-H^+ exchanger (NHE) on the basolateral membrane. Another source of intracellular HCO_3^- is a secretin-sensitive Na-HCO_3^- cotransporter expressed on the basolateral membrane. Following the ingestion of a meal, the GI hormone secretin activates the CFTR anion channels on the apical membrane via a cAMP-dependent mechanism. The chloride ions that enter the duct lumen via the open CFTR channels fuel the adjacent anion exchange (AE) pumps that secrete HCO_3^- into the duct lumen in exchange for Cl^-, because the rate of HCO_3^- secretion by the Cl^--HCO_3^- pump depends on the availability of luminal Cl^-. When luminal Cl^- concentration is low, the selectivity of the CFTR channels is shifted toward conducting HCO_3^-, rather than Cl^-. The net lumen negative potential that is created by these anion transport processes drives the secretion of Na^+ and K^+ through the paracellular channels, with water following via this pathway and through aquaporin (AQP) channels to produce an isosmotic pancreatic juice. Although a less powerful stimulant than secretin, acetylcholine also acts on ductal epithelial cells to enhance HCO_3^- secretion. The cholinergic neurotransmitter engages with muscarinic receptors on the duct cell to stimulate HCO_3^- secretion via Ca^{++}-activated chloride channels, which provide the luminal Cl^- needed to fuel the Cl^--HCO_3^-exchanger. Cholecystokinin (CCK), another weak secretory stimulant of ductal cell HCO_3^- secretion, also stimulates Ca^{++}-dependent signaling, which enables the GI hormone to potentiate the effects of secretin on pancreatic ducts. Table 5-1 summarizes the different endocrine, paracrine, and neurocrine mediators that have been reported to exert an influence on pancreatic duct secretion.

Trace amounts (<3 mEq per liter) of several other ionic species are present in pancreatic juice. These include calcium, magnesium, zinc, inorganic phosphate, and sulfate. Many stimulants of acinar cell (enzyme) secretion (e.g., vagal stimulation, cholecystokinin) also enhance the secretion of divalent cations. The concentrations of calcium and magnesium in the secreted

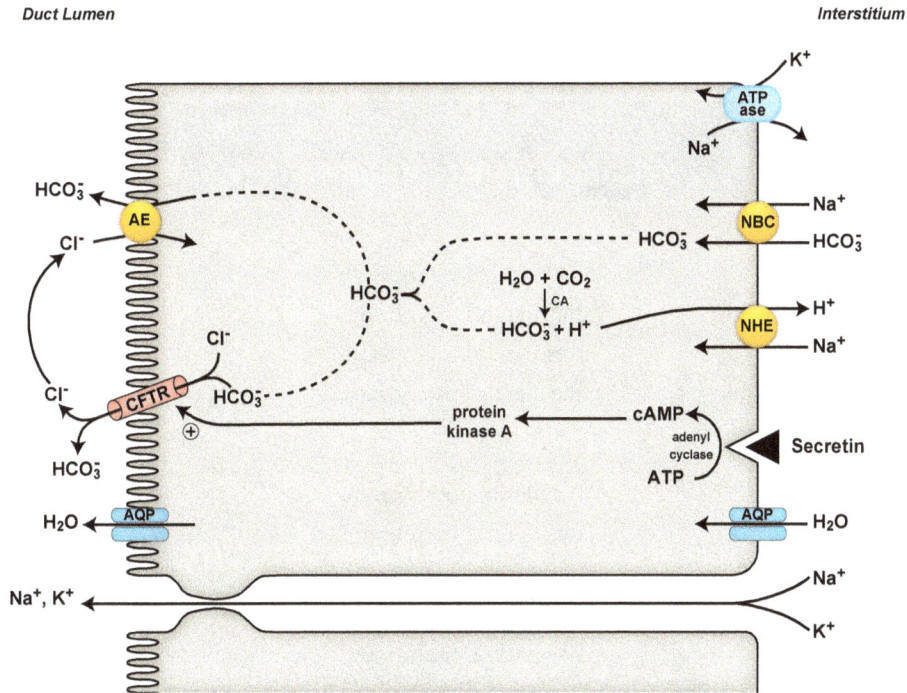

FIGURE 5-3: Cellular mechanisms involved in the secretion of HCO$_3$– by pancreatic duct cells. HCO$_3$– accumulates in the duct cell as a product of the reaction (catalyzed by carbonic anhydrase, CA) of water and carbon dioxide, and by cotransport with sodium on the basolateral membrane. Secretin increases intracellular cAMP, which activates CFTR on the apical membrane via protein kinase A. Cl$^-$ that enter the duct lumen via the open CFTR channels fuel the adjacent anion exchange (AE) pumps that secrete HCO$_3$– into the duct lumen in exchange for Cl$^-$ because the rate of HCO$_3$– secretion by the Cl$^-$-HCO$_3$– pump depends on the availability of luminal Cl$^-$. When luminal Cl$^-$ concentration is low, the selectivity of the CFTR channels is shifted toward conducting HCO$_3$–, rather than Cl$^-$. The net lumen negative potential drives the secretion of Na$^+$ and K$^+$ through the paracellular channels, with water following via this pathway and through aquaporin (AQP) channels to produce an isosmotic pancreatic juice.

juice are less than that of plasma. The divalent cations enter the juice by simple diffusion through paracellular pathways. As the rate of pancreatic secretion increases, the concentrations of Ca^{++} and Mg^{++} in the juice decrease. In addition to the paracellular route, calcium is released along with the enzymes contained in the zymogen granules of acinar cells. Calcium secretion during stimulation of enzyme release parallels the protein concentration in the juice and is independent of plasma levels of calcium.

TABLE 5-1: Endocrine, Paracrine, and Neurocrine Regulators of Pancreatic Duct Secretion		
MEDIATOR	**RECEPTOR LOCATION**	**SECRETORY RESPONSE**
Endocrines		
Secretin	Basolateral membrane	Increase
Cholecystokinin	Basolateral membrane	Increase
Insulin	Basolateral membrane	Increase
Somatostatin	Basolateral membrane	Decrease
Neurocrines		
Acetylcholine	Basolateral membrane	Increase
Norepinephrine	Basolateral membrane	Increase/decrease
VIP	Basolateral membrane	Increase
Substance P	Basolateral membrane	Decrease
CGRP	Basolateral membrane	Decrease
Peptide YY	Basolateral membrane	Decrease
*Paracrines**		
ATP	Luminal	Increase
Adenosine	Luminal	Increase
Guanylin	Luminal	Increase
Uroguanylin	Luminal	Increase

* Secreted by ductal cells into lumen

Proteins

Pancreatic juice contains a variety of proteins, the majority of which (> 90%) are hydrolytic enzymes that can be classified according to the specific substrates upon which they act (Table 5-2). The four major enzyme groups are proteolytic, amylolytic, lipolytic, and nucleolytic. Most of the enzymes of pancreatic juice are proteolytic and can be divided into two groups—endopeptidases (trypsin, chymotrypsin, elastase) and exopeptidases (carboxypeptidases). Endopeptidases break peptide bonds within the molecule (at non-terminal amino acids) to largely

TABLE 5-2: Enzymes and Modulating Factors in Pancreatic Juice

ENZYMES (PROENZYMES)	SPECIFIC HYDROLYTIC ACTIVITY OR ACTION
Proteolytic	
a) endopeptidases	
trypsin(ogen)	Peptide bonds near basic residues
chymotrypsin(ogen)	Peptide bonds near aromatic or other large Hydrophobic residues
(pro)elastase	Peptide bonds near small hydrophobic residues
b) exopeptidases	
(pro)carboxypeptidase A	All peptide bonds at carboxyl end except those near basic residues
(pro)carboxypeptidase B	Only peptide bonds at carboxyl end near basic residues
(pro)aminopeptidase*	All peptide bonds at amino end
Amylolytic	
$\propto$-amylase	Internal $\propto$-1,4 glucosidic bonds of glucose polymers
Lipolytic	
lipase	Ester bonds at 1- and 3-positions of triglycerides
(pro)phospholipase A_2	Ester bonds at 2-position of phosphoglycerides
carboxyl ester hydrolase (cholesterol esterase)	Ester bonds of water-soluble lipids; cholesterol esters
Nucleolytic	
ribonuclease	Phosphodiester bonds of nucleotides in ribonucleic acid
deoxyribonuclease	Phosphodiester bonds of nucleotides in deoxyribonucleic acids
Modulating factors	
trypsin inhibitor	Forms stable complex with & inactivates trypsin
colipase	Complexes with lipase & anchors it to hydrophobic surface of oil droplet
monitor peptide	Stimulates release of cholecystokinin

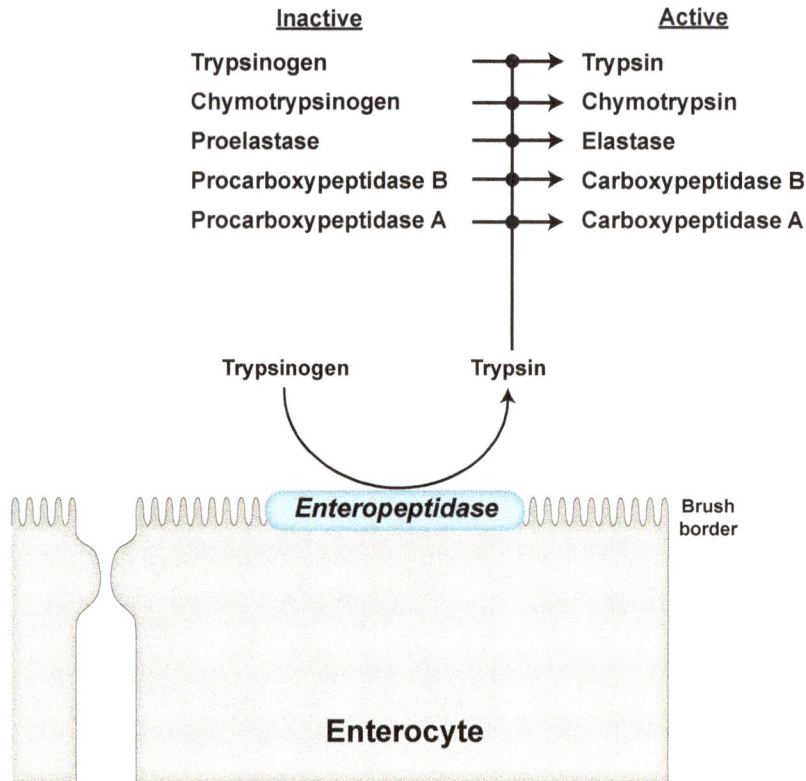

FIGURE 5-4: Mechanism of pancreatic enzyme activation in the intestinal lumen. Inactive proenzymes (zymogens) enter the duodenum (via the pancreatic duct) where the brush border enzyme enteropeptidase cleaves the activation peptide on trypsinogen to expose trypsin's catalytic domain. Trypsin converts other proenzymes, including trypsinogen itself, to their active forms via enzymatic cleavage.

generate oligopeptides, whereas exopeptidases cleave peptide bonds at the carboxyl terminus of proteins, largely yielding free amino acids and dipeptides. All of these proteases are synthesized, stored, and secreted as inactive precursors (called proenzymes or zymogens) that are activated in the duodenal lumen. Activation of the proteases involves hydrolytic cleavage of key peptide bonds, thereby inducing a conformational change in the enzyme that unmasks the "catalytic site" or the "binding site." The initial step in activation of pancreatic proteases is accomplished by an endopeptidase (sometimes inappropriately referred to as enterokinase), a brush border enzyme of the duodenal mucosa (Figure 5-4). This peptidase converts trypsinogen to its active form, trypsin. Trypsin, in turn, activates other trypsinogen molecules and the other pancreatic proteases. Amylase, ribonuclease, and the lipolytic enzymes, except phospholipase A_2, are secreted in their active forms. Pro-phospholipase A_2, like the proteases, is activated in the duodenal lumen by trypsin.

Pancreatic juice also contains three peptides that are not enzymes yet play an important role in regulating the activity or secretion of pancreatic enzymes. One of these, trypsin inhibitor, combines with trypsin in a one-to-one ratio to produce a complex that is inactive. Trypsin inhibitor also partially inhibits chymotrypsin. The presence of trypsin inhibitor in the gland serves to protect the tissue from autodigestion by the small amounts of trypsin that may become activated prematurely within the pancreas. The molar concentration of trypsin inhibitor in the juice, however, is much less than the molar concentration of trypsin and, therefore, activation of the pancreatic proteases in the duodenal lumen is not prevented. Colipase, another small polypeptide present in the secreted juice, enhances the lipolytic activity of pancreatic lipase. This cofactor prevents the bile salt-induced inhibition of lipase activity by anchoring the lipase molecule to the lipid-water interface of oil droplets. Colipase, in the presence of bile salts, lowers the pH optimum of lipase from 8.5 to 6.5, which is the average pH in the duodenum. The concentration of colipase in pancreatic juice is roughly equivalent to the concentration of lipase. Another peptide secreted into pancreatic juice is monitor peptide, which functions as an intraluminal CCK-releasing factor. This peptide, which is trypsin-sensitive, appears to modulate the release of CCK in response to protein ingestion, thereby providing an intraluminal mechanism for feedback regulation of pancreatic enzyme secretion. The regulatory role of monitor peptide is further discussed below.

The capacity for protein synthesis of the human pancreas exceeds that of all other organs and largely reflects the high rates of synthesis of digestive enzymes in acinar cells. The synthesis, packaging, and secretion of proteins (i.e., enzymes) is an orderly sequence of events that occurs in the endoplasmic reticulum (ER) and other organelles, proceeding from the basal to apical regions of the pancreatic acinar cell. The peptide chain (enzyme) is synthesized in the rough ER located in the basal portion of the cell, segregated away from lysosomal enzymes and processed in the Golgi apparatus. The digestive enzymes (and inactive precursors) exit the Golgi complex in condensing vacuoles, which provide an acidic environment for the stored and concentrated proteins. The condensing vacuoles mature into zymogen granules that ultimately reside in the apical region of the acinar cell. The low pH within the condensing vacuoles and granules inhibits enzyme activation and helps prevent pancreatic autodigestion. In response to a secretory stimulus (e.g., neural or humoral), the zymogen granule membrane

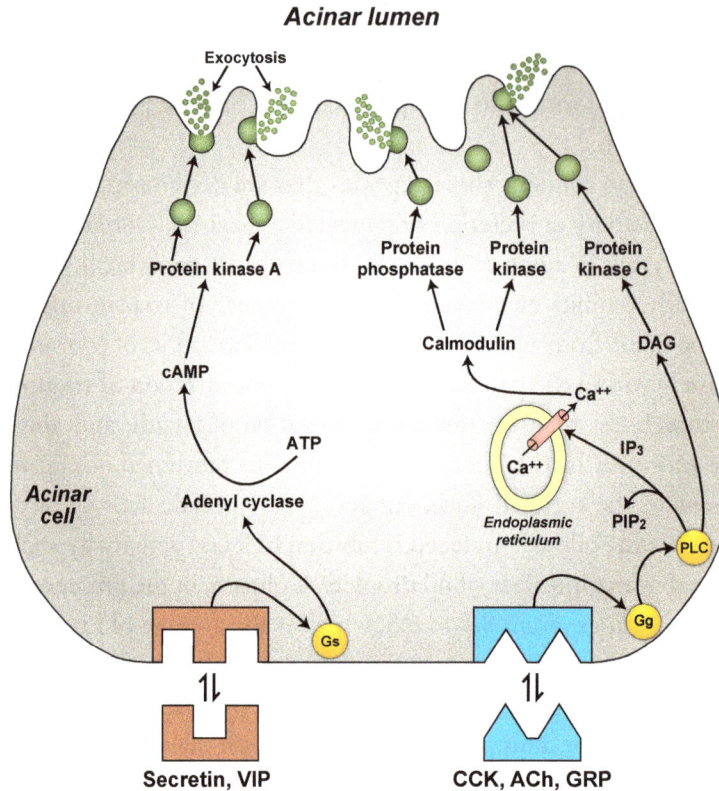

FIGURE 5-5: Mechanisms underlying the secretion of proteins from the pancreatic acinar cell. Two of the major pathways used by agonists that simulate the release of digestive enzymes from acinar cells are depicted. Cholecystokinin (CCK), acetylcholine (Ach) and gastrin releasing peptide (GRP) activate the G-protein, Gg, which stimulates phospholipase C (PLC) and ultimately activates protein kinase C (PKC) and the release of calcium from the endoplasmic reticulum. The elevated intracellular Ca++ activates calmodulin, which activates protein kinases and phosphatases. The resulting phosphorylation status of proteins leads to exocytosis and protein secretion. The second pathway, used by secretin and VIP, promotes exocytosis by Gs-dependent activation of adenyl cyclase, which results in the generation of cAMP, and the subsequent activation of protein kinase A.

fuses with the apical cell membrane and subsequently releases the granule contents into the duct lumen.

The regulated secretion of digestive enzymes occurs through modulation of zymogen granule exocytosis in acinar cells (Figure 5-5). These cells express different G-protein-coupled receptors (GPCRs) that can engage a specific mediator to elicit exocytosis. The GPCRs on acinar cells stimulate enzyme secretion via one of two signaling mechanisms that involve increased intracellular levels of either cAMP or Ca^{++}. Receptors for cholecystokinin, acetylcholine and GRP are coupled by the G-protein Gq to activation of phospholipase-C (PLC), which cleaves phosphatidylinositol 4,5 biphosphate (PIP_2) to produce inositol 1,4,5-triphosphate (IP_3). The IP_3 elicits the release of Ca^{++} from stores in the ER, resulting in a rapid increase in cytosolic Ca^{++} concentration. The rise in intracellular Ca^{++} activates calmodulin, which activates phosphatases (PP) and protein kinases (PK). The formation of diacylglycerol (DAG) and the subsequent activation of protein kinase C (PKC) also result from receptor-mediated PLC activation. The spikes in intracellular Ca^{++} and resultant alterations in phosphorylation of the cytoskeleton and other cell structures that are elicited in response to Gq-dependent mediators promote the movement of granules toward, and subsequent fusion with, the apical membrane, which results in a burst of zymogen granule exocytosis and secretion. The GPCR-dependent stimulation of pancreatic enzyme exocytosis and secretion that occurs in response to secretin and VIP involves the activation of Gs, which stimulates adenyl cyclase to generate cAMP from ATP (Figure 5-5). The elevated intracellular cAMP level leads to the activation of protein kinase A, which mediates the phosphorylation events that promote zymogen granule exocytosis and secretion. Of the two stimulus-secretion coupling pathways that mediate enzyme secretion, the Gq-Ca^{++}-dependent pathway is quantitatively more important. Nonetheless, secretagogues that increase intracellular cAMP potentiate the effects of secretagogues that elevate Ca^{++}, and vice versa.

Each zymogen granule is believed to contain the entire complement of pancreatic enzymes synthesized by the acinar cells and, therefore, the enzymes are secreted in a constant proportion when the pancreas is stimulated. Because the ratio of one enzyme to another is identical in pancreatic juice and zymogen granules, it appears that the relative concentrations of the different enzymes in the juice are determined at the time of synthesis. However, the relative concentration of the enzymes in the juice can be altered by the diet, that is, the concentration of a given enzyme can be increased when the level of its substrate in the diet is raised. For example, a carbohydrate-rich diet enhances the synthesis of amylase and increases the proportion of this protein to the total pool of pancreatic enzymes (Figure 5-6). Such adaptation of pancreatic enzyme secretion to dietary stimulants involves changes at the level of gene transcription and requires several weeks of dietary modification before it is fully established. Insulin appears to mediate the adaptive response of amylase production and release to a carbohydrate-rich diet, whereas CCK has been implicated in the adaptive changes in the synthesis of lipolytic and proteolytic enzymes that result from dietary changes in fats and proteins, respectively.

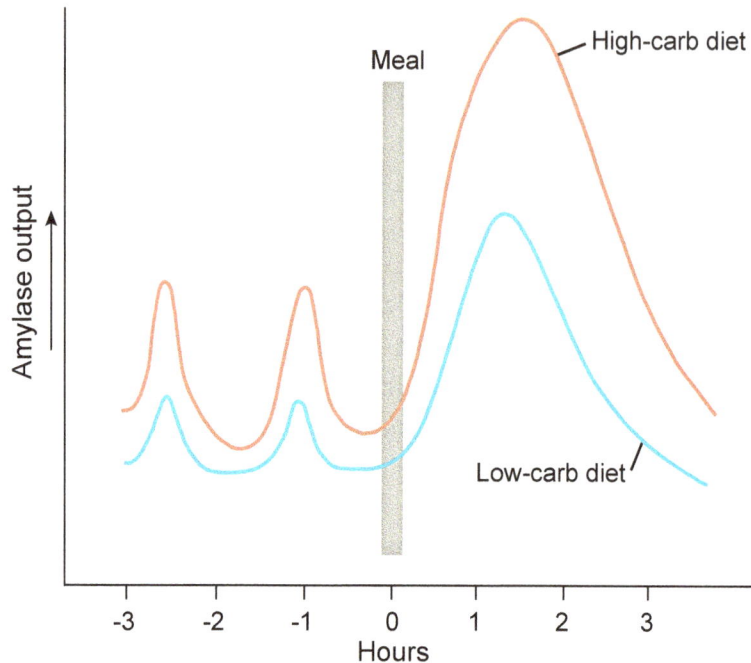

FIGURE 5-6: Effect of a long-term high carbohydrate diet on pancreatic amylase output. Both basal and postprandial output of amylase is enhanced with the high carbohydrate diet. The oscillations in amylase output in the interdigestive period (before the meal) reflect the influence of the migrating motility complex on enzyme release into the duodenal lumen.

REGULATION OF PANCREATIC SECRETION

Endocrine and Paracrine Regulation

The most potent humoral stimulants of pancreatic secretion are secretin and cholecystokinin. Although a variety of peptides and hormones are known to influence the secretion of bicarbonate and water by the duct cells (Table 5-1), secretin is considered to be the principal physiological agonist of this response. The pancreas is very sensitive to secretin, with small increases in the blood concentration of this hormone eliciting profound increases in fluid and bicarbonate output. Secretin also exerts a mild stimulatory effect on enzyme output by the acinar cells. Conversely, cholecystokinin is a potent stimulant of enzyme secretion, but a weak stimulant of fluid and bicarbonate output. Other peptides structurally related to these two agonists produce predictable effects on pancreatic secretion. Vasoactive intestinal polypeptide (VIP) is structurally similar to secretin and can stimulate bicarbonate and water secretion. VIP is a weaker agonist than secretin. When VIP is administered along with secretin, it competitively inhibits the secretory effects of the latter. Gastrin possesses the same carboxy-terminal structure as cholecystokinin and can stimulate pancreatic enzyme output. The effect of gastrin is considerably less than that of cholecystokinin, and gastrin can inhibit the cholecystokinin-induced enzyme output. Because of these structure-activity relationships, pancreatic cells are believed to have both secretin and CCK receptors that recognize structural analogues of each of these hormones. Other peptide hormones have been shown to alter pancreatic secretion by acting on receptors other than the secretin and cholecystokinin receptors that are expressed on the basolateral membrane of the acinar and/or duct cells. For example, somatostatin, pancreatic polypeptide, and glucagon released from islet cells inhibit secretion by both duct and acinar cells.

Although the actions of CCK on pancreatic enzyme secretion have long been attributed to direct engagement of the gut-derived hormone with its receptors on acinar and duct cells, there is now clear evidence that a large part of the CCK-mediated pancreatic response is linked to the activation of a vagovagal reflex (Figure 5-7). With this mechanism, CCK released from I cells in the duodenum interacts with CCK_1 receptors expressed on local sensory neurons to stimulate the afferent arm of a vagovagal reflex. The afferent signals are received and integrated in the dorsal vagal complex (DVC) of the brain. Vagal efferents from the DVC converge on cholinergic intrapancreatic neurons that release acetylcholine and VIP. The neurotransmitters bind to muscarinic and VIP receptors on acinar and duct cells to stimulate both enzyme and bicarbonate secretion. This mechanism offers an explanation for the observation that pancreatic enzyme output in response to CCK is significantly reduced (but not abolished) in normal human subjects treated with atropine and in patients after vagotomy. The incomplete inhibition noted following vagotomy and atropine suggests that blood borne CCK, which enters the circulation in the small bowel and is ultimately delivered to the pancreatic microcirculation, also contributes to the secretory response that is elicited by a meal (Figure 5-7).

Neural/Neurocrine Regulation

Parasympathetic innervation of the pancreas is derived from the dorsal vagal nucleus. Preganglionic axons emerge from the brain stem in the vagus nerve and synapse on cell bodies of

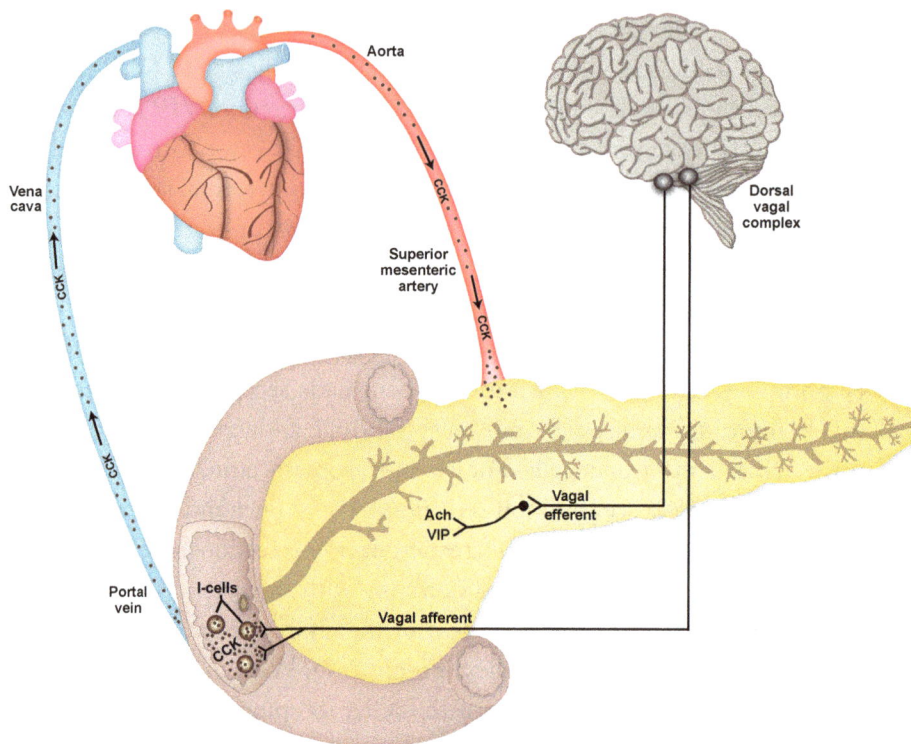

FIGURE 5-7: Neural and hormonal mechanisms of CCK-mediated pancreatic enzyme secretion. Following a meal, nutrients delivered to the duodenum elicit the release of CCK from I cells. Some of the CCK enters the portal vein and is delivered, via the circulatory system, to the pancreas, where it acts as a hormone to stimulate acinar cells to secrete digestive enzymes. An alternate, and quantitatively more important, mechanism of CCK action involves the activation of afferent vagal fibers in the intestinal mucosa. The activated afferent fibers send a signal to the dorsal vagal complex (DVC) in the brain, where the sensory information is integrated and vagal efferent fibers to the pancreas are activated. The vagal afferents synapse with neurons in the pancreatic ganglia and signals from the DVC elicit the release of acetylcholine and VIP, which stimulate the release of enzymes from the acinar cells.

postganglionic fibers within the substance of the pancreas. The intrinsic postganglionic fibers, in turn, innervate islet, ductular, and acinar cells. Although acetylcholine is the neurotransmitter released near the effector cells during vagal stimulation, VIP has also been identified in the postganglionic nerve terminals that surround pancreatic ducts. Activation of the efferent parasympathetic (vagal) fibers or administration of acetylcholine enhances pancreatic enzyme secretion to a far greater extent than bicarbonate and water output. The enzyme output in response to nerve stimulation or cholinergic agents can be abolished by atropine. The concomitant release of VIP from postganglionic fibers is likely responsible for the bicarbonate and water secretion induced by vagal stimulation. Although vagotomy has little or no direct effect on pancreatic bicarbonate and water output, it does blunt the pancreatic response to GI hormones. Vagotomy similarly blunts the enzyme output response to CCK. Some neurocrines (e.g., substance P, CGRP, peptide YY) released in the pancreas exert an inhibitory influence on acinar and duct secretions.

The preganglionic sympathetic fibers supplying the pancreas originate in the fifth to tenth thoracic segments, pass down to the level of the lower six thoracic and upper two lumbar ganglia, and synapse with postganglionic fibers in the celiac and superior mesenteric ganglia. The postganglionic fibers course with the arterial blood vessels to the gland. The postganglionic nerve terminals are in close association with vascular smooth muscle rather than the parenchyma. The neurotransmitter released at the sympathetic nerve terminal is norepinephrine. In general, stimulation of the splanchnic nerve has either no effect or inhibits pancreatic secretion. The inhibition of pancreatic secretion, which is largely manifested as a reduction in fluid and bicarbonate output, is believed to be secondary to the sympathetically mediated constriction of blood vessels supplying the pancreas.

Potentiation of Secretory Responses

Secretin can potentiate the effects of cholecystokinin on pancreatic enzyme secretion, whereas CCK can potentiate the effects of secretin on bicarbonate and water secretion. The magnitude of the change in pancreatic protein secretion in response to the two hormones exceeds the sum of the effects of each hormone alone. Potentiation of the bicarbonate and water responses to secretin and cholecystokinin is more dramatic than their potentiating effect on protein secretion. There are similar synergisms between secretin and gastrin and between VIP and cholecystokinin. At the cellular level, potentiation requires that the stimuli act on different membrane receptors and trigger different intracellular events, which subsequently lead to the same secretory response (Fig. 5-5). Because secretin and CCK stimulate different receptors and use different stimulus-secretion coupling mechanisms, they can potentiate each other's actions. Vagal stimulation (or acetylcholine) and cholecystokinin, which trigger identical intracellular events in acinar cells, are only additive, whereas acetylcholine and secretin result in a potentiated duct secretory response. Insulin has been shown to potentiate the effects of CCK and secretin on enzyme and fluid/bicarbonate output, respectively.

Feedback Regulation of Pancreatic Secretion

There is evidence that supports the existence of a negative feedback mechanism that enables signals within the lumen of the upper small intestine to modulate the rate of pancreatic secretion in response to the presence or absence of active proteolytic enzymes. This mechanism was proposed based on several key observations. First, the ingestion of trypsin inhibitor stimulates the release of CCK by I cells in the duodenum and results in enhanced pancreatic enzyme secretion. Second, the presence of active trypsin in the gut lumen reduces CCK release, with a concomitant reduction in pancreatic secretion. Third, surgical diversion of pancreatic juice from the gut is a potent stimulus for pancreatic secretion and increases plasma CCK concentration. Collectively, these observations, coupled to other evidence in the literature, are consistent with a feedback control mechanism that links pancreatic enzyme output to both CCK release and proteolytic activity in the small bowel. The existence of a trypsin-sensitive CCK-releasing factor (CCK-RF) that is secreted spontaneously into the bowel lumen explains how luminal proteolytic activity is able to regulate pancreatic secretion via CCK (Figure 5-8). When there is no food (proteins) in the proximal bowel to act as a substrate for pancreatic proteases, the CCK-RF is degraded and rendered inactive, thereby removing a stimulus for CCK release by I cells. However, when ingested proteins appear in the gut lumen, the pancreatic proteases bind to (and form a complex with) the proteins, ultimately leading hydrolysis of this alternative substrate. In this instance, the CCK-RF is protected from proteolytic degradation, allowing the releasing factor to stimulate I cells to elaborate CCK, which stimulates the pancreas to secrete more hydrolytic enzymes. The CCK-RF/protease feedback mechanism is presumed to fine tune the amount of active digestive enzymes that are present in the lumen and to ensure an optimal rate of digestion without excess output or wastage of pancreatic enzymes.

A protease-dependent feedback mechanism has also been described that links duodenal acidification and secretin output to a releasing factor. Although less is known about this feedback mechanism than that described for CCK, the available evidence indicates that phospholipase A2 (PLA_2) that is present in pancreatic juice functions as a secretin-releasing factor. This contention is supported by reports describing an increased pancreatic bicarbonate secretion and secretin release from the duodenum after luminal acidification, and inhibition of these responses following blockade of PLA_2 function with anti-PLA_2 serum. Pancreatic PLA_2 also stimulates the release of secretin from S cells in the duodenum. Like CCK-RF, PLA_2 is vulnerable to protease degradation when the relative levels of ingested proteins to proteases in the gut lumen are low (e.g., during fasting) and is protected from degradation when the protein to protease ratio is high (e.g., following a meal).

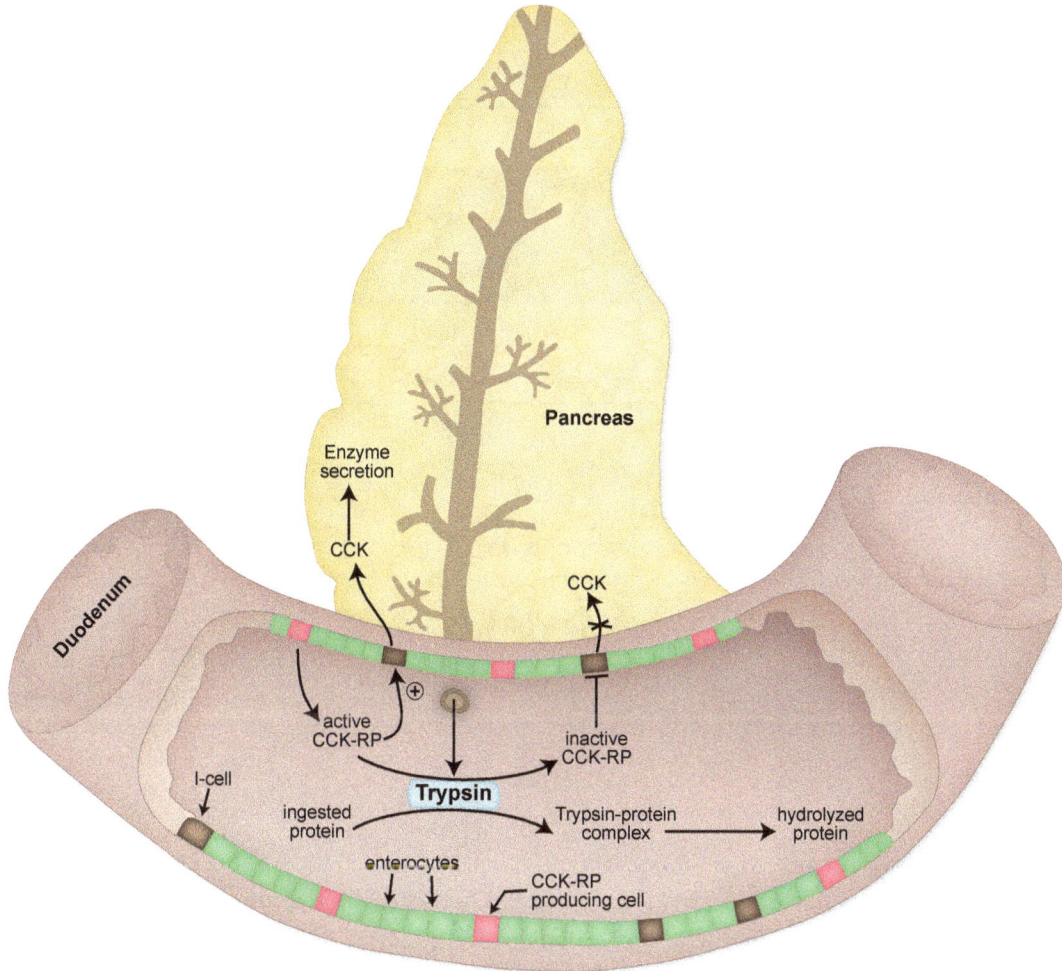

FIGURE 5-8: Regulation of pancreatic enzyme secretion by cholecystokinin-releasing peptide (CCK-RP). CCK-RF is secreted spontaneously into the lumen of the proximal small intestine where it can be cleaved and inactivated by trypsin. The fate of CCK-RP is determined by the relative abundance of ingested protein and proteases in the gut lumen. Following a meal, ingested proteins compete for the trypsin that would otherwise inactivate CCK-RP. With the prevention of CCK-RP inactivation, the CCK-producing I cells are stimulated (by CCK-RP) to release CCK, which gains access to the circulation and enhances pancreatic enzyme secretion. During fasting, the competition between CCK-RP and ingested proteins is diminished, CCK-RP is degraded, and the stimulus for pancreatic enzyme output is reduced.

OVERVIEW OF PANCREATIC SECRETORY RESPONSE TO MEALS

TABLE 5-3: Mechanisms that Elicit Pancreatic Exocrine Secretion During and After a Meal

PHASE	% MAXIMUM ENZYME SECRETION	STIMULI	MECHANISM/ MEDIATOR
Cephalic	25	Site, smell, taste, mastication	Vagal/acetylcholine VIP, GRP
Gastrics	10	Distension Protein digestion products	Vagovagal/acetylcholine Gastrin?
Intestinal	65	Duodenal acid (pH < 4.5) Lipids, amino acids, peptides Distension	Secretin CCK 5-HT, Vagovagal (enteropanceatic) reflex

Basal Secretion

Basal pancreatic secretion of bicarbonate is about 2% of the maximal secretory response, whereas the enzymes are secreted at roughly 10% of the maximal output. Transient increases in pancreatic enzyme and bicarbonate secretion occur every 60 to 120 minutes and are synchronous with phase III of the interdigestive migrating motility complex (MMC), which is regulated by cholinergic input (see Figure 5-6). Blocking motilin, a hormone linked to phase III of the MMC, abolishes the interdigestive pattern of pancreatic secretion. The intermittent release of pancreatic juice into the small bowel between meals is considered integral to the function of the MMC to clear the small bowel of debris and bacteria.

Cephalic Phase of Secretion

The cephalic phase of pancreatic secretion is initiated by the sight, smell, taste, chewing, and swallowing of food. The enhanced pancreatic secretion during the cephalic phase represents

10% to 15% of the volume and 25% of the enzyme response to a meal. The duration of the cephalic phase in response to sham feeding typically lasts about 60 minutes. This phase is associated with a pancreatic secretory response that is low in volume but rich in enzymes. The input from the sensory stimuli (e.g., smell, taste, etc.) is integrated in the dorsal vagal complex of the brain, with output signals transmitted to the pancreas via the vagus nerve. Both vagotomy and atropine block the enhanced secretion during the cephalic phase. Of the neurotransmitters that have been implicated in the efferent arm of this vagal response, acetylcholine (acting via muscarinic receptors) appears to be the most important, with VIP and GRP making smaller but significant contributions. Acetylcholine, VIP, and GRP are all known to stimulate the exocytosis of zymogens from pancreatic acinar cells (Figure 5-5).

Gastric Phase of Secretion

The gastric phase of pancreatic secretion, which accounts for about 10% of the enzyme secretion during a meal, is triggered by the entrance of food into, and distension of, the stomach. Distension of the stomach with a balloon results in a pancreatic secretory response similar to that observed during the cephalic phase, that is, an enzyme-rich, low volume/HCO_3^- secretion. The mechanoreceptors that mediate this response to stretch are located in the body of the stomach; antral distension does not result in enhanced pancreatic secretion. The gastric phase response, like the cephalic phase, is blocked by either vagotomy or atropine administration. These observations are consistent with a cholinergic (vagovagal) reflex that originates in the stomach and terminates in the pancreas. Gastrin has also been implicated in the gastric phase of pancreatic secretion, because it functions as an agonist on acinar cells and is released from G cells in the antral mucosa when protein-containing food enters the stomach. However, the quantitative significance of this mechanism in humans remains unclear.

Intestinal Phase of Secretion

The intestinal phase of pancreatic secretion is initiated by the entrance of chyme into the duodenum and accounts for 70% to 80% of the pancreatic response to a meal. The acidity of the chyme and the products of protein and lipid digestion contained in chyme are the principal determinants of the type of pancreatic response elicited.

Secretin, considered the most important physiological stimulant of pancreatic HCO_3^- secretion following a meal, is released from S cells in the intestinal mucosa in response to intraluminal acidity. The increases in blood secretin concentration and pancreatic HCO_3^-

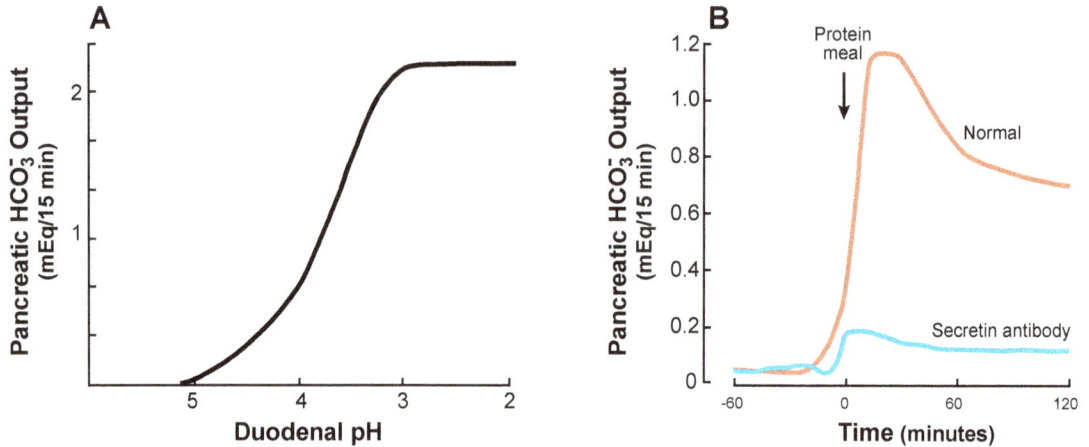

FIGURE 5-9: Dependence of pancreatic bicarbonate output on duodenal pH (panel A) and secretin (panel B). *Panel A* illustrates that once pH in the duodenal lumen falls to a threshold value, progressive increases in the acid load (reductions in pH) are accompanied by enhanced levels of pancreatic HCO3– secretion. The increasing HCO3– output at lower pHs is associated with corresponding increases in plasma secretin levels (not shown). Panel B illustrates the effect of immunoblockade of secretin (anti-secretin antibody treatment) on the response of pancreatic HCO3– output to ingestion of a protein meal [modified from Liddle, RA. In: Physiology of the Gastrointestinal Tract (LR Johnson, ed.), 2012, Elsevier Inc., used with permission].

secretion are related to the degree of luminal acidity and the length of the gut exposed to acid. The threshold for enhanced HCO3– secretion is a luminal pH of about 4.5 (Figure 5-9, panel A). When luminal pH falls below 4.5, the rate of HCO3– secretion and amount of secretin released are inversely related to pH, with HCO3– and secretin outputs reaching maximum values at pH 3.0. The postprandial pH of the upper small intestine rarely falls below 4.0 to 3.5, and chyme is normally neutralized before it enters the jejunum. The critical role of secretin in mediating the intestinal phase of pancreatic bicarbonate secretion during a meal is evidenced by the potent inhibition ($\geq$80%) of HCO3– output that is noted following secretin blockade using an anti-secretin antibody (Figure 5-9, panel B). Nonetheless, secretin alone cannot account for the pancreatic bicarbonate and water secretion that is observed after a meal. This apparent discrepancy has been explained by the involvement of a cholinergic enteropancreatic reflex that stimulates pancreatic duct cells to secrete HCO3– and water. This possibility is supported by the observation that blockade of cholinergic input to the pancreas by either atropine treatment or vagotomy partially inhibits the HCO3– secretion induced by duodenal acid, without altering blood secretin concentration. The ability of cholecystokinin to potentiate the effects of small amounts of circulating secretin on the duct cells may also account for the non-secretin component of HCO_3^- output after a meal.

Cholecystokinin, a potent stimulant of pancreatic enzyme secretion, is released from I cells of the intestinal mucosa by the products of fat and protein digestion (Figure 5-10). Undigested lipids or proteins do not stimulate the release of cholecystokinin. The circulating levels of cholecystokinin are directly related to the concentration of the luminal stimulants and the length of the gut exposed to them. The most potent lipid stimulants are the long-chain fatty acids (>8 carbons in length) and monoglycerides. Various peptides containing glycine and amino acids with aromatic residues (glycylphenylalanine, glycyltryptophan), as well as some essential amino acids (e.g., phenylalanine, methionine, valine) are also potent stimulants of CCK release.

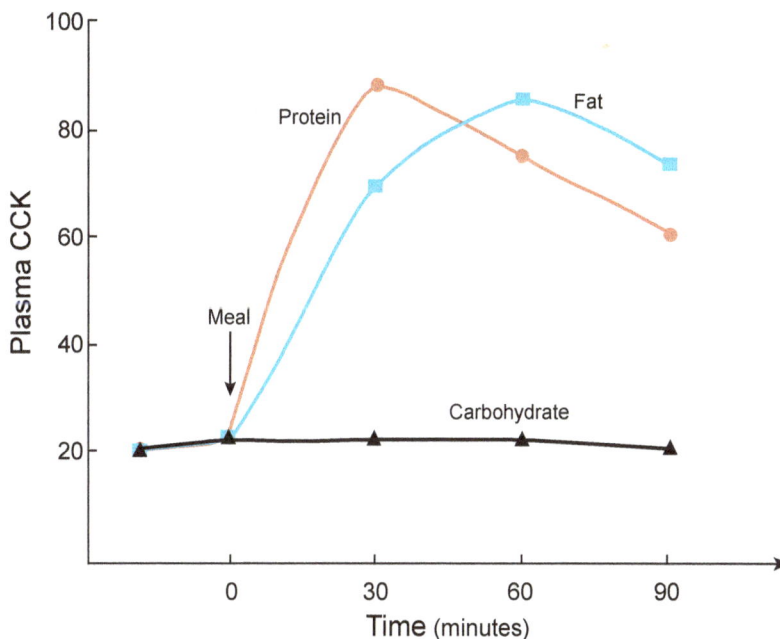

FIGURE 5-10: Changes in plasma CCK concentration in cats fed either a protein, fat or carbohydrate meal. Ingestion of either protein or fat elicits a robust increase in plasma CCK while no response is noted after a carbohydrate meal [based on data from Backus RC et al., *Regulatory Peptides* 57:123–131, 1995. https://doi.org/10.1016/0167-0115(95)00027-9. Used with permission of Elsevier].

Carbohydrates do not affect CCK secretion (Figure 5-10). Secretin, which is released after food ingestion due to duodenal acidification and, to a lesser extent, by the products of lipid and protein digestion, can potentiate the postprandial effects of CCK on pancreatic acinar cells.

Although CCK enhances pancreatic enzyme secretion in part by acting as a hormone, that is, direct stimulation of acinar cells by the elevated CCK in blood, its role in the activation of sensory afferents in the duodenum and eliciting a cholinergic enteropancreatic reflex (Figure 5-7) appears to be more quantitatively important in stimulating enzyme output. Another paracrine agent produced in the proximal small intestine that stimulates pancreatic secretion following a meal by way of a vagal cholinergic mechanism is serotonin (5-hydroxytryptamine, 5-HT). Mechanical stimulation of the mucosa as well as luminal factors, such as osmolarity and disaccharides, induce the release of 5-HT from mucosal enterochromaffin cells. The 5-HT activates receptors ($5-HT_2$ and $5-HT_3$) on vagal afferent fibers that originate in the duodenal mucosa and subsequently stimulates the DMV in the brainstem to mediate postprandial enzyme secretion through a cholinergic pathway. The quantitative importance of the CCK- and 5-HT-mediated enteropancreatic reflexes in mediating the intestinal phase of pancreatic enzyme secretion is evidenced by reports describing complete inhibition of this response in animals treated with a combination of CCK1 receptor and 5-HT receptor antagonists. The involvement of a cholinergic pathway in this response is supported by the observation that vagotomy or atropine treatment can completely block the pancreatic enzyme secretion following a meal.

It is apparent that the intestinal phase of pancreatic secretion is complex, being the result of additive and potentiating effects of hormones (secretin and cholecystokinin) and neural activity (vagovagal reflexes) on pancreatic secretory cells. Despite these multiple interactions, the enhanced pancreatic enzyme secretion produced by a meal peaks at levels about 70% of the maximal rates of enzyme secretion that are attainable with exogenous hormonal stimulation. The inability of luminal stimulants to elicit a maximal response may result from inadequate excitatory mechanisms or it may reflect the influence of the protease-dependent negative feedback mechanisms that come into play during the intestinal phase of digestion. The CCK (CCK-RF) and secretin (PLA_2) releasing factors produced by the intestinal mucosa (discussed above) are able to stimulate their target cells (I- and S-cells, respectively) during a meal because they escape degradation by pancreatic proteases that are engaged in the hydrolysis of ingested protein. As digestion proceeds however, the relative excess of pancreatic proteases leads to the degradation of the endocrine cell-stimulating peptides (CCK-RF, PLA_2), resulting in the loss of key signals (CCK, secretin) that mediate the intestinal phase of digestion. Hence, this mechanism to fine tune the amount of digestive enzymes and HCO_3^- secreted in the bowel lumen may limit the capacity of the pancreas to achieve the maximal rate of secretion that is attainable with exogenous hormonal stimulation.

FUNCTIONS OF PANCREATIC JUICE

Bicarbonate

During a meal, the stomach can secrete as much as 20 to 40 mEq of hydrogen ions per hour. Despite the buffering capacity of ingested proteins and peptides, the pH of chyme is lowered to 2. Because the duodenal mucosa does not possess the protective barrier present in the stomach, the entrance of acidic chyme (containing pepsins) can result in ulcerations of the intestinal mucosa. Furthermore, the digestion of nutrients is impaired by an acidic chyme, because pancreatic enzymes require a pH between 6.0 and 8.0 for optimum activity. Indeed, lipase can be irreversibly inactivated when lumen pH falls below 4.0. Fortunately, the secretin-stimulated pancreas can secrete bicarbonate at a rate sufficient to neutralize all of the titratable acid delivered to the duodenum from the stomach. The bicarbonate ions that enter the intestinal lumen from duodenal secretions and bile also dissipate some of the acid entering the duodenum. However, the neutralizing capacity of the non-pancreatic secretions is minimal compared with that of pancreatic bicarbonate. Without the pancreas, the entire duodenum is acidified, whereas under normal conditions, most of the acid is dissipated at the level of the duodenal bulb.

Enzymes

Pancreatic juice contains a variety of digestive enzymes capable of hydrolyzing the major constituents of ingested food (Table 5-1). The pancreas normally secretes about ten times more enzyme than is required to hydrolyze all of the ingested nutrients to products that can be handled by the digestive (brush border and intracellular) enzymes and transport processes of the intestinal mucosa. However, the pancreas has much larger reserves for enzymes that hydrolyze carbohydrates or proteins, than for the enzymes involved in the digestion of lipids, particularly triglycerides. Nonetheless, maldigestion of fats and fecal fat excretion (steatorrhea) is not manifested until 80% to 90% of the pancreas is removed. Despite alternative sources of digestive enzymes (saliva and gastric juice), significant maldigestion and malabsorption occurs in the absence of pancreatic enzymes. Although patients can survive without pancreatic enzymes, health is impaired: body weight and muscle mass are reduced, and the patient is incapacitated by diarrhea associated with steatorrhea.

THE PANCREATIC CIRCULATION

The arterial supply of the pancreas is derived from the superior mesenteric artery (via the inferior pancreaticoduodenal artery) and the celiac artery (via the superior pancreaticoduodenal artery and pancreatic branches of the splenic artery). Within the gland, arterioles supply three distinct capillary plexuses: the acinar, ductular, and islet vascular beds (Figure 5-11). The intrapancreatic circulation is also characterized by an extensive portal system. The acinar capillary plexus receives blood from vessels draining the islet plexus. The ductular plexus receives blood from vessels draining the acini and islets. All three vascular plexuses eventually drain into either the splenic vein or superior mesenteric vein; these, in turn, empty into the portal vein.

Functional Implications of a Portal Circulation

The intrapancreatic portal circulation illustrated in Figure 5-11 allows the hormones secreted by the endocrine portion of the pancreas (islets of Langerhans) to reach the exocrine portion of the gland (acinar and duct cells) in very high concentrations. The hormones released from islet cells are proteins of low molecular weight (insulin, 6000; glucagon, 3550; somatostatin, 1638; pancreatic polypeptide, 4250) that are small enough to filter across the fenestrated capillaries surrounding the ducts and acini in significant quantities. All of the pancreatic hormones are known to influence pancreatic exocrine secretion. Insulin enhances the synthesis and release of amylase and potentiates the secretory effects of cholecystokinin. Glucagon inhibits enzyme secretion, whereas somatostatin and pancreatic polypeptide inhibit bicarbonate and water secretion. In view of these observations, the islets of Langerhans may be viewed as a "controlling center" for the exocrine function of the pancreas.

Role of the Microcirculation in Pancreatic Secretion

The pancreas is capable of secreting approximately 0.4 ml of juice per min per 100 grams of tissue during maximal stimulation induced by secretin. This rate of fluid secretion is approximately 30 times higher than the normal rate of capillary filtration in the pancreas. There is evidence that the Starling forces governing fluid exchange across pancreatic capillaries are altered

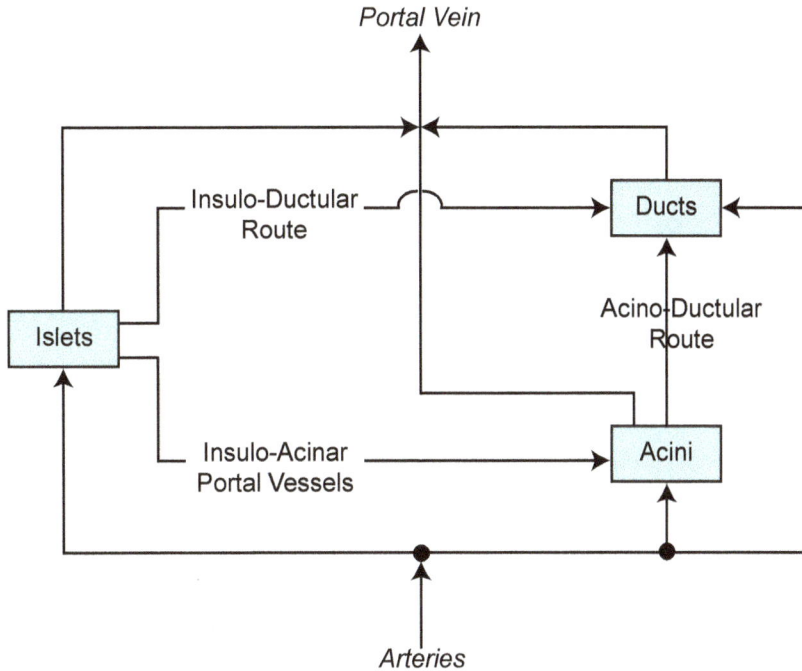

FIGURE 5-11: The microvascular arrangement in the pancreas. Arterioles perfuse distinct capillary plexuses that serve the acini, ducts, and islets of Langerhans. An extensive portal system is also evident, with the acinar capillary plexus receiving blood draining the islet plexus, and the ductal plexus receiving blood draining the acini and islets [modified from Ohtani et al., *Arch. Histol Jpn*; 46:1–42, 1983, used with permission].

during periods of enhanced secretion. After a meal, there is a doubling of pancreatic blood flow that is mediated, at least in part, by VIP. The arteriolar dilatation results in an increased capillary hydrostatic pressure that, in turn, provides the driving force for the movement of fluid from blood to ducts. Because, in some cases, pancreatic secretion can increase substantially without an accompanying increase in blood flow (and capillary pressure), the fluid for secretion can also be derived from the capillaries in response to changes in interstitial forces, that is, by increasing interstitial oncotic pressure and decreasing interstitial hydrostatic pressure. Such changes in the interstitial forces should result from the interstitial dehydration and reduced interstitial volume that occur when the ductal epithelium transports water from the interstitial spaces into the duct lumen. During a meal, it is likely that both capillary and interstitial forces serve to support the ongoing pancreatic secretion.

PANCREATIC PAIN

Pain is a prominent feature of pancreatic disease. Pancreatic pain is felt in the upper abdomen, centrally, and in both the right and left subcostal regions. The pain frequently radiates to the back and relief is often obtained by sitting up and bending forward into the fetal position.

PATHOPHYSIOLOGY AND CLINICAL CORRELATIONS

Pancreatitis is an inflammatory condition that can result in extensive damage and malfunction of the pancreas. Pancreatitis can be manifested as either an acute (e.g., obstruction of the pancreatic duct by a gallstone) or chronic (alcohol abuse) condition. In either instance, the premature activation of zymogens (proenzymes) may be an important initiating event. This occurs when the defense mechanisms (trypsin inhibitor, synthesis, storage and release of enzymes as inactive precursors) that normally exist to prevent premature zymogen activation are overwhelmed. A consequence of the intrapancreatic enzyme activation is autodigestion of pancreatic tissue, which perpetuates the inflammatory response and tissue injury. In addition to producing extensive local tissue damage, pancreatitis can result in injury and impaired function of distant organs, like the lungs, a response that has been attributed to the activated pancreatic enzymes.

Diagnosis of acute pancreatitis involves the identification of two of the three following criteria: 1) elevated serum lipase level (>three times the upper limit of normal value), 2) changes in pancreatic parenchyma detected by computed tomography, and 3) characteristic epigastric pain. Patients may also have associated nausea and vomiting. As serum amylase may be elevated in multiple conditions (renal failure), lipase is the primary pancreatic enzyme used for diagnostic purposes. Fluid resuscitation is critical in acute pancreatitis because of excess fluid loss from capillaries into the interstitial compartment that results from an inflammatory mediator-dependent increase in vascular permeability. Aggressive fluid (e.g., lactated Ringer's solution) administration helps to prevent complications that can be associated with acute pancreatitis, including excess fluid extravasation, systemic vasoconstriction, and disseminated intravascular coagulation. Patients with acute pancreatitis should be monitored frequently in the initial period following admission. Elevated hematocrit and/or blood urea nitrogen are indicators of potential worse injury (necrotizing damage) to the pancreas. Early enteral feeding may also be beneficial in acute pancreatitis patients.

The loss of exocrine and endocrine function that occurs in severe chronic pancreatitis produces a clinical picture that clearly illustrates the normal physiologic role of the pancreas. *Chronic pancreatitis* is often associated with alcoholism and with tobacco smoking; other cases are defined as idiopathic, with no defined causative agent. On rare occasion, chronic pancreatitis is associated with an autoimmune disease. Chronic pancreatitis results in progressive loss, first of exocrine and subsequently of endocrine tissue of the gland, and the affected cells are replaced by fibrous tissue. In advanced cases, calcification of the parenchyma occurs, and this can be recognized in radiologic imaging studies.

Severe loss of exocrine function results in malabsorption and weight loss. The malabsorption involves proteins, long-chain triglycerides, fat-soluble vitamins, and trace elements, with the exception of iron. Protein malabsorption may be manifest as "starvation" edema. Malabsorption of fat results in *steatorrhea*, that is, daily stool fat output of more than 5 g when 100 g of dietary fat is consumed daily. The stools are characteristically pale, bulky, and offensive. The action of colonic bacteria on unabsorbed lipids and proteins is responsible for the formation of malodorous compounds. Bacterial action on unabsorbed fats produces hydroxy-fatty acids similar to ricinoleic acid, the active principal of castor oil. These fatty acid derivatives cause water and electrolyte secretion by the colonic mucosa. In severe chronic pancreatitis, stool fat output may exceed 40 g per day, and this high output is because of the fact that triglyceride digestion by pancreatic lipase is a necessary prerequisite for fat absorption. However, only about 33% of patients with chronic pancreatitis have steatorrhea as defined above. This is explained by the fact that the pancreas has a large functional reserve. When more than 90% of exocrine function is lost, severe maldigestion and, consequently, malabsorption ensue.

As chronic pancreatitis progresses, endocrine tissue is also eventually lost with the development of frank diabetes mellitus; before reaching this stage, however, many patients will have already developed an abnormal glucose tolerance test.

The diagnosis of chronic pancreatitis can be established by intubation studies; a catheter is placed in the duodenal lumen and the pancreas is stimulated to secrete juice in response to a test meal or to intravenous secretin and cholecystokinin. Measurements of bicarbonate and enzyme secretion are obtained from the aspirates. Frequently, an impaired output of bicarbonate in response to secretin is the earliest sign of pancreatic insufficiency. These direct stimulation studies are usually reserved for expert centers and often combined with other endoscopic procedure.

Endoscopic cannulation of the pancreatic duct is used to demonstrate radiologically visible abnormalities of the duct system in chronic pancreatitis (irregularity, tortuosity, and focal dilatations) by *endoscopic retrograde pancreatography*. This invasive testing is the best criterion for diagnosis of chronic pancreatitis, but has risks of bleeding, infection, and inducing acute pancreatitis. Secretin stimulation with subsequent MRI scanning may also be employed to visualize, in a non-invasive fashion, the pancreatic ductal system for changes. More recently, *endoscopic ultrasonography* has been used to diagnose chronic pancreatitis with good accuracy, provided multiple criteria are met.

Cessation of alcohol and tobacco dependence should be encouraged for any patient with these aggravating factors to encourage pain relief and prevent further active injury to the pancreas. Pain management may be helpful for control of the chronic intermittent pain associated with chronic pancreatitis. In rare situations of chronic pancreatitis, resection of the head of the pancreas or *total pancreatectomy with auto-islet transplantation of pancreatic cells* may be considered to reduce severe pain. Enzyme supplementation with lipase containing compounds (either enteric coated pills or uncoated pills co-administered with proton pump inhibitors to prevent gastric acid degradation) may be used to reduce the incidence of steatorrhea.

REFERENCES

Crockett SD, Wani S, Gardner TB, Falck-Ytter Y. American Gastroenterological Association Institute Guideline on Management of Acute Pancreatitis. *Gastroenterology*. 2018; 154:1096–1101.

Forsmark C. Chronic Pancreatitis. In: Sleisenger and Fordtran's Gastrointestinal and Liver Disease. Chapter 59; pp. 985–1015, 2016. WB Saunders Co.

Liddle RA. Regulation of pancreatic secretion. In: Physiology of the Gastrointestinal Tract. (LR Johnson et al., eds.), Chapt. 52; pp. 1425–1457, 2012, Elsevier Inc.

Owyang C, Williams JA. Pancreatic secretion. In: Yamada's Textbook of Gastroenterology (DK Podolsky et al., ed.), 6th edition, Chapt. 25; pp. 450–474, 2016, John Wiley & Sons.

Pandol SJ. *The Exocrine Pancreas*. In: Colloquium Series on Integrated Systems Physiology: From Molecule to Function to Disease (Granger DN, Granger JP, eds.), Morgan and Claypool Publishers, 2011. https://doi.org/10.4199/C00026ED1V01Y201102ISP014

Williams JA. Regulation of normal and adaptive pancreatic growth. *Pancreapedia: Exocrine pancreas knowledge base*. March 15, 2017 [DOI: 10.3998/panc.2017.02]

* * * *

CHAPTER 6

The Liver and Biliary Tree

INTRODUCTION

Interposed between the digestive tract and other organ systems, the liver functions to store, metabolize, and distribute (via blood) the large amount of nutrients that are ingested in the form of food. The liver also interacts with the digestive system by secreting bile, a solution of detergent molecules called bile salts. Bile is concentrated and stored in the gallbladder during the interdigestive period. The presence of fat in intestinal chyme leads to gallbladder contraction and the delivery of bile into the gut lumen, an effect mediated by the hormone cholecystokinin. Although present in the gut lumen, bile promotes the efficient digestion and absorption of lipids. The bile salts and other constituents of bile are absorbed in the small bowel and returned to the liver via the portal vein for resecretion into bile. In addition to its role in fat digestion, bile excretion provides a major route for the elimination of endogenous compounds (e.g., cholesterol and bilirubin), drugs, and environmental pollutants that have been metabolized, detoxified, or inactivated by the liver.

FUNCTIONAL ANATOMY OF THE LIVER AND BILIARY TREE

Macroscopic Anatomy

The human liver is the largest solid organ in the body, weighing between 1.3 and 1.7 kg, depending on gender and body size. It is divided into two principal lobes, a large (50–70% of liver mass) right lobe and a smaller left lobe. Two additional lobes (caudate and quadrate) are contiguous with the left lobe. Each lobe is covered by a thin connective tissue sheath, the Glisson's

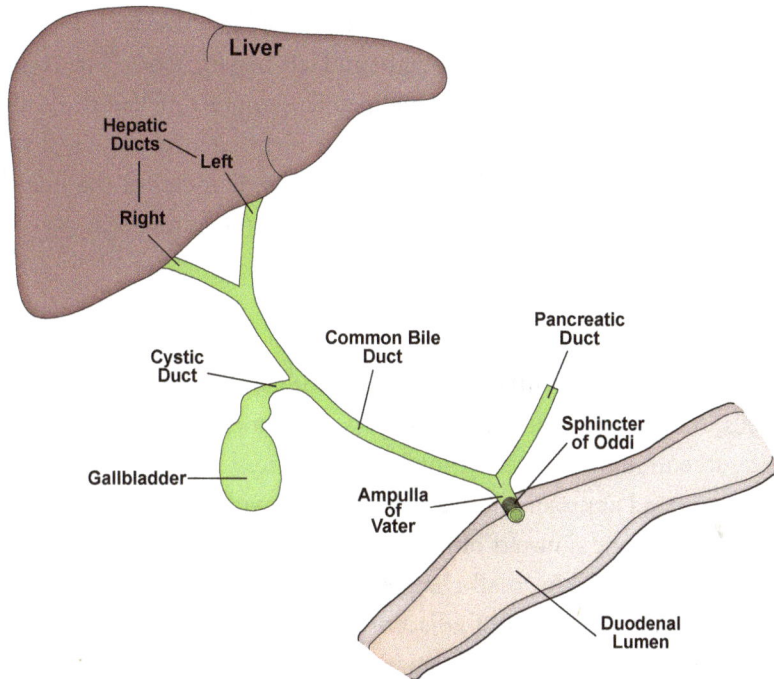

FIGURE 6-1: The biliary system. Bile formed in the canaliculi between adjacent hepatocytes flows into intrahepatic ducts, which in turn empties into extrahepatic ducts and the gallbladder. The flow of bile into the duodenum occurs via the common bile duct and the rate of bile entry into the duodenal lumen is regulated by the sphincter of Oddi.

capsule, over which lies the visceral peritoneum. There are two sources of blood supply to the liver, the portal vein and the hepatic artery. About 70% of liver blood flow and 40% of oxygen delivery is derived from the portal vein, with the hepatic artery providing the remaining 30% of blood flow and 60% of oxygen. The smaller contribution of the portal vein to liver oxygenation reflects the fact that portal blood is derived from (drains) the stomach, intestines, pancreas, and spleen, all of which extract oxygen from blood to meet their own metabolic needs. The hepatic artery and portal vein enter the liver at the porta hepatis to form branches that supply the left and right sides. Venous drainage of the liver microcirculation occurs through the hepatic veins, which converge before emptying into the vena cava.

The extrahepatic biliary system, which arises from small tubular structures called canaliculi, consists of the (1) hepatic bile ducts (right and left), (2) cystic duct, (3) common bile duct, and (4) gallbladder (Figure 6-1). The two hepatic ducts join with the cystic duct from the gallbladder to form the common bile duct. The pear-shaped gallbladder is located on the infe-rior surface of the liver at the junction of the right and left hepatic lobes. The common bile duct empties into the small intestine at the duodenal papilla (ampulla of Vater). At the entrance of the bile duct into the duodenum, there is a thickening of the circular muscle of the duct, which is called the sphincter of Oddi. The sphincter governs the diameter of the bile duct orifice in the ampulla and, thereby, controls the rate of bile flow into the gut lumen.

Microscopic Anatomy

The human liver is comprised of different cell populations, including hepatocytes, endothelial cells, Kupffer cells, stellate cells, and cholangiocytes. Hepatocytes, the polarized epithelial cells that produce bile, account for over 80% of total liver volume. These cells are arranged in parallel plates or cords that extend approximately 20 cell lengths, with each cell in the row exposed to blood flowing in the adjacent sinusoid or capillary (Figure 6-2). The biliary tree starts blindly with the bile canaliculi, which are small channels approximately 1 μm in diameter formed by the fusion of the plasma membranes of adjacent hepatocytes. The canaliculi have no epithe-lium, yet their walls are elongated by the formation of fingerlike microvilli that greatly increase the area available for bile secretion (Figure 6-2). The flow of bile in the canaliculi is countercur-rent to the flow of blood in hepatic sinusoids. Bile canaliculi drain into the ducts of Herring (also known as preductules), which in turn connect to ductules and ducts. The biliary ductules and ducts are lined by small cuboidal epithelial cells called cholangiocytes that exhibit charac-teristic features of a secretory cell, such as microvilli and micropinocytotic vesicles. Within the liver, the bile ducts merge and increase in diameter as they travel with branches of the hepatic artery and portal vein. The epithelium gradually changes from a cuboidal to a columnar type as the ducts increase in diameter.

Within the liver, the portal vein and hepatic artery give rise to numerous smaller vessels called terminal portal venules and hepatic arterioles, respectively (Figure 6-3, panel A). These terminal blood vessels supply a small mass of parenchyma called the liver acinus. The terminal portal venule and hepatic arteriole course along with a bile ductule (collectively called the portal tract or triad) and enter the acinus at its center. The hepatic arterioles empty either 1) directly into the sinusoidal capillaries that lie between rows of hepatocytes or 2) into the peribiliary capil-laries (supplying the bile ducts), which subsequently drain into the sinusoids. Because the portal venules also drain directly into the sinusoids, the hepatocytes are perfused with a mixture of venous and arterial blood. The sinusoids radiate toward the periphery of the acinus, where they drain into the larger central veins and ultimately into the hepatic veins and inferior vena cava.

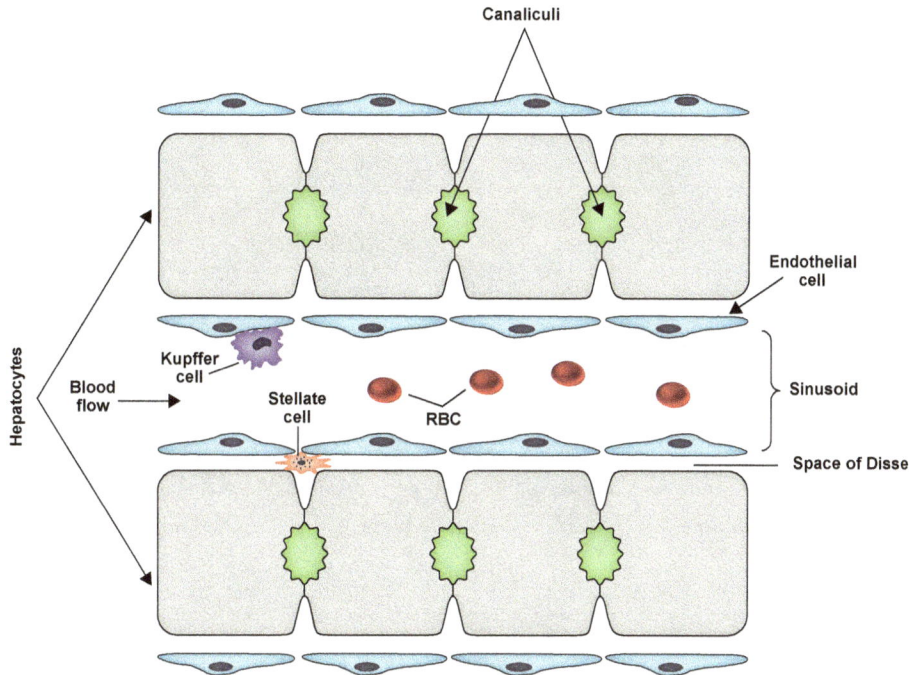

FIGURE 6-2: Anatomical arrangement of the major cell types in the liver. Plates of hepatocytes form bile canaliculi at the apical cell membranes. Endothelial cells separate the blood in capillaries (sinusoids) from the hepatocytes. The space of Disse, which is populated by stellate cells, lies between the endothelial cells and hepatocytes. Kupffer cells are attached to the endothelial cell surface in sinusoids.

The liver sinusoids are highly porous capillaries that allow easy access of plasma to the basolateral membrane of hepatocytes. Large discontinuities (fenestrations) in the endothelial cells that line the sinusoids and the absence of a basement membrane account for the ready passage of large molecules (e.g., plasma proteins), but not blood cells, into the space (called the space of Disse) between the sinusoid and hepatocytes (Figure 6-2). Kupffer cells, which are macrophages within the sinusoidal lumen that are attached to endothelial cells, perform functions related to the phagocytosis and clearance of particulate matter, including bacteria, endotoxins, immune complexes, and senescent (or cellular fragments of) blood cells. Stellate (fat-storing) cells, located in the space of Disse, are a major site of vitamin A storage and the production of hepatocyte growth factor (HGF). During chronic inflammation, stellate cells undergo transformation to myofibroblasts, which produces extracellular matrix materials (such as collagen) that are deposited in the space of Disse and impair hepatocyte function.

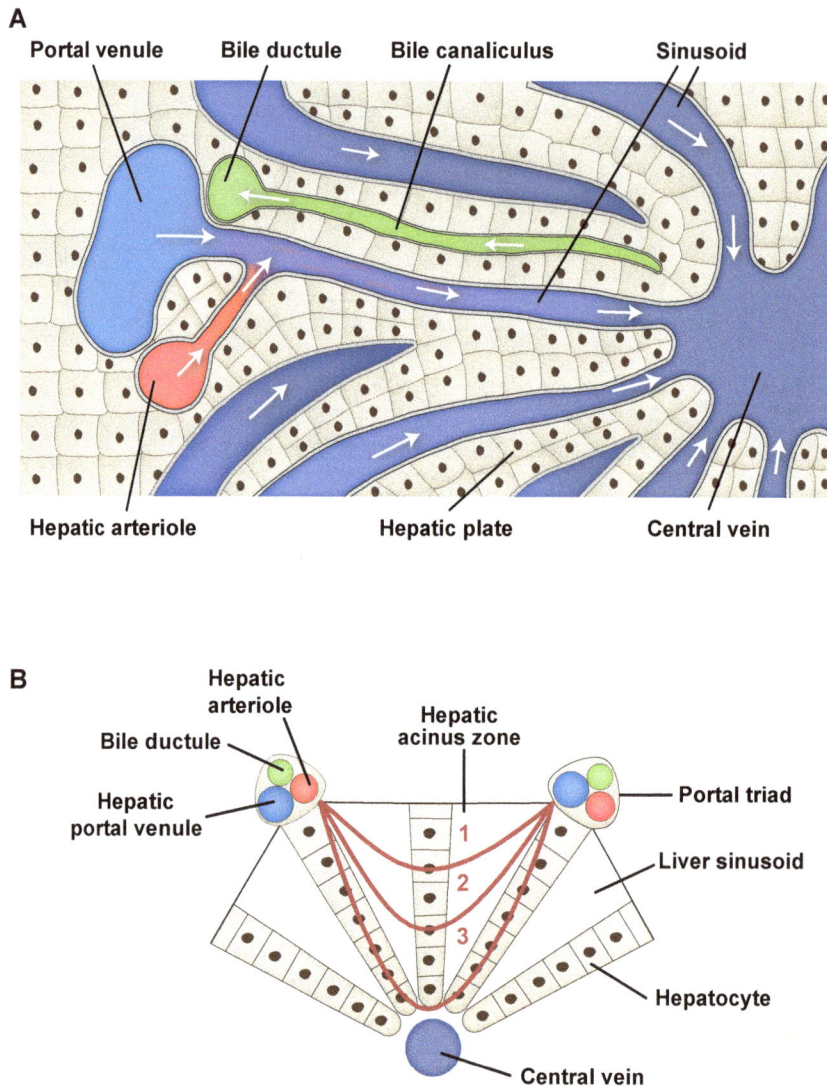

A

Portal venule   Bile ductule   Bile canaliculus   Sinusoid

Hepatic arteriole   Hepatic plate   Central vein

B

Hepatic arteriole

Bile ductule

Hepatic portal venule

Hepatic acinus zone

Portal triad

Liver sinusoid

Hepatocyte

Central vein

FIGURE 6-3: *Panel A.* Hepatocytes, sinusoids, and the intrahepatic biliary system. Sinusoidal blood is derived from microscopic branches of the portal vein (venules) and hepatic artery (arterioles) and empties into central vein. Blood flow is countercurrent to bile flow. *Panel B.* Zonal organization of hepatocytes that emphasizes the axial gradient of blood oxygen concentration and cell metabolism from the portal triad to the central vein [modified from Wang DQH, Neuschwander-Tetri BA, Portincasa P. *The Biliary System* (2nd Edition). In: Colloquium Series on Integrated Systems Physiology: From Molecule to Function to Disease (Granger DN, Granger JP, eds.), Morgan and Claypool Publishers, 2017. https://doi.org/10.4199/C00147ED2V01Y201611ISP071].

A functional model of the unit structure of the liver—analogue to the nephron in the kidney—that can explain the functions of the entire liver is the liver acinus. The liver acinus can be subdivided into three circulatory zones that surround the portal axis as layers (Figure 6-3, panel B). Zone 1 (periportal zone) encompasses the hepatocytes situated close to the portal triad and the origin of the sinusoid; hence, these hepatocytes are bathed by blood rich in nutrients and oxygen. The cells in zone 3 (perivenular or pericentral zone) are situated at the periphery of the acinus, near the central vein. Hepatocytes in zone 3 receive blood that has already exchanged nutrients and oxygen with cells in Zones 1 and 2 (the arbitrary intermediate zone). Consequently, although blood $pO2$ in periportal blood is 60 to 65 mm Hg, it is 30 to 35 mm Hg in pericentral blood. As a result of this microvascular organization, the metabolic requirements and transport functions of hepatocytes vary along the length of the sinusoid. Hepatocytes in zone 1 are exposed to the highest blood oxygen tension and the highest concentrations of nutrients and bile acids. These cells exhibit the highest mitochondrial density and are most active for both oxidative metabolism and the uptake of bile acids. Zone 1 cells exhibit significant resistance to injury caused by hypoxia and toxic substances. Conversely, the cells in zone 3 are more geared for anaerobic metabolism, and the rate of bile acid uptake is low. Zone 3 cells are the most sensitive to damage due to ischemia, nutritional deficiency, and toxic substances (e.g., acetaminophen, carbon tetrachloride). The enzymes (e.g., cytochrome P-450) that are involved in the detoxification and biotransformation of the majority of drugs and toxins in the liver are localized in zone 3 hepatocytes, which render these cells more vulnerable to the toxic effects of drugs like acetaminophen. The importance of adaptation to the microenvironment created by the microcirculation in the establishment of the gradient of hepatocyte function along the length of the sinusoid is evidenced by reports describing a reversal of the zone 1 to zone 3 gradient when blood supply and nutrient flow is reversed.

Hepatocyte Growth and Regeneration

The liver has an enormous capacity for regeneration after partial resection or damage, as little as 25% of the liver can regenerate into a whole liver. Mitotic figures in hepatocytes of normal liver are scarce, but in the regenerating liver every hepatocyte will divide in 24 hours. HGF, which stimulates mitogenesis and inhibits apoptosis in hepatocytes, is an important initiator of liver regeneration. In addition to the actions of HGF and other locally produced mitogens, hormones (insulin and glucagon from the pancreas) and growth factors (e.g., epidermal growth factor from the salivary glands and Brunner's glands in the duodenum) liberated from the digestive system exert trophic influences on the liver. The importance of insulin and glucagon in maintaining normal liver mass is underscored by the hepatic atrophy (to one third its normal size) that results from diversion of pancreatic venous blood away from the liver. Because they are delivered through the portal vein, the liver is the first recipient of insulin and glucagon produced by the pancreas. With the exception of neoplastic cells, and possibly hematopoietic stem cells, there is no other example of a cell with the proliferative capability of the hepatocyte. This unique characteristic of hepatocytes allows the liver to recover lost mass while simultaneously restoring all of the hepatic functions necessary to maintain body homeostasis.

COMPOSITION OF BILE

Hepatic Bile

Like most other exocrine secretions, bile is an isosmotic aqueous solution of organic and inorganic compounds (see Table 6-1). The solution is yellow, brownish, or olive-green in color and has a slightly alkaline pH. The concentrations of electrolytes in hepatic bile resemble those in

TABLE 6-1: Comparison of Hepatic and Gallbladder Bile Composition		
CONSTITUENT	HEPATIC BILE (mM)	GALLBLADDER BILE (mM)
Na^+	165	280
K^+	5	10
Ca^{++}	2.5	12
Cl^-	90	15
HCO_3-	45	8
Bile Acids	35	310
Lecithin	1	8
Bile pigments	0.8	3.2
Cholesterol	3	25
pH	8.2	6.5

plasma, with the exception that bicarbonate concentration may be twice as high. Bile acids are the major organic constituent of bile. The other major organic solutes include phospholipids, cholesterol, bilirubin, and a small amount of plasma proteins.

Bile acids in human hepatic bile can be divided into primary and secondary (Figure 6-4). Primary bile acids are synthesized from cholesterol by hepatocytes. The liver converts cholesterol to bile acids through a series of 14 reactions, beginning with the 7-α hydroxylation of cholesterol, the rate-limiting step in bile acid synthesis. This series of reactions primarily involves the addition of hydroxyl groups and a carboxyl group to the steroid nucleus, which increases the water solubility of bile acids relative to cholesterol. The major primary bile acids in man are cholic acid (α-trihydroxycholanic acid) and chenodeoxycholic acid (α-dihydroxycholanic acid).

Secondary bile acids are formed by dehydroxylation (removal of hydroxyl group at 7-α position) of primary bile acids by bacteria in the digestive tract. Bacterial metabolism of bile salts requires the presence of obligate anaerobes in concentrations above 10^4 per mm^3. Such conditions normally occur in the distal ileum, cecum, and colon. The secondary bile acids are absorbed in the distal bowel and returned to the liver via the portal vein, where they are secreted into bile. The major secondary bile acids in man are deoxycholic acid and lithocholic acid. Cholic acid, chenodeoxycholic acid, and deoxycholic acid normally appear in a ratio of 4:4:2 in hepatic bile, whereas lithocholic acid is usually present only in small quantities (10%–20% of the deoxycholic acid concentration). The low concentration of lithocholic acid results from its insolubility and consequent poor absorption in the distal bowel and excretion in stool.

Both primary and secondary bile acids are conjugated in the liver with the amino acids glycine or taurine (Figure 6-4), which makes the bile acid more water-soluble in the relatively acidic environment (the pH is lower than the pKa of unconjugated bile acids) of the duodenum. Conjugation lowers the pKa of the bile acids to between 4.0 (glycine conjugates) and 2.0 (taurine conjugates). When pH falls below 4.0, glycine-conjugated bile salts precipitate out of solution; however, the alkaline nature of biliary and pancreatic secretions prevents this from occurring in the biliary tract and intestine, respectively. The concentration of glycine conjugates is approximately three times higher than that of taurine conjugates in human bile, presumably due to the limited availability of taurine in the diet. Conjugated bile acids ionize more readily than free bile acids and therefore exist as salts of cations such as Na$^+$ and K$^+$, resulting for example in sodium taurocholate. Conjugated bile salts are rarely present in the feces because bacteria in the distal ileum and colon deconjugate the bile salts.

Three-dimensional analyses of conjugated and unconjugated bile acid molecules reveal that the hydroxyl groups are all on the same side of the molecule, whereas the water-insoluble portion of the molecule is on the opposite surface (Figure 6-5). This configuration imparts

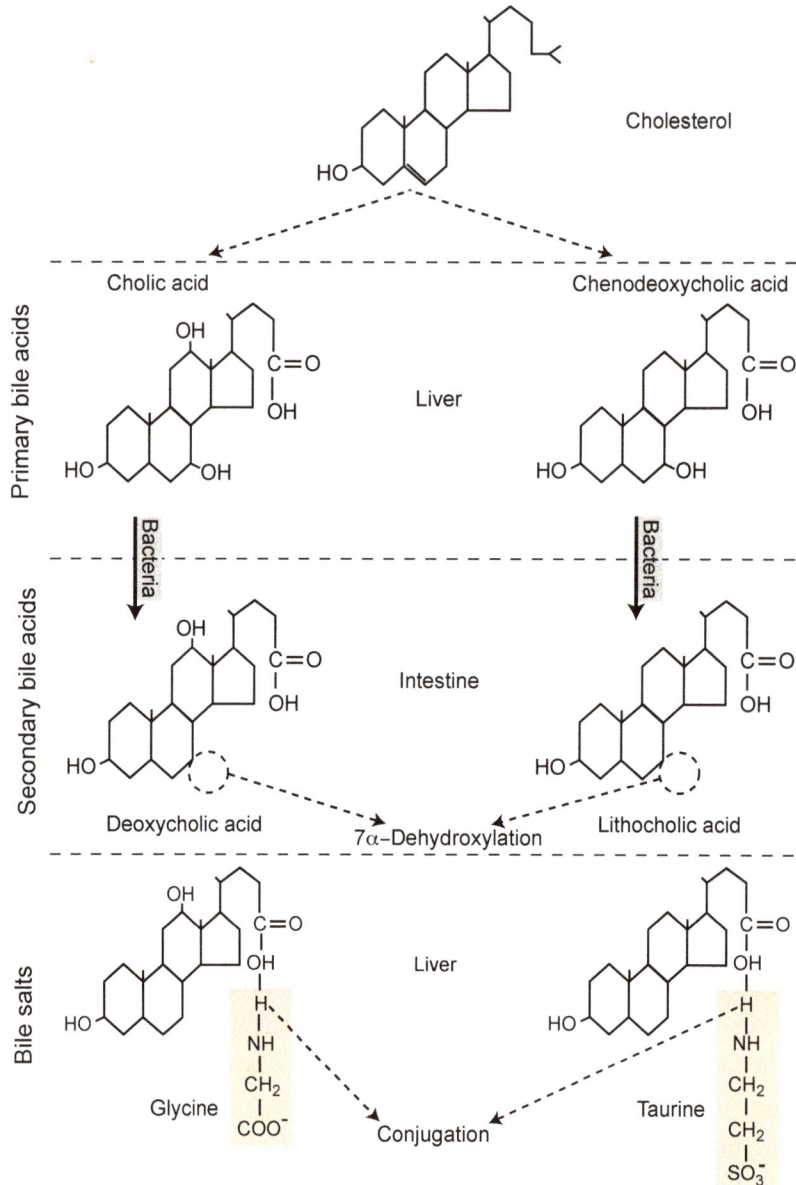

FIGURE 6-4: Formation of bile acids. The primary bile acids (cholic and chenodeoxycholic acids) are synthesized from cholesterol in the liver. Secondary bile acids (deoxycholic and lithocholic acids) are formed by the dehydroxylation of primary bile acids by bacteria in the intestine. The primary and secondary (when returned to the liver) bile acids are conjugated to small molecules, like glycine and taurine, by hepatocytes before secretion into bile.

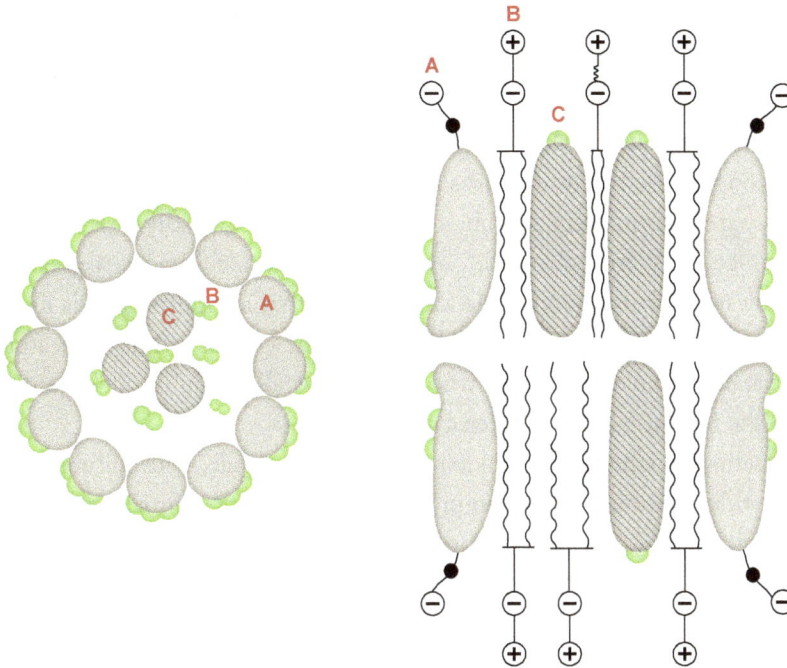

FIGURE 6-5: Structure of a mixed micelle in longitudinal (right) and cross (left) section. The micelle consists of bile acids (A), lecithin (B), and cholesterol (C). The water-soluble aspect of the bile acid covers the outside of the micelle (hydrophilic domain), whereas cholesterol and the nonpolar portion of the phospholipid are solubilized in the hydrophobic interior of the micelle [modified from Brooks FP, Gastrointestinal Pathophysiology, 2nd edition, New York, Oxford University Press, 1978. Used with permission].

amphipathic properties to the bile acid, that is, it has both hydrophilic and hydrophobic domains. Because of their amphipathic properties, bile acids tend to form molecular aggregates known as micelles when their concentration exceeds a critical level (called the critical micellar concentration or CMC). In micelles, the bile acid molecules are oriented so that the hydrophilic portion of the molecule faces the aqueous phase of the solution, whereas the hydrophobic portion faces inward, away from the aqueous phase. Therefore, the interior of the micelle becomes a lipid-soluble environment where lipids, such as cholesterol, which are very water-insoluble, can be maintained in a solution such as bile. In normal bile, bile acids, cholesterol, and phospholipids exist in the form of "mixed micelles" (Figure 6-5). The concentration of bile acids in hepatic bile is normally well above the CMC (1.5 to 5 mM), thereby ensuring an appropriate environment for solubilization of cholesterol and phospholipids.

Phospholipids and cholesterol comprise approximately 20% and 4% of the total solutes in hepatic bile, respectively. Lecithin accounts for 98% of the phospholipids, with the remainder consisting of sphingomyelin, phosphatidylethanolamines, and lysolecithin. Both lecithin and cholesterol are essentially insoluble in water; however, they are solubilized in bile acid micelles. Cholesterol, a nonpolar molecule, dissolves in the center of the micelle, whereas lecithin, which is amphipathic, partitions its fatty acyl chains into the center and leaves its polar head near the surface of the micelle. A "mixed micelle" of bile acids and lecithin has a greater capacity for solubilizing cholesterol than a simple bile acid micelle, that is, phospholipids like lecithin help to solubilize cholesterol. When the concentration of cholesterol in bile is greater than can be solubilized, crystals of cholesterol are formed. This crystallization of cholesterol plays an important role in the formation of cholesterol gallstones.

Bile pigments account for the golden yellow color of bile, and bilirubin is the principal pigment of bile. Bilirubin is a degradation product of heme, with approximately 85% of bilirubin derived from hemoglobin in erythrocytes and the remaining 15% from other heme proteins, like myoglobins and cytochromes. After the release of iron and globin from the hemoglobin molecule by macrophages in the mononuclear phagocyte system (residing in spleen, bone marrow, liver), heme is converted to biliverdin, which, in turn, is converted to bilirubin (Figure 6-6). When released into extracellular fluid by the macrophage, the bilirubin binds tightly to albumin and is delivered to the liver, where hepatocytes extract the bilirubin from blood and conjugate it with glucuronic acid (80%) and, to a lesser extent (10%), with sulfate. The increased water solubility that accompanies the conjugation of bilirubin allows for its excretion into bile. Upon reaching the bowel lumen, conjugated bilirubin is converted by bacterial action to highly soluble substances such as urobilinogen, which can be reabsorbed by the intestine or oxidized in the feces to stercobilin. Some urobilinogen is reabsorbed by the intestine, returned to the liver, conjugated and secreted into bile, whereas some of the urobilinogen is excreted into urine, where it is oxidized to form urobilin. The end-products of bile pigment degradation by bacteria such as stercobilin are responsible for the brown color of the stool, whereas urobilin is responsible for the color of urine.

A variety of other organic and inorganic compounds normally appear in bile in minute quantities. These compounds include lipids, steroid hormones (e.g., estrogens), and end-products of drug and hormone metabolism. Small amounts of heavy metals, such as lead and arsenic (and other environmental pollutants), also appear in bile, and the elimination of these substances by the biliary system may provide an important level of protection against their toxic effects. Copper is primarily excreted in humans via the biliary system and abnormalities in this process may result from Wilson's disease, a hereditary disorder that impairs copper transport into bile and is associated with very high (toxic) levels of hepatic copper.

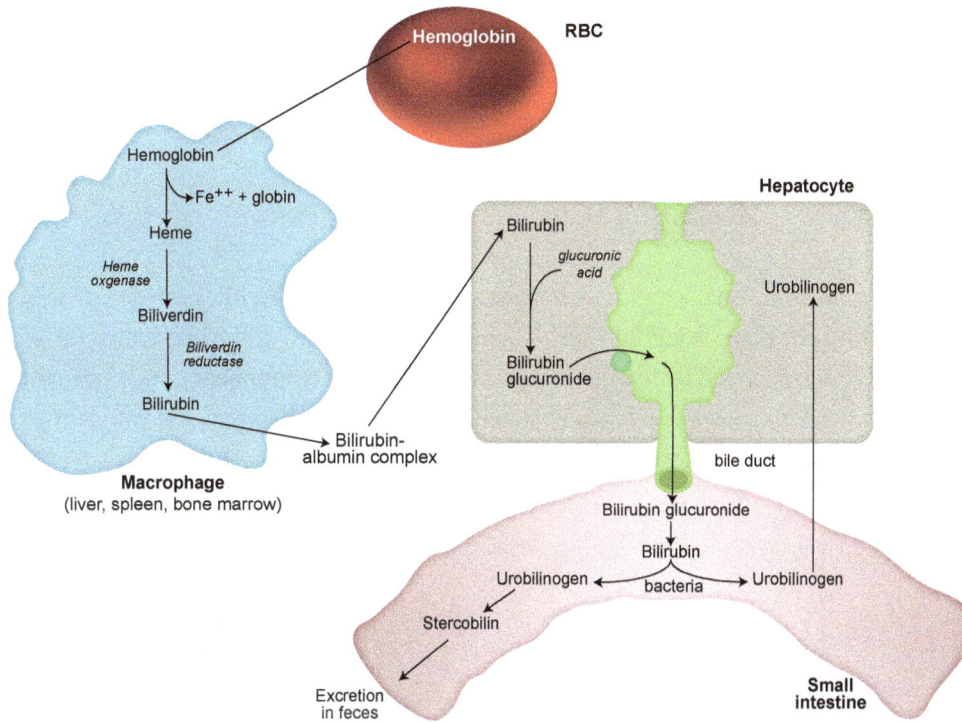

FIGURE 6-6: Production, secretion, and metabolism of bilirubin. Tissue resident macrophages (mononuclear macrophage system) generate bilirubin from hemoglobin. The bilirubin binds to albumin in extracellular fluid, allowing its transport to the liver. Hepatocytes conjugate bilirubin and secrete it into bile. After expulsion into the intestine, bilirubin is converted by bacteria to urobilinogen, which is either excreted in feces or reabsorbed and transported to the liver.

Gallbladder Bile

Table 6-1 compares the composition of gallbladder bile to that of hepatic bile. There are several major differences in the bile compositions, most of which are related to the absorptive functions of the gallbladder. The concentrations of cations, bile acids, lecithin, bile pigments, and cholesterol are much higher in gallbladder bile than hepatic bile. This difference results primarily from the reduction in the volume of bile in the gallbladder caused by water absorption. The concentrations of Cl^- and HCO_3^- in gallbladder bile are significantly less than in hepatic bile. These differences are attributed to Cl^- and HCO_3^- absorption. Because of HCO_3^- absorption, the pH of gallbladder bile is lower (slightly acidic) than that of hepatic bile (slightly alkaline) as well as plasma. Gallbladder bile remains isotonic with plasma despite the dramatic rise in electrolyte and bile acid concentrations. The fact that micelles have minimal osmotic activity, coupled to their ability to sequester counterions, such as sodium, explains the maintenance of isotonicity even though there is a larger concentration of solutes in gallbladder bile.

MECHANISMS OF HEPATIC BILE FORMATION

The liver of adult man normally produces between 800 and 1000 ml of bile each day, at an average rate of 30 to 40 ml per hour. Roughly, 75% of hepatic bile is formed at the bile canaliculi, whereas the remaining 25% is secreted by the bile ducts. The canalicular component of bile formation can be divided into two fractions: a bile acid-dependent fraction, and a bile acid-independent fraction, with the former accounting for most of the bile entering the canaliculi (Figure 6-7).

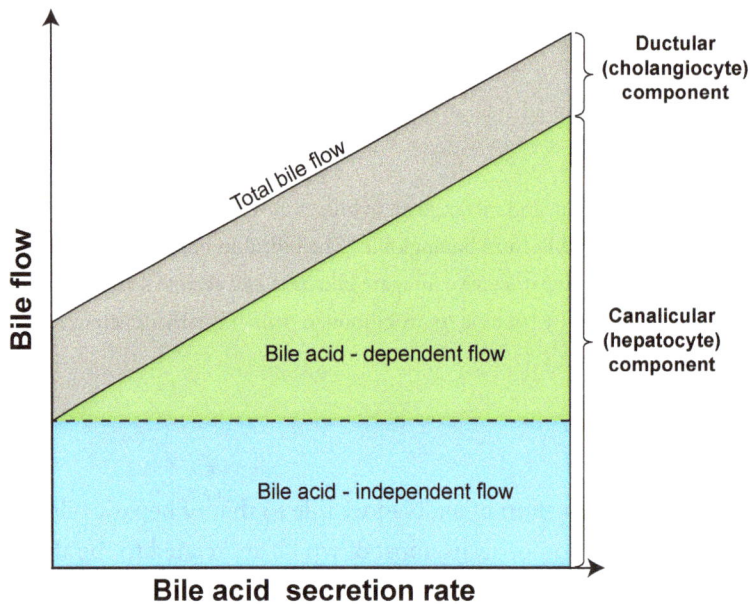

FIGURE 6-7: Components of hepatic bile flow. Total bile flow consists of bile acid-dependent canalicular secretion, bile acid-independent canalicular secretion, and ductular secretion. Total canalicular bile flow represents the difference between total bile flow and ductular secretion. Bile acid-independent canalicular secretion is defined as the canalicular bile flow when bile acid excretion rate is zero. [From Scharschmidt BF. In: Hepatology (Zakim D and Boyer T, eds.), Philadelphia, WB Saunders, 1982. Used with permission of Elsevier.]

Canalicular Bile Formation

Bile Acid-Dependent Secretion

Canalicular bile formation is generally regarded as an osmotic water flow in response to active solute transport. Because of the excellent correlation between bile flow and bile acid secretion (Figure 6-7), bile acids are considered to be one of the major solutes generating bile flow. Bile acids are secreted against a concentration gradient; the biliary concentration is 100 to 1000 times higher than the sinusoidal (plasma) concentration. The secretion of bile acids into the canalicular lumen is mediated by a transport protein called bile salt export pump (BSEP), which is enriched in the canalicular membrane and requires ATP as an energy source (Figure 6-8). An osmotic gradient is established between the canalicular lumen and the hepatic interstitial space owing to the osmotic activity of the bile acids (or of their associated counter-ions) secreted into the canaliculus. The osmotic pressure gradient draws water through the hepatocyte and the intercellular spaces (tight junctions) and into the lumen, thereby leading to the formation of isotonic bile (see Figure 1-13 for discussion of standing osmotic gradient theory).

The bile acids that are actively secreted into the canaliculi are derived either by de novo synthesis from cholesterol or by extraction from portal venous blood. The rate-limiting reaction in the conversion of cholesterol to primary bile acids is the hydroxylation step catalyzed by the enzyme 7-α hydroxylase. Feedback inhibition of bile acid synthesis by the concentration of bile acids in portal blood occurs at this step. Hepatocytes have a high extraction efficiency for bile acids, that is, 50% to 90% (depending on the bile acid) of the bile acids in portal venous blood is taken up by the liver in a single pass. This highly efficient uptake of bile acids on the sinusoidal membrane of hepatocytes involves different transport mechanisms that can extract both conjugated and unconjugated bile acids from blood (Figure 6-8). Uptake of conjugated bile acids, which account for the majority of bile acids returned to the liver via portal blood, is mediated in large part by a transporter called the Na^+ taurocholate co-transporting polypeptide (NTCP) that is dependent on sodium and derives its energy from the Na^+ concentration gradient maintained by the Na^+-K^+ ATPase. A family of Na^+-independent, non-specific anion transporters known as organic anion transporter proteins (OATPs) contribute to the uptake of unconjugated, and to a lesser extent conjugated, bile acids by hepatocytes. Glutathione (GSH) and HCO_3^- are the primary anions involved (exchanged) in OATP-mediated bile acid transport. Secretion of bile acids into the canaliculi by BSEP, rather than hepatocellular uptake by NTCP and OATP, is the rate-limiting step in overall transport of bile acids from blood to bile, inasmuch as the velocity of bile acid uptake by the hepatocyte exceeds the secretory transport maximum by about six-fold.

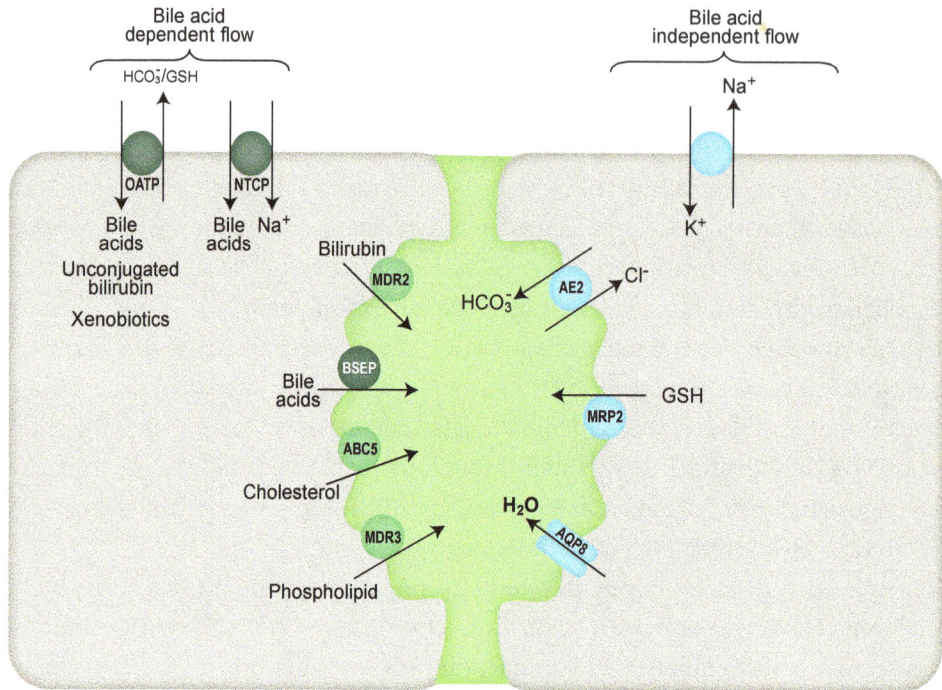

FIGURE 6-8: Mechanisms involved in the formation of bile acid-dependent and -independent canalicular secretions. Different transport proteins (NTCP, OATP) enable hepatocytes to take up bile acids from sinusoidal blood. Other transport proteins on the canalicular membrane allow for the secretion of bile acids (BSEP), cholesterol (ABC5), phospholipids (MDR3), and bilirubin (MRP2) into the canalicular lumen. Bile acid-independent bile flow across the canalicular membrane is mediated by HCO_3^- exchange via AE2, GSH secretion via MRP2, and water efflux via aquaporin-8 (AQP8).

The secretion of bilirubin, cholesterol, and phospholipid across the canalicular membrane and into bile also involves ATP-dependent transport proteins, most of which belong to the ATP-binding cassette (ABC) transport protein family (Figure 6-8). Bilirubin uptake by hepatocytes is mediated by OATP and is actively extracted from blood by hepatocytes in proportion to its free (unconjugated) concentration in plasma. Upon entry into the hepatocyte, bilirubin is conjugated with glucuronic acid, which markedly increases its water solubility. The conjugated bilirubin is then transported across the canalicular membrane by the multidrug resistance protein transporter, MDR2. Because of the broad specificity of this transport protein, it is also involved in the secretion of conjugated drugs into bile. Similar protein transporters are also engaged in the secretion of cholesterol (ABC5) and phospholipids (MDR3) into bile;

however, these molecules do not require conjugation for secretion. The rate(s) of secretion of phospholipids and cholesterol into bile is directly related to bile acid secretion, which serves to maintain appropriate solubility ratios between the bile acids and both hydrophobic lipids.

Bile Acid-Independent Secretion

There is considerable evidence suggesting that bile is also formed at the canaliculus by a mechanism that is independent of bile acid secretion. Inhibitors of the enzyme Na^+-K^+ATPase significantly reduce bile acid-independent secretion, indicating that water movement coupled to active Na^+ secretion into the canaliculus makes a major contribution to this fraction of bile formation. There is also evidence implicating HCO_3^- exchange with chloride (via the anion-exchange protein AE2), reduced glutathione (GSH) excretion (via multidrug-resistance protein-2, MRP2), and water efflux through aquaporins (AQP8) in the elaboration of bile acid-independent secretion (Figure 6-8). In the human liver, bile acid-independent mechanisms contribute much less than bile acid-dependent processes to total canalicular bile formation.

Ductal Component of Bile Flow

Bile leaving the extensive network of canaliculi drain into small intrahepatic bile ducts that are lined with cholangiocytes, which are polarized epithelial cells that can modify both the composition and volume of canalicular bile. The most important physiological consequences of the activities of the apical and basolateral channels and transporters of the cholangiocytes are: (1) the movement of water via aquaporins (AQP1 anf AQP4) into the duct that is driven by chloride secretion, and (2) the alkalinization of bile via the secretion of HCO_3^- into the duct lumen (Figure 6-9A). Hence, the influence of cholangiocytes on bile formation is similar to that described in chapter 5 for pancreatic ductal cells. In cholangiocytes, the Na/K/2Cl cotransporter loads the cells with Cl^-, which is transported into the duct lumen by two channels: the CFTR and a Ca^{++}-stimulated Cl^- channel. The ductal accumulation of Cl^- produces a lumen negative potential that favors Na^+ entry into the duct lumen via paracellular pathways and water movement via aquaporin channels. Bicarbonate ions enter the cholangiocyte by a Na^+-HCO_3^- cotransporter, then it enters the duct lumen via an anion-exchange protein (AE2) that exchanges Cl^- for HCO_3^-. When regulatory peptides and hormones, most notably secretin, activate their receptors on cholangiocytes, intracellular cAMP level is elevated, and this stimulates the apical Cl^- channels (CFTR) and the Cl^--HCO_3^- exchanger, resulting in a more dilute and alkalinized bile.

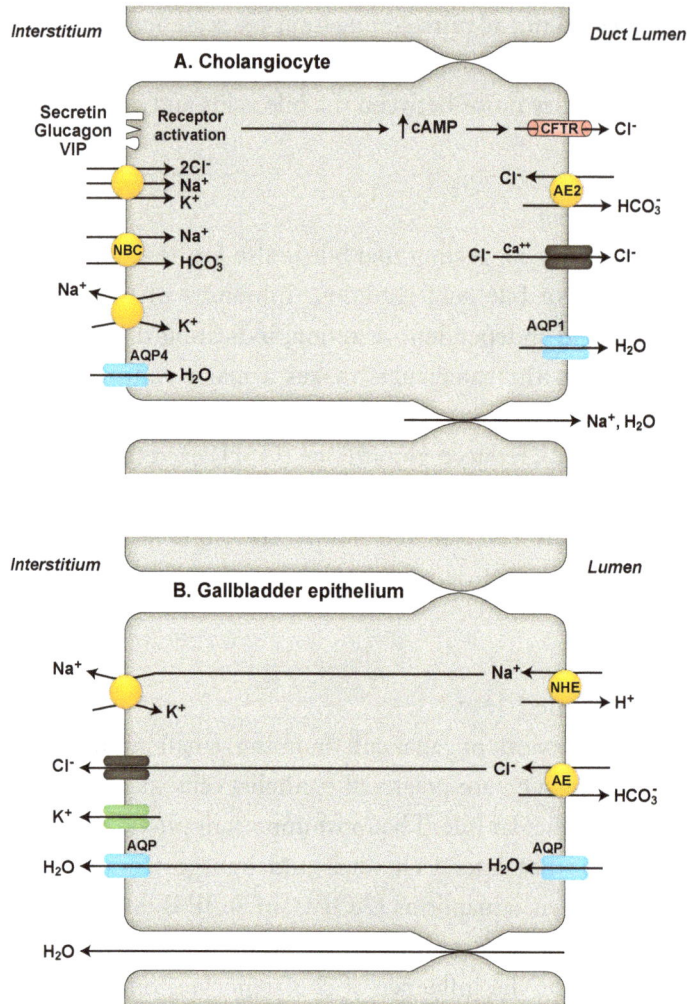

FIGURE 6-9: Major transporters and channels in cholangiocytes (Panel A) and gallbladder epithelium (Panel B). Secretin-induced stimulation of cholangiocytes increases intracellular cAMP concentration and elicits chloride secretion via the cystic fibrosis transmembrane regulator (CFTR). The transmembrane chloride gradient drives Cl^--HCO_3^- exchange via AE2, whereas transepithelial water flow is mediated through aquaporin channels (AQP) on the basolateral and apical membranes. In the gallbladder, parallel Na^+-H^+ (NHE) and Cl^--HCO_3^- (AE) exchangers on the luminal (apical) membrane of epithelial cells mediate the electroneutral uptake of Na^+ and Cl^-, whereas Na^+ exits the cell via the Na^+-K^+ ATPase activated pump located on the basolateral membrane. The accumulation of Na^+ and Cl^- and hypertonicity in the lateral intercellular space provides a driving force for osmotic water flow from the lumen to the interstitium. NBC, sodium-bicarbonate co-transporter.

The quantitative significance of ductal secretion to total bile flow is highly species-dependent. Species that feed intermittently, such as man, have an important ductal contribution, whereas the ductal contribution is negligible in continuous eaters, such as rodents. The epithelium of the biliary ducts and ductules also has the capacity to absorb electrolytes and water. Although little is known concerning the mechanisms involved in water absorption by ductal epithelium, it appears that the common bile duct can take over a significant fraction of the gallbladder's concentrative capacity after removal of the gallbladder (cholecystectomy).

THE GALLBLADDER: ANATOMY AND FUNCTION

The gallbladder is a thin muscular organ that forms a blind appendage of the common bile duct. The inner wall of the gallbladder is directly exposed to bile and is lined by a single layer of tall, columnar epithelium with tight junctions that are highly resistant to the permeation of bile salt molecules. Interspersed among the epithelial cells of the gallbladder mucosa, which exhibits folds analogous to intestinal villi, are goblet cells that secrete mucus and protect the epithelium from the potentially toxic effects of bile salts. Lying beneath the mucosal layer are smooth muscle cells that express receptors for, and can contract in response to, hormonal (e.g., cholecystokinin) and/or neural (acetylcholine) stimuli. The gallbladder smooth muscle also receives input from branches of the vagus nerve.

The human gallbladder is not a large organ; when relaxed it can accommodate 40 to 60 ml of fluid. With an appropriate stimulus (e.g., a fatty meal), the gallbladder will contract and eject up to 70% of its contents due to contraction. The gallbladder is not essential to life, and its removal does not significantly compromise the effective assimilation of a meal.

Once bile is formed at the canaliculi and ducts, it flows down the biliary tree and drains into either the gallbladder or the intestine. The force responsible for propelling bile down the biliary tree is the secretory pressure generated by the hepatocytes and ductal epithelia. Biliary secretion pressures as high as 20 to 30 mm Hg can be generated owing to active secretion of bile acids and electrolytes, and possibly as a result of the contraction of stellate cells and myofibroblasts surrounding bile ductules. During the interdigestive period, there is an increased resistance to bile flow in the common bile duct due to constriction of the sphincter of Oddi, which can generate a pressure as high as 30 mm Hg. The increased sphincteric tone, coupled with a relaxed gallbladder, allows the biliary secretion pressure to propel about half of hepatic bile (~450 ml) preferentially into the cystic duct and gallbladder. The other half of daily hepatic bile output (~450 ml) will drain into the duodenum in the interdigestive period. Storage of bile within the gallbladder during the interdigestive periods serves two important physiologic functions: (1) to concentrate bile, and (2) to store the bile and deliver it into the duodenum at the appropriate time after ingestion of food.

Water and Electrolyte Transport

As illustrated in Table 6-1, there are major differences in the composition of hepatic and gallbladder bile, which can be largely explained by water absorption in the gallbladder. In the fasted state, the human liver can secrete a volume of bile (~900 ml/day) that is several times the maximum fluid capacity of the gallbladder (40–60 ml). The ability of the gallbladder to concentrate bile some 5 to 20 times accounts, in part, for the discrepancy between the amount of bile secreted by the liver and the amount stored by the gallbladder. Part of the discrepancy is explained by the fact that as much as 50% of daily hepatic bile output (~450 ml) will drains into the duodenum in the interdigestive period.) Figure 6–10 illustrates the time required for conversion of hepatic bile into gallbladder bile. Within 4 hours, up to 90% of the water in hepatic bile is removed by the gallbladder, resulting in a more concentrated bile.

Active sodium transport is primarily responsible for the ability of the gallbladder to concentrate bile (Figure 6-9B). On the apical (luminal) membrane of gallbladder epithelium, parallel Na^+-H^+ (NHE) and Cl^--HCO_3^- (AE) exchangers allow for the electroneutral uptake of Na^+ and Cl^-, whereas Na^+ exits the cell via the Na^+-K^+ ATPase activated pump located on the basolateral membrane. The resulting accumulation of Na^+ and Cl^- in the lateral intercellular space creates a region of hypertonicity that provides a driving force for osmotic water flow from the gallbladder lumen to the mucosal interstitium. Aquaporin (AQP) channels on the apical and basolateral membrane also contribute to water flow from lumen to interstitium. Because of the high water permeability of gallbladder epithelium, only slight hypertonicity is required to account for the high rate of water absorption.

The large volumes of water absorbed by gallbladder epithelium have a "concentrating" effect on bile constituents, which are either impermeable or only slightly permeable at the epithelial barrier. The resulting increase in bile concentrations of K^+, lecithin, conjugated bilirubin, and unconjugated bile salts creates a steeper gradient for passive diffusion of these "permeable" substances across gallbladder epithelium. This process allows for up to 30% of bile lecithin and lysolecithin, and a smaller amount (<5%) of bile salts and bilirubin, to be absorbed in the gallbladder. Unesterified long chain fatty acids (e.g., oleic acid) are also absorbed and subsequently metabolized by the gallbladder.

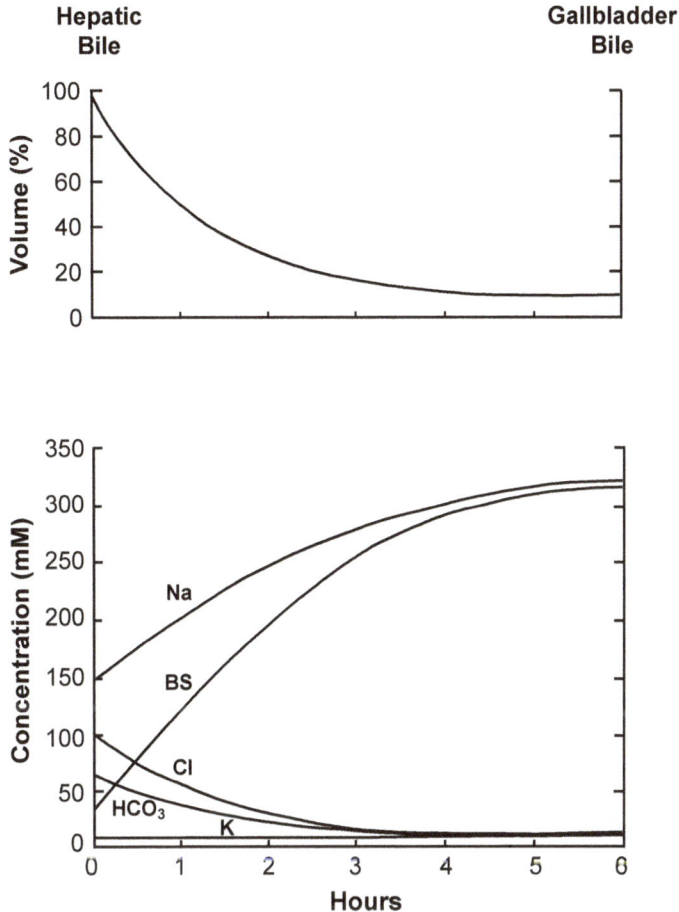

FIGURE 6-10: Time course of hepatic bile modification in the gallbladder. Ninety percent of water is removed from hepatic bile by the gallbladder in only 4 hours [modified from Makhlouf GM, *Viewpoints Digestive Diseases* 11:1–4, 1979].

An important modification of hepatic bile that takes place in the gall bladder is acidification. The pH of hepatic bile is slightly alkaline, but it becomes acidic in the gallbladder. This has been explained by a faster exchange of Na^+-H^+ than Cl^--HCO_3^- on the apical membrane of gallbladder epithelium (Figure 6-9), which results in the net secretion of H^+ ions in the lumen, the neutralization of HCO_3^-, and acidification of bile. Acidified bile is physiologically advantageous because it enhances the solubility of calcium salts and diminishes the likelihood of calcium salt precipitation and gallstone formation.

Motor Functions

In the interdigestive period, hepatic bile is diverted into the gallbladder where it is concentrated and stored until feeding. Shortly after a meal, the gallbladder contracts and the sphincter of Oddi relaxes. The resulting rise in biliary pressure and reduction in resistance to bile flow lead to the ejection of gallbladder bile into the duodenal lumen. Rhythmic contractions of the gall bladder musculature begin within 10 to 30 minutes after a meal and can persist for up to 90 minutes after the meal. The time of onset, intensity, and duration of gallbladder ejection is influenced by meal composition. With a fatty meal for example, contraction of the gallbladder can occur within 20 minutes of ingestion. A fatty meal elicits two phases of gallbladder ejection, with an early rapid phase completed within 30 minutes of the test meal which accounts for about two thirds of the total ejection volume, followed by a slow ejection phase that expels the remaining volume. During the entire contraction period, the gallbladder can discharge 70% to 85% of its contents into the duodenum.

The sphincter of Oddi regulates bile flow into the duodenum. It exhibits an intrinsic electrical activity that is not synchronous with that of the longitudinal muscle of the duodenum. The sphincter exhibits several high-pressure phasic contractions per minute that are superimposed on its basal tone. Between meals, a contracted sphincter prevents the reflux of duodenal contents into common bile duct. Bile flow from the common duct in humans is well correlated with the electrical activity of sphincteric muscle but not the duodenal muscle, indicating that the sphincter itself exerts primary control over the rate of bile excretion into the duodenum. Nonetheless, the contractile activity of the duodenum does influence bile flow. When the sphincter of Oddi is open, bile enters the bowel in spurts between periods of duodenal relaxation and contraction.

Ejection of bile into the duodenal lumen will occur during gallbladder contraction even in the absence of neural or hormonal (CCK) relaxation of the sphincter of Oddi. This occurs as a result of the tendency for the sphincter to yield when the pressure in the common bile duct exceeds approximately 10 mm Hg. Contractions of the gallbladder can generate transient pressures exceeding 20 mm Hg in the biliary tree. Therefore, when the sphincter yields to such high pressures, bile spurts into the duodenal lumen. Although this mechanism serves to ensure bile secretion and prevents marked elevations in biliary pressure during gallbladder contraction, it cannot generate the high rates of bile flow that are associated with a relaxed sphincter of Oddi.

THE ENTEROHEPATIC CIRCULATION

Fate of Bile Salts in the Intestine

When a meal is ingested, the gallbladder contracts and bile salts are secreted into the duodenal lumen. The concentration of bile salts in the duodenum during the early stage of gastric emptying is approximately half that of gall bladder bile. However, after gastric emptying, the luminal bile salt concentration is further reduced to approximately 5 to 10 mM owing to dilution with chyme. Even with the ten-fold dilution of gallbladder bile, the concentration of bile salts in the intestinal lumen always remains above the critical micellar concentration. Pancreatic lipolysis of lecithin and triglycerides changes the composition of mixed micelles in the duodenal lumen. Here, the mixed micelles are maximally enlarged with the products of lipid digestion, and larger spherical aggregates composed of the same lipids (called liposomes) are also formed.

An important function of the intestinal micelle is to solubilize the nonpolar products of luminal fat digestion within the polar environment of the gut lumen. The micelle facilitates the passive entry of the products of lipid digestion into the intestinal epithelial cell. Once the lipids move by simple diffusion across the brush border membrane, the bile salts remain within the bowel lumen, where they are driven toward the ileum by intestinal motility.

Bile salts are absorbed in the small intestine by both passive diffusion and active transport. Passive diffusion of bile salts occurs along the entire length of the small intestine and colon. Estimates based on studies in primates and humans suggest that as much as 50% of the bile salts in the intestines are absorbed passively. Both conjugated and unconjugated bile salts undergo passive absorption. However, the unconjugated bile acids more readily undergo passive absorption in the small bowel than glycine conjugates, with taurine-conjugates exhibiting the slowest rate of passive absorption. The more lipophilic bile acids (e.g., deoxycholate, chenodeoxycholate) diffuse more readily into intestinal epithelial cells than the hydrophilic bile acids (e.g., cholic acid). Deconjugation and dehydroxylation of bile acids by bacteria in the distal ileum and colon enhance their lipid solubility and facilitate passive absorption. Deconjugation increases passive bile salt absorption nine-fold, wheras dehydroxylation increases it fourfold. The rate of deconjugation of glycine conjugates is over twice that of taurine conjugates.

Active absorption of bile salts occurs exclusively in the distal ileum. Like the absorption of most water-soluble organic substances, active bile salt transport is dependent upon the presence of sodium ions (see Figure 1-10 in Chapter 1). The Na^+-dependent transporter that mediates the apical membrane step of this process is known as the apical sodium-bile salt transporter (ASBT). Once the bile salt enters the ileal enterocyte via ASBT, they leave the cell through the basolateral membrane by way of a Na^+-independent anion channel. A reciprocal relationship exists between active and passive transport rates of specific bile salts. The most polar bile salts (e.g., taurine conjugates), which are poorly absorbed by passive diffusion, have the highest affinity for ASBT and exhibit the highest active transport rates in the ileum. The active transport process for bile salts exhibits saturation kinetics, and there is competition among bile salts for the ASBT sites.

Between 7% and 20% of the total bile salt pool, that is, the total amount of bile salts in the body at any given moment (approximately 3 grams), escapes absorption in the small and large bowel and is excreted in the feces. The fecal bile acids in health consist of deoxycholate and lithocholic acids (70% to 80%) and unmetabolized primary bile salts, cholic and chenodeoxycholic acids (7% to 8%). The daily fecal loss (600 mg/day) of these bile salts is equal to their rates of hepatic synthesis, thereby maintaining a constant total bile salt pool.

The rate of spillage of certain bile salts into the colon appears to play an important role in governing daily fecal water loss. The two major dihydroxy bile acids of human bile, deoxycholic acid and chenodeoxycholic acid, modulate this process. An increased luminal concentration (above 1.5 mM) of either of these bile acids in the colon causes diarrhea, whereas decreased concentrations are associated with constipation. Although the enhanced fecal water loss noted in patients with a surgically resected ileum has long been linked to bile acid loss in stool, recent evidence indicates that bile acid diarrhea is a more common occurrence, affecting 1 in every 100 people (without ileal resection).

Transport of Absorbed Bile Salts to Liver and Resecretion into Bile

Once bile acids are absorbed in the small intestine, they enter the blood circulation, that is, the portal vein. In blood, bile acids are tightly bound to albumin (50% to 75%) and specific lipoproteins (20% to 50%). Free bile acids are more tightly bound to plasma proteins than are conjugated salts. In the fasted state, the bile salt concentration in portal venous blood is approximately 20 μM. After a meal, the portal venous blood concentration increases to two to seven times the fasting level.

The bile acids absorbed from the intestine are transported by the portal vein to the liver, where the hepatocytes are extremely efficient at clearing portal venous blood of bile acids. It is estimated that up to 90% of the bile salts in the portal vein are extracted in a single pass through the liver. Hepatocellular uptake of bile acids is mediated by a variety of transporters that are specific for conjugated (e.g., NTCP) and unconjugated (e.g., OATP) bile acids (Figure 6-8). Once the bile acids enter the hepatocyte, the deconjugated bile acids are reconjugated and a portion of the secondary bile acids are rehydroxylated. Then, the bile salts are resecreted across the canalicular membrane into bile, along with newly synthesized bile acids, by the BSEP transport (see Figure 6-8).

The flow of bile salts from liver to intestine and back to the liver again is known as the enterohepatic circulation (Figure 6-11). The major physiologic significance of the enterohepatic circulation is that the recycling and reuse of bile salts allows for efficient lipolysis and fat absorption with a limited total body pool of bile salts. An amount of bile salts twice the total body pool is required for efficient lipolysis of an average meal, whereas five times the total body pool is required after a meal with a very high fat content. Thus, the total amount of body bile

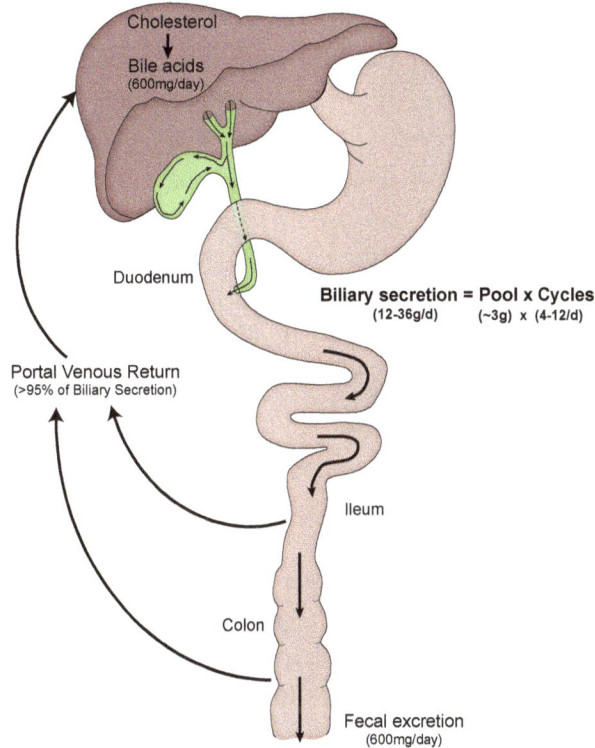

FIGURE 6-11: The enterohepatic circulation of bile acids in healthy humans, with typical kinetic values. The total body bile acid pool of man undergoes 4 to 12 cycles per day in the enterohepatic circulation. The recycling and reuse of bile acids allow for efficient lipolysis and fat absorption with a limited total body pool of bile acids [modified from Wang DQH, Neuschwander-Tetri BA, Portincasa P. *The Biliary System* (2nd Edition). In: Colloquium Series on Integrated Systems Physiology: From Molecule to Function to Disease (Granger DN, Granger JP, eds.), Morgan and Claypool Publishers, 2017. https://doi.org/10.4199/C00147ED2V01Y201611ISP071].

salts must be cycled two to five times through the enterohepatic circulation after a meal. The rhythm of the enterohepatic circulation, its acceleration during eating and deceleration during fasting, is determined by the pattern of dietary intake. Most individuals eat three meals a day and sleep at night. Accordingly, the total body bile salt pool of man undergoes 4 to 12 cycles per day in the enterohepatic circulation (Figure 6-11).

A number of other substances that are secreted into bile also undergo recycling in the enterohepatic circulation. Endogenous substances that are recycled include fat-soluble vitamins.

Non-steroidal anti-inflammatory drugs (NSAIDs), cardiac glycosides, and antibiotics are examples of exogenous compounds, which undergo recycling. Enterohepatic cycling may influence drug pharmacokinetics by prolonging elimination and may have a therapeutic advantage. For example, sulindac, an NSAID, is recycled through the biliary system to a greater degree than its active sulfide metabolite, thereby sustaining higher plasma concentrations of the prodrug and minimizing exposure of the intestine to active drug. Some substances, such as lithocholic acid and cholesterol, are poorly recycled in the enterohepatic circulation, thereby facilitating the elimination of these potentially deleterious compounds. With liver disease, the biliary excretion of drugs may be reduced, which increases the risk of toxicity.

REGULATION OF THE BILIARY SYSTEM

Bile Salt Concentration in Portal Blood

The major physiologic determinant of the rates of bile acid secretion and synthesis is the amount of bile acids in portal venous blood. As previously discussed, the concentration of bile acids in the hepatocyte exerts negative feedback control over the rate-limiting enzyme in bile acid synthesis (7 α-hydroxylase), and it is the primary regulator of the bile acid-dependent fraction of bile formation. The increased amount of bile acids in portal venous blood after a meal leads to an increased intracellular concentration of bile acids in hepatocytes. The intracellular bile acids bind to cytosolic transcription factors that regulate genes that are involved in both the synthesis and transport of bile acids. This allows an increased intracellular bile acid concentration to decrease their synthesis and rate of uptake into the hepatocyte while enhancing their canalicular secretion. Conversely, interruption of the enterohepatic circulation (following ileal resection or ingestion of bile acid-binding agents such as cholestyramine) leads to a reduction in bile acid secretion (due to inadequate absorption) and a five- to tenfold increase in bile acid synthesis. A similar relationship between synthesis and secretion is observed long after a meal, that is, bile acid synthesis is maximal yet their rate of secretion is low, whereas the gallbladder fills slowly with bile.

Gastrointestinal Hormones

Secretin

Secretin is released from the duodenal mucosa in response to acidic chyme and subsequently exerts some influence over the transport functions of the bile ducts/ductules and gallbladder. Secretin stimulates the formation of a HCO_3-rich fluid by the bile ducts and ductules through engagement with receptors expressed on the basolateral membrane of cholangiocytes. The effects of secretin on biliary epithelium are mediated by cyclic AMP, which activates the Cl^--HCO_3^- exchanger (AE2) and enhances Cl^- transport into the duct lumen through activation of CFTR while increasing water flux via aquaporin channels (Figure 6-9A). The net result is an increased formation and flow of HCO_3^--rich bile in the ducts and ductules. The responsiveness of the ductal system to secretin is species-dependent, with intermittent feeders (dog, man) more responsive than continuous eaters (rat, rabbit). Bile flow increases three-fold and bicarbonate output four-fold when secretin is administered to humans. The ductal responses to secretin are not mediated by vagal stimuli, because cholinergic blockade does not abolish the response to endogenous or exogenous secretin. Although secretin can elicit gallbladder smooth muscle contraction, supraphysiologic concentrations of the hormone are required to elicit this response. However, secretin may act to antagonize the actions of CCK.

Cholecystokinin

Cholecystokinin is released from the duodenal mucosa in response to the products of lipid digestion. It is the primary stimulus for gallbladder contraction and relaxation of the sphincter of Oddi following a meal. The peptide can exert direct (endocrine) effects on the smooth muscle of both the gallbladder and the sphincter of Oddi, which are more sensitive to CCK compared to visceral smooth muscle in other regions of the G.I. tract. However, CCK also appears to mediate the gallbladder and sphincter responses to a meal through a neural mechanism (similar to that described for CCK-mediated pancreatic secretion in Figure 5-7) wherein CCK interacts with vagal afferent fibers in the duodenum to elicit a vagovagal reflex that culminates in the release of acetylcholine from vagal efferents that impinge on gallbladder smooth muscle to cause contraction. Similarly, the vagovagal reflex results in the release of the inhibitory neurotransmitters nitric oxide and VIP from efferent fibers on sphincter of Oddi smooth muscle to produce relaxation. The neurocrine role of CCK in mediating gallbladder contraction and sphincter of Oddi relaxation following a meal is believed to be more important than the endocrine actions of the peptide.

Other Hormones

Other hormones can also influence the rate of bile formation. For example, VIP is known to stimulate cholangiocytes to liberate a HCO_3^--rich secretion via the same cAMP-dependent mechanism that is activated by secretin. Somatostatin and gastrin, on the other hand, counter the effects of secretin by decreasing cAMP levels in cholangiocytes. Some steroid hormones, particularly estrogens and 19-alkylated androgens, inhibit bile formation by suppressing both bile acid-dependent and independent secretions. A reduction of bile formation is a well-known side effect of the therapeutic use of estrogen. The significant depression of bile acid secretion during pregnancy may be explained by high estrogen levels.

Nervous Influences

There are both vagal (parasympathetic) and sympathetic fibers supplying the biliary system. Vagal fibers travel in the hepatic branch of the anterior (right) vagal trunk and terminate on bile ducts, as well as the myenteric plexus located in the submucosal and subserosal layers of the gall-bladder. Cholinergic innervation of hepatic parenchymal cells is sparse. Sympathetic innervation of the biliary system is derived from the celiac plexus. Both efferent and afferent fibers are present. Sympathetic neurons supply the smooth muscle and myenteric plexus of the gallbladder and the bile ducts. Adrenergic innervation of hepatocytes is sparse; however, the intrahepatic blood vessels (portal venules, hepatic arterioles, and sinusoids) are richly innervated.

Parasympathetics

Although there is no evidence that hepatic secretion of bile salts is under neural control, parasympathetic nerve activity does appear to exert a significant influence on gallbladder emptying and, to a lesser extent, regulate bile formation. Cholinergic fibers from the vagi appear to mediate the gallbladder contraction and sphincter of Oddi relaxation that occur during the cephalic and gastric phases of digestion. Acetylcholine administration and electrical stimulation of the parasympathetic nerves to the biliary tract produce both gallbladder contraction and relaxation of the sphincter of Oddi. Furthermore, gallbladder emptying is impaired in patients who have undergone a vagotomy. Acetylcholine administration and vagal stimulation also both produce an increase in bile flow, which can be blocked by atropine. Cholangiocytes, which express muscarinic receptors, produce a HCO_3^--rich biliary secretion when stimulated with acetylcholine. The increase in bile flow observed after a meal is attenuated after bilateral vagotomy, suggesting that enhanced parasympathetic nerve activity accounts for a portion of the secretory response to feeding.

Sympathetics

Intravenously administered catecholamines (e.g., norepinephrine) and stimulation of sympathetic nerves both cause a reduction in bile formation. However, it is unclear whether the effects of adrenergic stimulation on bile flow result from a direct influence on the biliary system, from indirect effects mediated by G.I. hormone release, or from alterations in blood flow. Sympathetic nerve stimulation causes relaxation of the gallbladder, yet the response of the sphincter of Oddi is variable. A physiologic role for sympathetic nerve activity in the regulation of biliary function appears unlikely.

Table 6-2 summarizes major aspects of control of the biliary system during the various phases of digestion.

TABLE 6-2: Major Factors Affecting Gallbladder Emptying and Bile Synthesis and Secretion in Response to a Meal			
PHASE OF DIGESTION	STIMULUS	MEDIATING FACTOR	RESPONSE
Cephalic	Taste and smell of food; food in mouth and pharynx	Impulses in branches of vagus nerve (acetylcholine)	Increased rate of gallbladder emptying
Gastric	Gastric distension Nutrients	Impulses in branches of vagus nerve (acetylcholine)	Increased rate of gallbladder emptying; increased bile formation by cholangiocytes
Intestinal	Fat digestion products in duodenum	Cholecystokinin (direct & vago-vagal responses)	Increased rate of gallbladder emptying
	Acid in duodenum	Secretin	Increased bile formation by cholangiocytes
	Absorption of bile acids in the distal part of ileum	High concentration of bile acids in portal blood	Stimulation of bile acid-dependent secretion; inhibition of bile acid synthesis
Interdigestive period	Low rate of release of bile into duodenum	Low concentration of bile acids in portal blood	Stimulation of bile acid synthesis; inhibition of bile acid-dependent secretion

THE HEPATIC CIRCULATION

Under normal resting conditions, total hepatic blood flow accounts for approximately 25% of cardiac output in humans, that is, 1250 to 1500 ml/min. Hepatic blood flow increases following the ingestion of a meal, largely owing to an increased flow in the portal vein. The enterohepatic circulation of bile salts may contribute to the postprandial blood flow response, inasmuch as an increased bile salt concentration in the portal vein causes hepatic arterial blood flow to increase.

The liver is probably the most important blood reservoir in man, with blood accounting for 25% to 30% of total liver volume. Hepatic blood volume can expand to as much as 60% of total liver volume with cardiac failure. The capacitance function of the liver plays an important role during hemorrhage. Hemorrhage can result in the expulsion of up to 50% of the entire blood content of the liver, which is equivalent to an infusion of 6% to 7% of total blood volume. Sympathetic nerve activation represents the most powerful regulator of hepatic blood volume. Fifty percent to 60% of this liver blood volume can be expelled by sympathetic nerve action within 90 seconds without impairment of liver function.

LIVER AND BILIARY PAIN

Pain arises from the liver only as a result of stretching the liver capsule and is experienced in a diffuse area over the right hypochondrium. Distention of either the common bile duct or the gallbladder results in pain experienced in the midepigastrium or the right subcostal region. Occasionally, pain from the biliary tract is experienced in the area below the scapula.

PATHOPHYSIOLOGY AND CLINICAL CORRELATIONS

Gallstones

Gallstones, which afflict about 20 million Americans (with a million new cases per year), represent a major medical and surgical problem. Gallstones result from the precipitation of insoluble material in bile. Cholesterol accounts for nearly 90% of gallstone material and bile pigment (calcium bilirubinate) for the remainder. Stones with varying mixtures of these two compounds are common. Despite the fact that it contains considerable amounts of cholesterol, which is essentially insoluble in water, normal bile is quite translucent. Cholesterol is held in solution in mixed phospholipid/bile salt micelles. The micelles allow cholesterol to remain "solubilized" during its passage through the biliary system. Stone-forming (lithogenic) bile exists when the concentration of bile acids and/or phospholipids is too low to solubilize cholesterol in micelles, and the excess cholesterol appears as vesicles in the aqueous phase. These vesicles, which are prone to aggregation, create a nidus from which cholesterol crystals nucleate and eventually lead to the formation of gallstones. Either excessive cholesterol secretion or reduced secretion of either bile salts or phospholipids can initiate stone formation. Cholesterol stones exhibit a yellow-green color.

Pigment stones result from deconjugation of water-soluble bilirubin diglucuronide by β-glucuronidase; insoluble bilirubin precipitates as its calcium salt. Stones comprised of pure calcium bilirubinate are black, whereas stones are composed of calcium salts of unconjugated bilirubin are brown. The β-glucuronidase present in bile is normally inhibited by glucaric acid, also a bile constituent, and the precipitation of calcium bilirubinate may result from an imbalance between the enzyme and its inhibitor. This balance can be upset by additional β-glucuronidase derived from bacteria in an infected biliary tract. A large load of bilirubin, resulting from hemolysis and excretion into bile, predisposes the gallballer to pigment stone formation.

Women have much higher predisposition to the development of gallstones in addition to individuals who are obese or have diabetes mellitus, hyperlipidemia, Crohn's disease, pregnancy, or an ethnic association with Pima Indians. Drugs, such as estrogens, oral contraceptive agents, and hyperlipidemic agents such as clofibrate increase cholesterol secretion and predispose to stone formation. In patients with a resected terminal ileum or with small bowel involvement in Crohn's disease, the pool of bile salts falls due to failure of reabsorption. The lack of bile salts results in the development of cholesterol gallstones.

Although most individuals with gallstones do not exhibit associated symptoms, others present with right upper quadrant abdominal pain, fever, nausea, and vomiting. Diagnostic strategies suggesting gallbladder origin include a Murphy's sign (stop in breathing upon palpation of the right upper quadrant). Ultrasonography is a diagnostic test that uses acoustical waves to identify gallbladder wall changes suggestive of acute cholecystitis. Surgical treatment for gallstones (excision of the gallbladder and removal of stones from the biliary tree) is reserved for patients that are symptomatic and have evidence of obstruction of the cystic duct by a gallstone, with resulting inflammation of the gallbladder wall (*cholecystitis*). This surgical procedure now involves laparoscopic or robotic removal of the gallbladder. The risks associated with gallbladder removal (cholecystectomy) are small because the surgical procedures are safe and simple. Cholecystectomy does not alter life expectancy or appreciably affects quality of life. However, in the absence of a gallbladder and the inability to expel concentrated bile in response to a meal, patients are less able to tolerate large fatty meals.

Conditions Associated with Hepatocellular Injury

The metabolically diverse functions of the liver render it vulnerable to the toxic effects of drugs, infectious agents, industrial chemicals and natural products. Viral infections, such as viral *hepatitis A, B, or C*, may cause hepatocellular injury and elicit the release of liver enzymes, such as alanine aminotransferase or aspartate aminotransferase, into blood. The presence of these enzymes in blood is indicative of an active or ongoing insult to hepatocytes. Liver injury can also result from other viruses (herpes simplex virus or cytomegalovirus) and drugs. The injuries to the liver can be acute, subacute, or chronic in nature. The synthetic functions of the liver may be affected, with the injury manifested as a reduction in albumin concentration (a measure of protein synthesis) and/or disorders in coagulation (represented by increases in the prothrombin time). With acute hepatitis A viral infection or the acute phases of infection with hepatitis B or C, supportive care is the primary treatment modality. In the case of chronic viral hepatitis B or C, recent advancements in drug therapies have resulted in control (hepatitis B) or eradication (hepatitis C) of these viral infections before progression to *cirrhosis* (fibrosis and inflammatory injury to the liver). With the recent development of directly acting antiviral agents, patients with chronic hepatitis C and cirrhosis are very likely to experience eradication (greater than 90%) of the 6 most common genotypes of hepatitis C virus.

Acetaminophen (called paracetamol in the United Kingdom), the most widely utilized analgesic and antipyretic drug in the United States, accounts for more than 50% of overdose-related acute liver failure and approximately 20% of the liver transplant cases. The hepatotoxic effects of acetaminophen are not due to the drug itself but to a reactive metabolite (*N*-acetyl-*p*-benzoquinone imine, NAPQI) produced by cytochrome P-450 enzymes in the liver. Like many other hepatotoxins, acetaminophen causes zone-specific damage that is explained by the high expression and preferential induction cytochrome P450 isoforms in hepatocytes of the pericentral (perivenous) region of the liver acinus. The depletion of reduced glutathione (GSH) by NAPQI is a critical step in acetaminophen-induced hepatotoxicity. Normally, GSH detoxifies NAPQI by forming a stable protein-adduct of the reactive metabolite. However, with an overdose of acetaminophen, NAPQI concentration exceeds the detoxification capacity of GSH. *N*-acetylcysteine (NAC), a GSH precursor, is currently the only approved antidote for acetaminophen toxicity. NAC treatment restores GSH levels in hepatocytes, detoxifies NAPQI and helps to promote recovery of the liver. For patients with acetaminophen-induced fulminant hepatic failure, defined as hepatic encephalopathy (confusion) in the setting of jaundice (elevated bilirubin) without underlying liver disease, a transplant may be required due to the failing state of the liver.

The liver is also subject to an autoimmune process called autoimmune hepatitis. This condition was originally known as lupoid hepatitis due to its serological resemblance to systemic lupus erythematosus. It is characterized by a female preponderance (approximately 3 times as many women to men) beginning in the 5th and 6th decades of life. The classic presentation of autoimmune hepatitis includes the development of jaundice (elevated bilirubin) and/or elevated levels of alkaline phosphatase and transaminases. Although elevated bilirubin often reflects hemolysis, with this disorder, it likely results from cholestatic injury to the liver. Patients can present with the development of serologic markers with antinuclear antibodies, anti-smooth muscle antibodies, or anti-liver-kidney microsome-1 antibodies. Total immungobulins (IgG) is also commonly elevated. A liver biopsy may show a plasma cell infiltrate with an interface hepatitis (piecemeal necrosis). Because of the variability in presentation, a scoring system has been developed to standardize the diagnosis. Patients may develop cirrhosis in a natural history setting without treatment. Treatment is directed at immunosuppression in patients with significantly elevated aminotransferases and biopsy findings of interface hepatitis. Glucocorticoids are often chosen as initial therapy with concomitant or subsequent steroid-sparing agents such as azathioprine. With normalization of aminotransferases and a reduction in inflammation noted on a liver biopsy, approximately 80% of patients can be weaned from immunosuppression.

REFERENCES

Bernal W and Wendon J. Acute Liver Failure. *N Engl J Med* 2013;369:2525–2534.

Dawson PA. Bile formation and the enterohepatic circulation. In: Physiology of the Gastrointestinal Tract (Johnson LR, ed.), Chapt. 53, pp 1461–1484, Elsevier Inc., 2012.

Feranchak AP. Bile secretion and cholestasis. In: Yamada's Textbook of Gastroenterology, 6th edition, Chapt. 26, pp. 474–496, (Podolsky DK, Camilleri M, Fitz JG et al., eds.), 2016.

Greenberger N, Paumgartner G. Diseases of the Gallbladder and Bile Ducts. Harrison's Principles of Internal Medicine, 18th Ed. E. Last Accessed August 25, 2014.

Kanel GC. Liver: anatomy, microscopic structure, and cell types. In: Yamada's Textbook of Gastroenterology, 6th edition, Chapt. 10, pp. 145–160 (Podolsky DK, Camilleri M, Fitz JG et al., eds.), 2016.

Manns MP, Cjaza, AJ, Gorham JD, Krawitt EL, Giorgina M, Vergani D, Vierlin JM. Diagnosis and Management of Autoimmune Hepatitis. *Hepatology.* 2010; 51:1–31.

Wang DQH, Neuschwander-Tetri BA, Portincasa P. *The Biliary System (2nd edition).* In: Colloquium Series on Integrated Systems Physiology: From Molecule to Function to Disease (Granger DN, Granger JP, eds.), Morgan and Claypool Publishers, 2017. https://doi.org/10.4199/C00147ED2V01Y201611ISP071

CHAPTER 7

The Small Intestine

Introduction

Anatomy

Serosa

Muscularis Externa

Submucosa

Mucosa

Muscularis Mucosae

Lamina Propria

Epithelium

Mucosal Growth and Adaptation

Water and Electrolyte Absorption

Water Absorption

Electrolyte Absorption

Sodium

Potassium

Chloride

Bicarbonate

Calcium

Magnesium

Iron

Secretion of Water and Electrolytes

Carbohydrate Digestion and Absorption

Dietary Carbohydrates

Digestion

Transport of Monosaccharides

INTRODUCTION

The small intestine is a tubular structure extending from the stomach to the colon. Its functions include mixing and propulsion of lumen contents, digestion and absorption of nutrients, regulatory peptides/amines (e.g., hormones), and participation in the immune response. The first three of these functions play important roles in the overall process of assimilation of ingested food. The chyme entering the small intestine is thoroughly mixed with pancreatic and biliary secretions by contractions of the muscularis externa. The coordinated activity of the smooth muscle coat assures that chyme is gradually moved analward. Specialized endocrine cells in the mucosa secrete various peptides and hormones that regulate the amount and composition of digestive secretions delivered to the small intestine. The absorption of electrolytes, nutrients, and water are facilitated by the large surface area of the mucosa. Every 24 hours approximately

8 liters of water, partially hydrolyzed nutrients, and various secretions enter the small intestine, where most of it is absorbed. The efficient assimilation of nutrients that occurs in the small intestine allows the human body to meet the daily caloric requirement of ~30 kcal/kg body weight in sedentary individuals.

ANATOMY

The small intestine is about 6 to 7 meters in length when relaxed (post mortem) and can be as short as 3 meters if tension is present (in vivo). Its diameter is about 4 centimeters but varies along the length of the gut, being larger in caliber in the proximal segments. The small bowel is divided into three parts: the duodenum, jejunum, and the ileum. The duodenum, which is mostly retroperitoneal, extends from the gastric pylorus to the ligament of Treitz and represents the first 20 to 30 cm of the small bowel. The jejunum originates at the ligament of Treitz, where the small intestine reenters the peritoneal cavity, and extends for approximately 2.5 meters. The ileum accounts for the remainder of the intraperitoneal small bowel and is roughly 3.5 meters in length. The luminal diameter of the small bowel is highest in the duodenum and lowest in the ileum. The differences between the segments of small intestine are not as obvious as are the similarities. At every level of the gut, the wall can be clearly partitioned into four basic layers: serosa, muscularis externa, submucosa, and mucosa (Figure 7-1).

Serosa

The serosa is the outermost layer of the intraperitoneal small intestine (jejunum and ileum). This thin lining is an extension of the visceral peritoneum and consists of simple squamous epithelial cells (mesothelium) that rest on connective tissue. While the jejunum and ileum are completely covered by serosa, the duodenum is invested with serosa only on its anterior surface; the posterior surface is covered by loose connective tissue.

Muscularis Externa

The muscularis externa is composed of two well-defined layers of smooth muscle: a thinner outer layer that surrounds a thicker inner layer. Although all of the smooth muscle cells are spindle-shaped (200 μm long and 6 μm wide), their orientation differs in the two layers of smooth muscle. In the outer layer, the long axes of the cells are oriented in the longitudinal

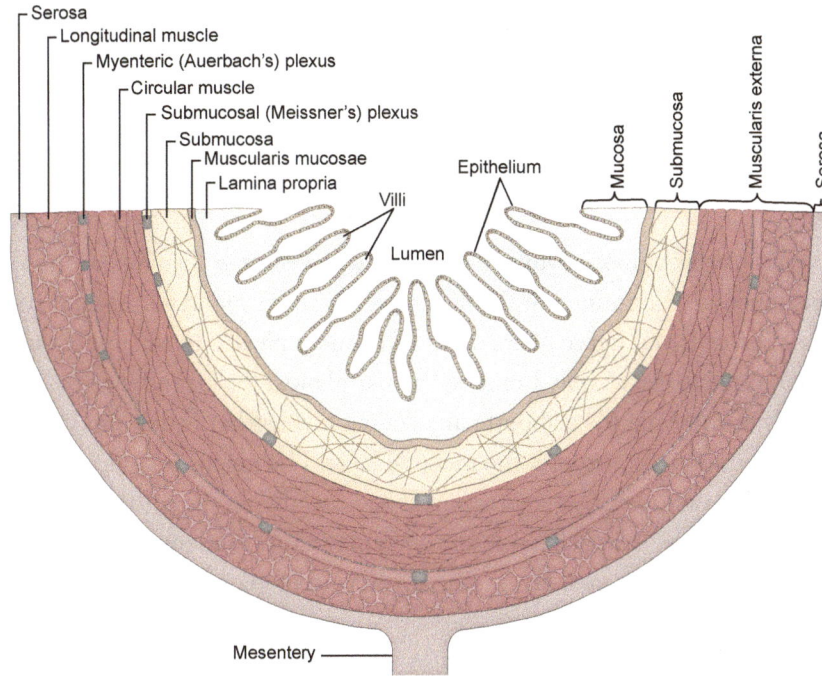

FIGURE 7-1: Cross-section of the small intestine, illustrating the different layers of the bowel wall and key structures localized in these layers that contribute to the motor, digestive, and transport functions of this tissue.

direction, while in the inner layer the long axes of the cells are oriented in a circular direction. Within each layer, the smooth muscle cells are densely packed (180,000 cells per mm³). The cells are separated mainly by small spaces (10 nm) that are occupied by collagen fibers and by a few larger spaces containing nerves, capillaries, interstitial cells, and connective tissue. In many cases, especially in the circular muscle layer, the plasma membranes of adjacent smooth muscle cells appear fused at various points, forming nexi. These nexi, occupying approximately 5% of the cell surface, allow for electrical coupling among the cells and muscle bundles. The presence of these nexi is particularly important since bundles of smooth muscle cells (containing 200 to 400 cells), not individual cells, represent the smallest electrically excitable unit of visceral smooth muscle.

The longitudinal and circular muscle layers are separated by a connective tissue septum that varies greatly in thickness and contains the myenteric (Auerbach's) plexus and parasympathetic ganglion cells. The contractile and electrical activities of these two layers are coupled by means of connective tissue bridges and the myenteric plexus, respectively. Another plexus of nerve fibers (Meissner's plexus) is located in the submucosa adjacent to the circular smooth muscle layer. The Meissner's plexus may also play a role in coordinating the activity of the longitudinal and circular muscle layers.

Submucosa

The submucosa is a layer of loose connective tissue lying between the circular muscle layer and the muscularis mucosae. Contained in this layer are the larger blood vessels and lymphatics, as well as the ganglion cells and nerve fibers of the Meissner's plexus. In the upper duodenum, the submucosa also contains elaborately branched acinar glands, the Brunner's glands. The serous and mucous cells of these glands elaborate an alkaline mucus that is delivered to the duodenal lumen via short ducts passing through the mucosal layer. The mucus provides a protective coating while the alkaline nature of this secretion aids in neutralizing the acidic chyme entering the upper duodenum from the stomach. In lower portions of the duodenum and the remainder of the small intestine, the submucosal layer is largely devoid of glands.

Mucosa

The mucosal layer, comprising the luminal surface of the small intestine, is designed to provide a large surface area for contact with lumen contents. Three anatomic modifications amplify the mucosal surface above that predicted for a simple cylinder (Figure 7-2). Spiral or circular concentric folds called plicae circulares (or circular folds of Kerckring) contribute to this amplification. They are absent from the duodenal bulb (uppermost region of duodenum), more prominent in the duodenum and jejunum, and gradually become less conspicuous in the terminal ileum. The mucosal surface area is further extended by finger-like projections called villi and depressions called crypts. There are 3 to 20 crypts surrounding each villus. The crypt to villus ratio varies from 14:1 in the duodenum to 6:1 in the ileum. The villi are 0.5 to 0.8 mm in height in the duodenum and jejunum but rarely exceed 0.5 mm in the ileum. Crypt depth usually ranges between 0.3 and 0.6 mm. Finally, each villus and crypt is lined by columnar epithelial cells, which are covered with numerous closely packed microvilli. The microvilli on the cells at the tip of the villus (1.0 to 1.5 μm) are longer than those on the cells lining the crypts (0.5 μm). The combination of the plicae circulares, the villi, and the microvilli increases the mucosal surface area by about 600-fold (Fig. 7-2), creating a total surface area of 200 square meters. However, the effective area available for absorption is less than this value owing to the closeness of adjacent villi and the inaccessibility of the lower two thirds of the villi and crypts to luminal contents.

Structure	Relative surface increase (x cylinder)	Surface area (m²)

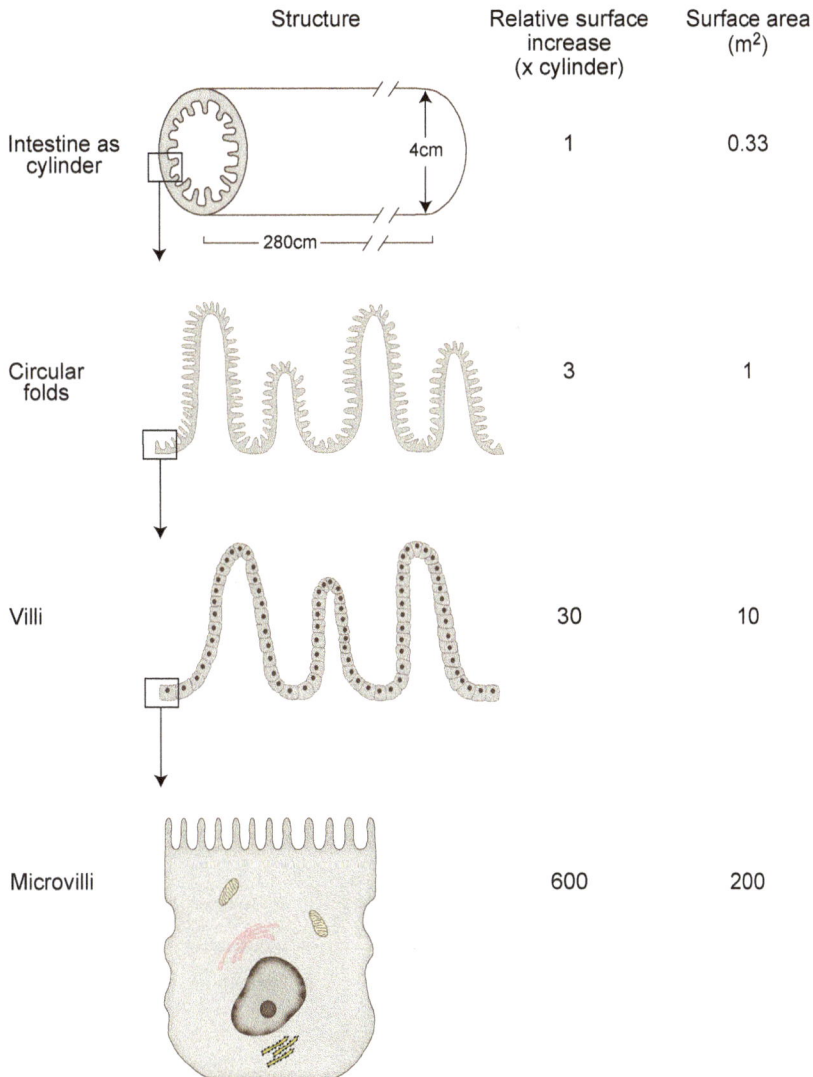

Intestine as cylinder — 1 — 0.33

Circular folds — 3 — 1

Villi — 30 — 10

Microvilli — 600 — 200

FIGURE 7-2: Anatomic features of the small intestine that amplify the mucosal surface above that predicted for a simple cylinder. Collectively, the circular folds (plicae circularis), villi, and microvilli increase the mucosal surface area by about 600-fold, creating a total surface area of ~200 square meters.

Muscularis Mucosae

The mucosa of the small intestine can be subdivided into three distinct layers: the muscularis mucosae, the lamina propria, and the epithelium (Figure 7-3). The base of the mucosa is lined by a thin continuous sheet of smooth muscle cells about 3 to 10 cells thick, separating the mucosa from the submucosa. This sheet of muscle is called the muscularis mucosae. Its physiologic functions remain uncertain; however, it may control the movement of villi.

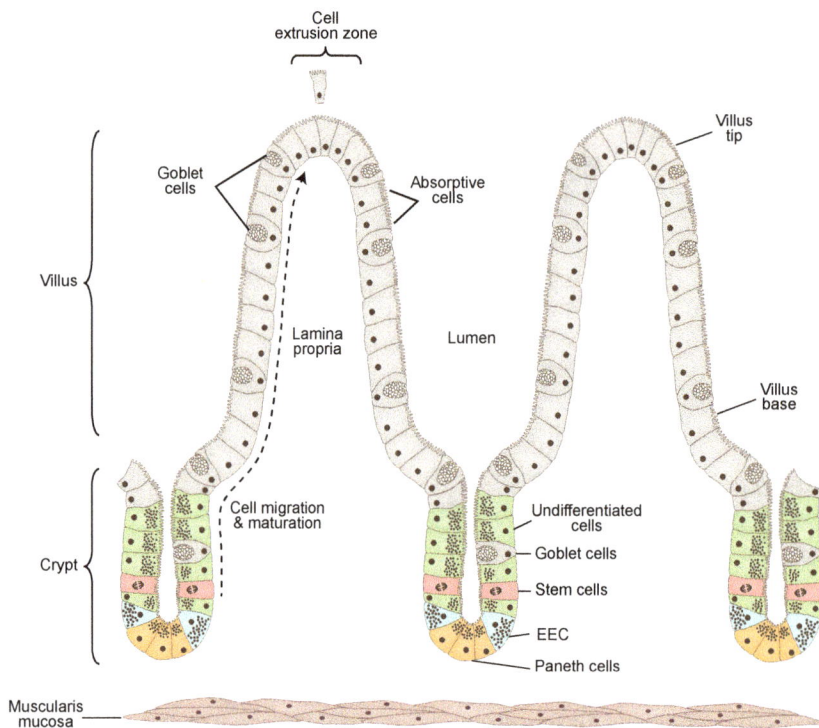

FIGURE 7-3: Histologic organization of the mucosa of the small intestine, illustrating the types of epithelial cells that line both the villi and crypts. The absorptive cells of the villus epithelium are engaged in digestion and transport, while the goblet cells secrete mucus as a protective coating of the epithelium. An important feature of the crypt epithelium is the presence of stem cells that undergo mitosis and migrate up to the villus tip, differentiating into mature epithelial cells along the way. At the villus tip epithelial cells are extruded at a rate matching the rate of migration along the villus.

Lamina Propria

The lamina propria consists primarily of connective tissue. It supports the epithelial lining, forming the core of each villus and surrounding the crypts. In addition to providing structural support, the lamina propria also contains blood vessels that nourish the epithelium and serve as conduits for transport of water-soluble substances absorbed by the epithelium. Within the core of the villus there is a large centrally located lymphatic vessel (the lacteal) that serves as a channel for the transport of chylomicrons (lipids) and absorbed water to the systemic circulation. Some small unmyelinated nerve fibers, fibroblasts, and smooth muscle cells are also present in the lamina propria. The smooth muscle cells are oriented along the longitudinal axis of the villus and likely play a role in villus contractions.

Immune cells that reside within the lamina propria provide an important line of defense against luminal contents. Lymphocytes, macrophages, granulocytes, mast cells, and plasma cells populate the lamina propria. Distinct lymphoid structures that detect and respond to antigens and microbes are also found in the lamina propria. Small lymphoid nodules are present in the upper small intestine, while large organized aggregates of lymphoid tissue (Peyer's patches) in the ileum extend as deeply as the submucosa. Peyer's patches contain both T-lymphocytes, as well as B cell precursors, that are destined to populate the lamina propria of the intestine. The lymphoid cells, nodules, and Peyer's patches of the lamina propria, along with the intraepithelial lymphocytes, constitute the so-called "gut-associated lymphoid tissue" or GALT, a major subdivision of the immune system. Peyer's patches are considered to be the initiating sites of mucosal immunity, since they contain a precursor population of B-lymphocytes that migrate to and populate the lamina propria. These cells terminally differentiate to become plasma cells in the lamina propria, where they synthesize and secrete a unique form of antibody, IgA, which is abundant in mucosal secretions. Secretory IgA serves as the first line of defense in protecting the intestinal epithelium from enteric toxins and pathogenic microorganisms.

Epithelium

A continuous sheet of epithelial cells (one cell thick) covers the surfaces of the villi and lines the crypts. The cell types that have been identified in the epithelial lining of the small intestine are listed in Table 7-1. Some of the cell types (e.g., goblet cells) are common to both villus and crypt epithelium, while others are exclusively found either in the villus region (e.g., enterocytes) or crypt region (e.g., stem cells).

The most prevalent cell type on the villus surface is the simple columnar absorptive cell, or enterocyte (Figure 7-4). The most striking feature of the enterocyte is the presence of 3000 to 7000 microvilli on its apical surface. The plasma membrane covering the microvilli is coated with a filamentous glycoprotein-containing structure, the glycocalyx. The microvillus and its

TABLE 7-1: Epithelial Cells Lining the Villi and Crypts of the Small Intestine.		
CELL TYPE	LOCATION	FUNCTION
Enterocytes	Villus	digestion & absorption
Goblet cells	Villus & crypt	mucus secretion
Endocrine cells	Villus & crypt	hormone/peptide secretion
Microfold (M) cells	Villus (overlying Peyers patches)	uptake of luminal antigens
Intraepithelial Lymphocytes	Villus	preserve mucosal barrier prevent pathogen entry/spreading
Paneth cells	Crypt	host defense maintain stem cells
Undifferentiated (stem) cells	Crypt	cell division & differentiation

ABSORPTIVE CELL — Glycocalyx, Microvilli, Basolateral Membrane

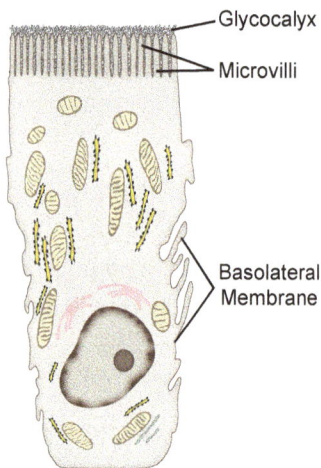

GOBLET CELL — Microvilli, Mucous Granules

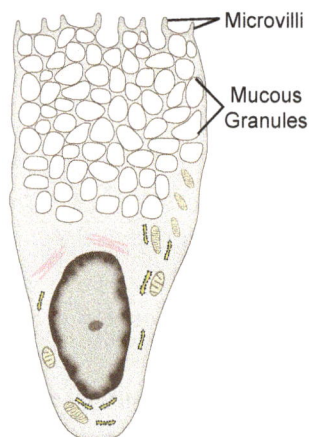

ENTEROENDOCRINE CELL — Microvilli, Secretory Granules

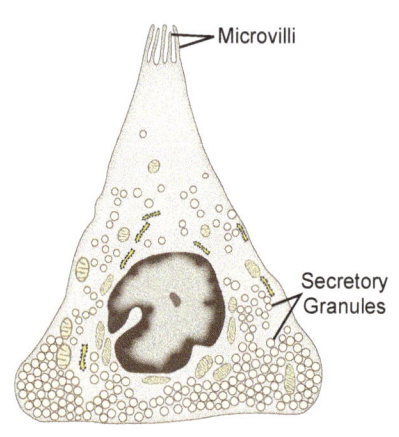

FIGURE 7-4: Ultrastructural features of enterocytes, goblet cells, and endocrine cells of the small intestine.

glycocalyx (the brush border) are viewed as the digestive-absorptive unit of the enterocyte. The plasma membrane of microvilli has an unusually large protein-to-lipid ratio, owing to the presence of specialized proteins possessing enzymatic, receptor, and transport properties. A number of enzymes involved in the hydrolysis of peptides (peptidases) and carbohydrates (disaccharidases) to transportable moieties (e.g., amino acids and glucose) are found in the microvillus membrane. In addition, proteins that are responsible for the co-transport of Na^+ and various organic solutes are localized in the microvillus membranes. Finally, receptor proteins specific for certain substances have been localized to the microvilli of enterocytes in different regions of the small intestine. For example, the receptors for the intrinsic factor-vitamin B_{12} complex are present in the microvilli of ileal, but not jejunal, enterocytes, providing an explanation for why vitamin B_{12} absorption occurs exclusively in the ileum. The specific functions of the glycocalyx remain poorly understood; it may provide a binding surface for adsorption of pancreatic enzymes and offer support for large hydrophilic portions of intrinsic enzymes of the microvilli.

The basolateral membrane of the enterocyte is strikingly different from the apical "brush border" membrane. Its protein-to-lipid ratio is lower, and it is thinner and more permeable than the brush border membrane. Many nutrients (e.g., monosaccharides, amino acids) that are transported into the enterocyte via the apical membrane leave the cell through specific transporters that reside in the basolateral membrane. The enzyme, Na^+ K^+-ATPase, is also present at the basolateral membrane but not at the brush border. The fact that the Na^+ K^+-ATPase is limited to the basolateral membrane underscores the importance of this membrane in electrolyte transport.

In addition to the enterocytes, the mucosal epithelium contains secretory cells such as goblet cells and endocrine cells. These secretory cells are not limited to the villus (as are the enterocytes) but are also present in the crypts. The goblet cells secrete their product (mucus) into the lumen of the gut, whereas the secretions of the enteroendocrine cells (EEC) are directed into the lamina propria. These two secretory cells have different distribution frequencies along the gut. The goblet cells are more prevalent in the lower than in the upper small intestine, whereas the EEC are generally more numerous in the upper portion of the gut.

Goblet cells are mucus-secreting cells that have a characteristic "wine goblet" appearance (Fig. 7-4). The apical portion of the cell is distended by many mucin-containing granules, while the nucleus and remaining cytoplasm are displaced to the narrow basal region. The luminal brush border is less developed than those of enterocytes, the microvilli being shorter and more sparsely distributed. The mucin granules are discharged into the lumen in response to noxious luminal stimuli (i.e., irritation) to form a protective coat of mucus over the epithelium.

The apical portion of the intestinal EEC is narrow and contains tufts of microvilli, which protrude into the lumen. These tufts are chemoreceptors that are sensitive to specific (meal-related) luminal stimuli. Some EEC have a highly developed system of cytoplasmic filaments that serve to release endocrine/paracrine factors in response to mechanical or chemical stimuli. Individual EEC secrete either a peptide (e.g., cholecystokinin), or an amine (e.g., serotonin), or both (e.g., serotonin and substance P). Some EEC exhibit true endocrine function, releasing their hormones to act at distant sites. Examples of these cells are the EEC that store and secrete cholecystokinin, secretin, and GIP. Other EEC appear to have a paracrine function, secreting substances (somatostatin and serotonin) that modulate local events. At least 10 different EEC types producing over 15 gut hormones/regulatory peptides have been identified in the small intestine (see Table 1-1 in Chapter 1).

The mucosal epithelium contains two types of specialized cells that are part of the GALT system, the microfold (M) cells and the intraepithelial lymphocytes (IEL). The M cells are interspersed among the enterocytes in the mucosal epithelium overlying the Peyer's patches. They are shaped like the letter **M,** with a thin apical portion bridging adjacent enterocytes and cytoplasmic extensions along the lateral margins of the enterocytes. The M cells usually envelop several lymphocytes and macrophages. Their apical portions are devoid of microvilli, covered instead by a series of ridges one third to one fifth the height of the microvilli of enterocytes. The numerous pinocytotic vesicles located in the apical region of M cells suggest that these cells function to transport antigenic substances from the lumen to the intestinal lymphoid system. The IEL are more ubiquitously distributed, accounting for about one sixth of the villus epithelium. They are larger than the simple lymphocytes of the lamina propria and normally reside in the paracellular space between adjacent epithelial cells. They are activated by specific luminal antigens and function to preserve the integrity of the mucosal barrier, and to prevent pathogen entry and spreading.

The diverse group of cells that line the intestinal villi and crypts are derived from a population of pluripotent stem cells that reside near the base of the crypt. These undifferentiated cells are the most abundant cell type in the crypts and are extremely proliferative. Because of their high mitotic rate, the stem cells are extremely sensitive to radiation, ischemia, and antimitotic agents (methotrexate and colchicine). As they divide, the cells migrate (at a rate of ~10 μm/hour) in a conveyor belt fashion upwards from the crypts onto the villi, exit the cell cycle, and differentiate into the different mature specialized cells that populate the villi. This process enables the proliferative cells in the crypts to supply about 300 cells to each villus per day. The best recognized aspect of this maturation process is the acquisition of the full complement of enzymes characteristic of the mature enterocyte, a gradual process that occurs as these cells migrate to the villus tip.

The absorptive, goblet, and enteroendocrine cells continue to migrate up the villus until they arrive at the tip, where they undergo apoptosis and exfoliate into the gut lumen by a process known as *anoikis*. Normally, the villus cell population is maintained constant owing to a balance between cell exfoliation at the villus tips and the proliferation and maturation of the cells in the crypts. The process of maturation, migration, and extrusion requires approximately 3 to 6 days. The cell turnover time is greater in the proximal than in the distal bowel, presumably owing to the greater height of villi in this region. The Paneth cells never leave the crypts; they eventually degenerate and are phagocytosed by cells lining the crypts. The turnover time for Paneth cells is several weeks.

In summary, the epithelium of the small intestine is a dynamic sheet of specialized cells that govern the movement of materials between the lumen and the lamina propria. The most important function of the villus epithelium is absorption of ingested nutrients and water, whereas the crypt epithelium serves as a source for the mature specialized cells of the villi and is an important site for fluid and electrolyte secretion in the small intestine.

Mucosal Growth and Adaptation

The high cell turnover rate of the small intestinal mucosa is closely regulated to maintain a balance between cell loss and cell production. Growth of the mucosa occurs when the turnover rate decreases, that is, increased cell proliferation or decreased cell loss or both. Atrophy occurs when the turnover rate increases, that is, decreased cell proliferation, increased cell loss or both. Mucosal growth is affected by several factors, all of which are linked to the presence of food in the gastrointestinal tract. It is well documented that food deprivation leads to a reduction in villus height and crypt depth. The starvation-induced atrophy can be reversed by oral, but not parenteral, feeding. The physiologic stress of lactation is associated with hypertrophy and hyperplasia of the intestinal mucosa. An increased food intake in response to the nutritional demands of lactation is believed to be the stimulus for mucosal growth. Intestinal resection induces an adaptive hyperplasia of the remaining intestinal mucosa. For example, resection of the jejunum leads to changes in the ileum, which now takes on the structural and functional characteristics of the jejunum. The postsurgical increase in mucosal mass occurs only if oral feeding, rather than parenteral feeding, is instituted. The mechanisms by which food intake induces adaptive growth in response to physiologic and surgical stresses are shown in Figure 7-5. The presence of food in the intestinal lumen can directly influence mucosal growth at the site of absorption by providing nutrients for the mucosal cells, that is, the enterocytes can derive nutritional support from absorbed material before the nutrients enter the circulation. Ingested amino acids, disaccharides and polyamines are examples of luminal factors that directly stimulate mucosal growth. Polyamines supplied either via the lumen or synthesized by epithelial cells are required for normal intestinal growth. Luminal nutrients can also influence mucosal growth by eliciting the release of gastrointestinal peptides, which regulate mucosal growth either by a paracrine or endocrine action. Cholecystokinin, secretin, and gastrin-releasing peptide that

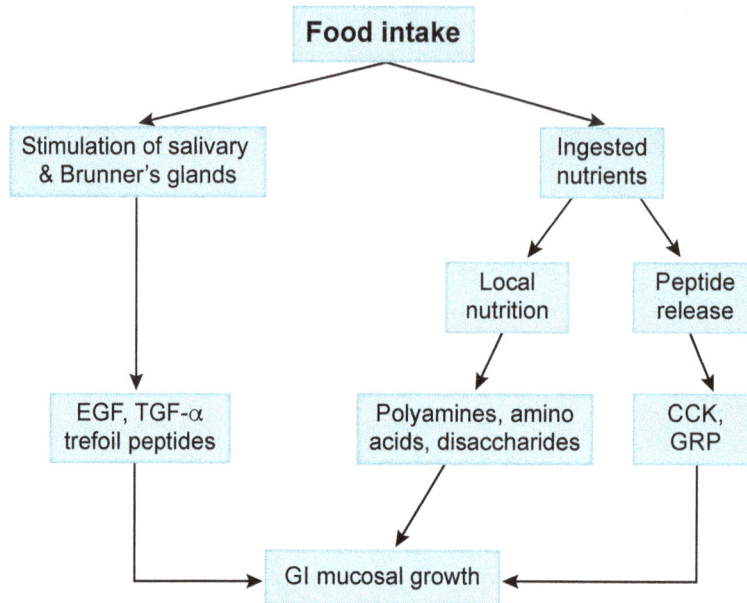

FIGURE 7-5: Mechanisms the link food intake to gastrointestinal mucosal growth. The presence of ingested nutrients in the bowel lumen both directly and indirectly (via GI hormones & peptides) stimulate growth, as well as trophic factors (EGF, trefoil peptides) released by salivary and Brunner's glands in response to ingestion of a meal.

have been implicated in the regulation of mucosal growth in the small intestine and in the functional development of the intestinal mucosa. A variety of peptide growth factors produced in the GI tract also demonstrate trophic actions in the small bowel. Notable examples in this regard are epidermal growth factor (EGF) and trefoil peptides, which are produced in the submaxillary glands and by Brunner's glands in the duodenum. EGF promotes the proliferation and differentiation of stem cells.

WATER AND ELECTROLYTE ABSORPTION

Water Absorption

The small intestine of normal healthy individuals absorbs large quantities of water each day. This water is derived from ingested food and drink (approximately 1.5 to 2.0 liters per day) and from the secretions of the salivary glands, stomach, pancreas, liver, and intestine (approximately 6.0 to 7.0 liters per day). A total daily water load of 8–9 liters is presented to the small intestine, where approximately 80% of it is absorbed. The maximal absorptive capacity of the small intestine is unknown but may be as high as 15–20 liters per day. In some animals (e.g., rat, cow), water absorption can proceed at such a rapid rate that intravascular hemolysis occurs.

Water transport in the small intestine is passive, with the rate of absorption varying with the location along the bowel (duodenum or ileum), the rate of active solute transport, and luminal osmolality. The duodenum and jejunum are the major sites of water absorption, which can be explained by the relatively large pore size (8 Å, radius) and consequent low resistance to water movement across their mucosae. Water movement across the ileal mucosa is more restricted (mucosal resistance is higher) due to the lower permeability of the tight junctions, which behave like 4 Å radius pores.

Water can easily move either into or out of the intestinal lumen, depending upon the total osmotic activity of intestinal contents. Thus, the direction and magnitude of water movement in the proximal bowel are largely influenced by the nature of the ingested meal. Ingestion of a *hypotonic* meal (approximately 200 mOsm) leads to net water flow from intestinal lumen to blood, with the blood-lumen osmotic pressure difference acting as the driving force (Figure 7-6). Water leaves the lumen more rapidly than electrolytes and absorbable nutrients; therefore, luminal osmolality rises until it equals plasma osmolality. Ingestion of a *hypertonic* meal (approximately 600 mOsm) leads to net water movement from blood to bowel lumen. The accumulation of water in the lumen (and the movement of ions and nutrients out of the lumen) leads to dilution of chyme and brings the luminal contents to isotonicity. As illustrated in Figure 7-6, the distal small bowel plays a more important role in the absorption of water following a hypertonic meal, while the proximal bowel is more important following hypotonic meals. The important physiologic aspect of water fluxes in the duodenum is that net movement of water is in the direction that maintains isotonicity of the intestinal contents with plasma. The transfer of water by osmosis to make chyme isosmotic with plasma requires only a few minutes (50% is absorbed in 3 minutes). Thereafter, the chyme remains isosmotic during its passage through the small intestine and, once isotonicity is achieved, water absorption is largely dependent on solute absorption.

The net flow of water across the intestinal mucosa is the result of two large unidirectional fluxes. One flux (absorptive) is directed from the intestinal lumen to the interstitial fluid and then to blood or lymph vessels. The second flux (secretory) occurs in the opposite direction, that is, from interstitial fluid to lumen. In the presence of isosmotic chyme, the absorptive flux is believed to occur at the villus tips, while the secretory flux is generated by crypt epithelium.

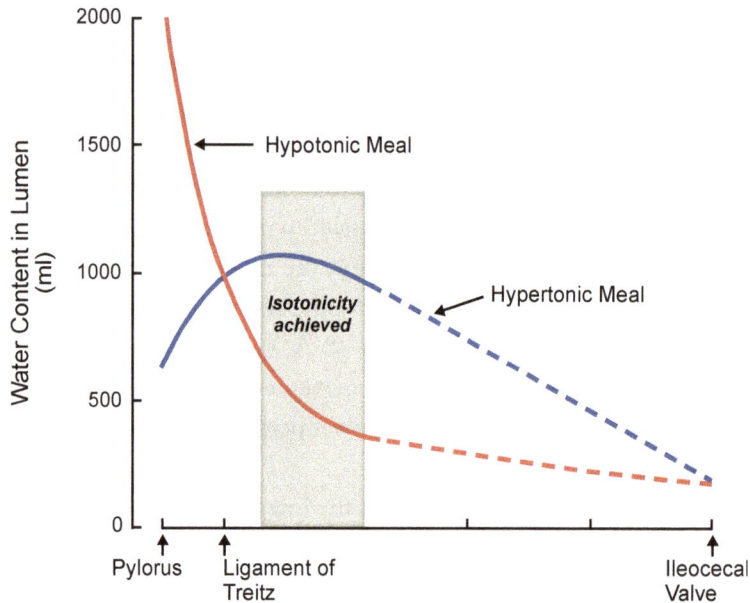

FIGURE 7-6: Water content in the intestinal lumen after ingestion of either a hypotonic (steak) or hypertonic (doughnut-milk) meal. In both instances, the meal achieves isotonicity in the proximal jejunum. Thereafter (broken lines), the movement of water across the mucosa is coupled to sodium transport. [Modified from Fordtran JS and Ingelfinger FJ. Intestinal absorption. In: HANDBOOK OF PHYSIOLOGY. Section 6: Alimentary Canal. Vol. III. Section Editor C. F. Code. Baltimore, Maryland: Williams & Wilkins Co. 1967. Pp. xii+1099–1570.

The difference between the blood-to-lumen (secretory) and lumen-to-blood (absorptive) fluxes represents the net water flux. The magnitude of the net water flux in the small bowel is one fifth to one tenth that of the two unidirectional fluxes. For example:

Net water flux (ml/min) = (Lumen-to-blood flux) – (Blood-to-lumen flux)
0.20 = 2.20 – 2.00

As a result of the large difference between net and unidirectional fluxes, a relatively small change in one of the unidirectional fluxes has a great impact on net water movement. Alterations of the unidirectional water fluxes account for the rapid attainment of osmotic equilibrium of chyme entering the duodenum. Hypotonic chyme enhances the lumen-to-blood flux of water, thereby inducing net water absorption. Conversely, hypertonic chyme can cause net water secretion by enhancing the blood-to-lumen flux. When luminal osmolality exceeds plasma osmolality by 100 to 130 mOsm, the two unidirectional fluxes are equal, and the net flux of water is zero. Further increments in lumen osmolality lead to net water secretion. An alternative stimulus for net water secretion is evidenced in disease states such as cholera, where a specific bacterial toxin elicits water secretion across crypt epithelium that is coupled to active chloride transport into the gut lumen.

While the rate of water absorption in the duodenum is highly dependent upon lumen osmolality, water movement across the mucosa of distal segments of the bowel is more closely coupled to solute transport because here the lumen is in osmotic equilibrium with plasma. The coupling of solute and water fluxes in the jejunum and ileum can be explained by the standing osmotic gradient theory (see Chapter 1, Figure 1-11). According to this model, an osmotic gradient is established between the lumen and lateral intercellular space by active transport of electrolytes into the latter compartment. The osmotic gradient draws water across the tight junction and through the cell into the intercellular space. Therefore, as ions are absorbed so is an isosmotic equivalent of water. Because the mucosal membrane of the ileum offers more resistance to water movement (due to the smaller pores), the composition of absorbed fluid is slightly hypertonic in this region of the small bowel.

Electrolyte Absorption

Sodium

The small intestine functions as a highly efficient sodium-conserving organ. The average adult intake of sodium is approximately 250 to 300 mEq per day. A similar amount of sodium enters the intestinal lumen from salivary, gastric, biliary, pancreatic, and intestinal secretions. Of the ~600 mEq of sodium entering the small intestine each day, approximately 75% is absorbed there. Over half of the sodium is absorbed in the jejunum and, of the remainder, half is absorbed in the ileum and the rest in the colon. While the rates of sodium absorption in the human jejunum and ileum are comparable, the predominant mechanism of transport is different.

There are two mechanisms that are largely responsible for sodium absorption in the small intestine: (1) nutrient-coupled sodium absorption, and (2) electroneutral sodium-hydrogen exchange (Figure 7-7). For both mechanisms, the sodium-potassium pumps located on the basolateral membrane create the steep chemical (140 mM to 15 mM) and electrical (−40 mv to +3 mv) gradients for sodium that drives its entry into the cell across the apical membrane. Nutrient-coupled sodium absorption is the primary mechanism for sodium absorption after a meal and it largely occurs in the duodenum and jejunum. Glucose and amino acids are the major nutrients that drive Na^+ entry into the cell via substrate-specific transport proteins

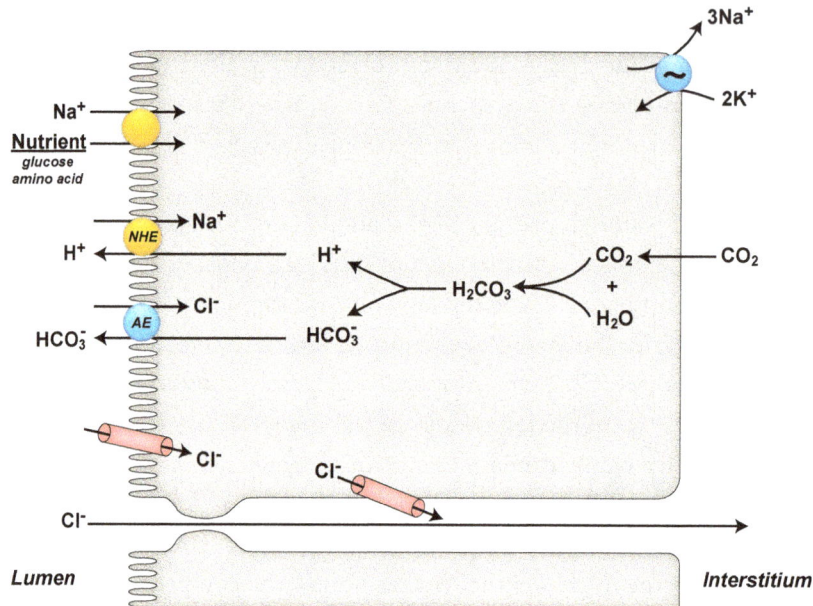

FIGURE 7-7: Major mechanisms of sodium and chloride absorption in the small intestine. Nutrient-coupled sodium transport and electroneutral sodium-hydrogen exchange are largely responsible for the sodium absorption. Cl^-–HCO_3^- exchange and passive movement via the paracellular pathway account for chloride absorption; a minor contribution to Cl^- absorption is via diffusion through channels.

located on the apical membrane. The "downhill" movement of sodium along its electrochemical gradient indirectly provides energy for the movement of the nutrient against its concentration gradient, that is, from gut lumen into the cell. While the absorbed Na^+ is extruded from the epithelial cell via the Na^+-K^+ pump on the basolateral membrane, the nutrient (glucose, amino acid) exits the cell by carrier-mediated facilitated diffusion.

Electroneutral Na^+-H^+ exchange is another important mechanism for sodium absorption in the small bowel. The Na^+-H^+ exchange protein (NHE) is located in the apical membrane throughout the small intestine. This protein couples the uptake of Na^+ into the enterocyte with the extrusion of H^+ into the lumen. NHE is stimulated by a reduction in intracellular pH or an increased luminal pH caused by HCO_3^- derived from duodenal, pancreatic and biliary secretions. The alkalinity of the luminal contents in the duodenum and jejunum is the principal stimulant for Na^+-H^+ exchange in this region of the small intestine. In the ileum, intracellular pH exerts a controlling influence on NHE and here, Na^+-H^+ exchange occurs in parallel with

Cl⁻-HCO₃⁻ exchange, which results in electroneutral NaCl absorption. The Na⁺-H⁺ and Cl⁻-HCO₃⁻ exchangers are coupled via the shared regulation by intracellular pH. Thus, activation of NHE promotes the extrusion of H⁺, which alkalinizes the cell and elicits the extrusion of HCO₃⁻ in exchange for Cl⁻. In much the same way that the steep electrochemical gradient for Na⁺ drives sodium into the cell and provides energy for the "uphill" movement of hydrogen ions into the lumen, the "downhill" transport of bicarbonate ions out of the cell energizes the "uphill" entry of chloride into the cell. The hydrogen and bicarbonate ions that enter the lumen rapidly form water and carbon dioxide. Therefore, the net outcome of activating these coupled transporters is maintenance of cell pH and electroneutral Na⁺ absorption.

Potassium

The average daily adult intake of potassium is 40 to 60 mEq. Most of the ingested potassium, as well as K⁺ entering the bowel lumen from salivary, gastric, pancreatic, and biliary secretions, is absorbed in the jejunum, with little absorption taking place in the ileum. Potassium absorption in the jejunum is largely passive, and likely involves both diffusion and convection. Diffusion occurs down a concentration gradient (lumen-to-plasma) of approximately 10 mEq per liter (14 mEq/liter in lumen and 4 mEq/liter in plasma), while potassium is also pulled along by water movement (convection). Both diffusion and convection of potassium across the jejunal mucosa occurs primarily through paracellular pathways.

Chloride

Intestinal absorption of chloride ions involves both active and passive processes and occurs through both cellular and paracellular pathways (Figure 7-7). In the jejunum, chloride absorption is primarily passive, with both electrical and concentration gradients acting as driving forces. This "voltage-dependent" chloride absorption results from the small (3 to 5 mv) positive electrical potential between the gut lumen and plasma created by active sodium (coupled and uncoupled) transport. Concentration-dependent chloride absorption occurs when the luminal chloride ion concentration exceeds 100 mEq per liter (plasma). The diffusional flow (down both concentration and potential difference gradients) of chloride ions occurs largely through the paracellular pathways (diffusion of Cl⁻ through apical and basolateral membrane channels also occurs but to a more limited extent) and serves to maintain electrical neutrality across the jejunal mucosa. Chloride absorption in the ileum also occurs via Cl⁻-HCO₃⁻ exchangers (described above) that exist and function either in the presence or absence of a parallel Na⁺-H⁺ exchanger. The net result in both instances is the 1:1 exchange of luminal (apical membrane)

Cl$^-$ for intracellular HCO$_3$-. The ileum can absorb chloride ions against a concentration gradient of 40 to 50 mEq per liter.

Bicarbonate

Bicarbonate is rapidly absorbed in the jejunum by a process that involves the formation of carbon dioxide from HCO$_3$- in chyme and H$^+$ secreted by the mucosa (e.g., by NHE). The product, H$_2$CO$_3$ dissociates into CO$_2$ and H$_2$O. The reaction gives rise to a markedly elevated luminal Pco_2 (~100 mm Hg), thereby providing a steep gradient for carbon dioxide diffusion into the mucosal epithelium. Within the cell, carbon dioxide reacts with water in the presence of carbonic anhydrase to form carbonic acid, which dissociates into hydrogen and bicarbonate ions. In the proximal small intestine, HCO$_3$- ions diffuse into the blood, whereas the H$^+$ ions are secreted into the lumen. In the ileum, on the other hand, HCO$_3$- is actively secreted into the lumen in exchange for Cl$^-$ (Fig. 7-7). As a result of this anion exchange (AE) process, the ileal contents become alkaline (pH between 7 and 8). Bicarbonate ions are also secreted into the lumen of the duodenum by the Brunner's glands. The alkaline secretion of the duodenal glands aids in the neutralization of acidic chyme from the stomach. The maximal bicarbonate output of the Brunner's glands is equivalent to basal pancreatic bicarbonate secretion. Secretin, which promotes bicarbonate delivery into the duodenum from both the pancreas and biliary system, also stimulates Brunner's gland secretion.

Calcium

Adult man ingests approximately 1.0 gram of calcium per day, with up to half derived from milk and milk products. An additional 200 to 300 mg is secreted into the intestine in various digestive juices. Most calcium is ingested in a nonionic form, that is, as salts formed with oxalate, phytate, carbonate, and other molecules. While these salts are insoluble at a neutral pH, the low luminal pH that results from gastric acid secretion enhances solubility and permits their absorption in the intestine. Approximately 30% of the calcium entering the small bowel is absorbed. While calcium is absorbed along the entire length of the small intestine, the mechanism(s) that contribute to this process differs among the bowel segments. Both transcellular (active) and paracellular (passive) transport processes are engaged in intestinal Ca^{++} absorption (Figure 7-8). Transcellular mechanisms contribute to Ca^{++} absorption in the duodenum and jejunum, while the paracellular mechanism is available along the entire length of the small bowel and is largely responsible for Ca^{++} absorption in the ileum.

The active (transcellular) absorption of calcium involves three distinct steps that relate to: 1) Ca^{++} entry/uptake into the cell, 2) solubilization and mobilization of Ca^{++} within the cytosol,

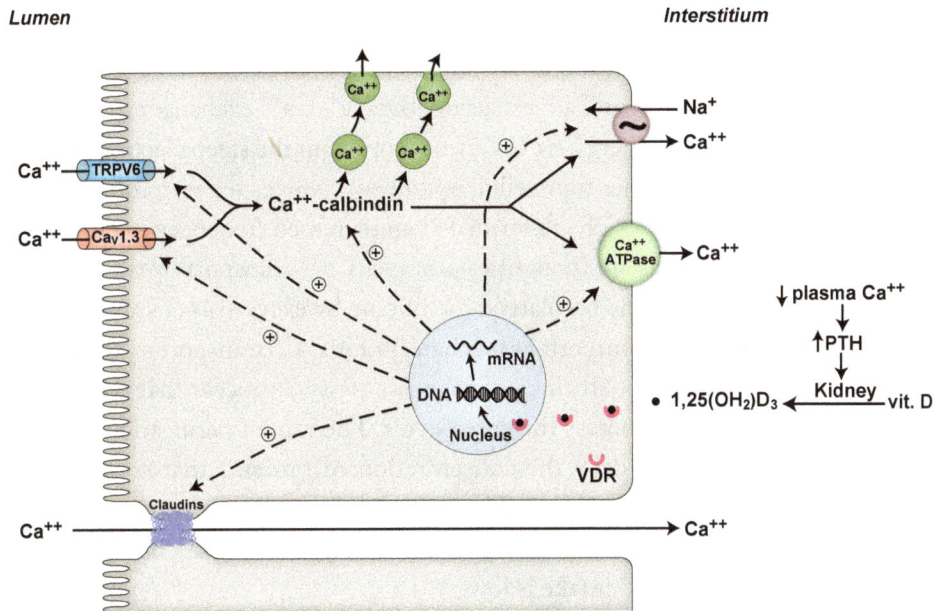

FIGURE 7-8: Mechanism of calcium absorption in enterocytes of the small intestine. Ca^{++} entry/uptake into the cell occurs via Ca^{++}-specific channels (e.g., TRPV6, $Ca_v1.3$). Once Ca^{++} enters the cytosol it binds to calbindins, which solubilize the cation and carries Ca^{++} to the basolateral membrane for presentation to two transport proteins (Ca^{++}-ATPase and Na^+-Ca^{++} exchanger) that extrude Ca^{++} from the cell. Absorbed Ca^{++} is also sequestered within and extruded from the enterocyte via membrane vesicles. Changes in plasma Ca^{++} concentration influence the rate of Ca^{++} absorption by promoting the formation of 1,25 dihydroxyvitamin D3, which enters the enterocyte and engages with its nuclear receptor (VDR) to subsequently transcribe genes encoding proteins that contribute to Ca^{++}.

and 3) Ca^{++} extrusion at the basolateral membrane (Figure 7-8). for Ca^{++} to enter the enterocyte at the brush border, it must engage with proteins that function as Ca^{++}-specific channels. Two Ca^{++} channels that have been implicated in this process are TRPV6 (transient receptor potential vanilloid 6) and $Ca_v1.3$ (an L-type channel). The expression of TRPV6 is highest in the duodenum and decreases through the jejunum to the ileum, while $Ca_v1.3$ exhibits the opposite expression pattern, that is, highest in the ileum and lowest in the duodenum. Once Ca^{++} enters the cytosol via one or both of these channels, it binds to proteins (calbindins) with high Ca^{++} affinity. Calbindin molecules can bind either 2 or 4 calcium ions. These binding proteins serve to minimize the rise in intracellular free Ca^{++}, prevent the formation of insoluble salts, and avert premature cell death via apoptosis during Ca^{++} absorption. Calbindin also carries

Ca^{++} to the basolateral membrane for presentation to two transport proteins (Ca^{++}-ATPase and Na^+-Ca^{++} exchanger) that function to extrude Ca^{++} from the cell. While the Ca^{++}-ATPase uses ATP to generate the energy for Ca^{++} extrusion, the Na^+-Ca^{++} exchanger (which exchanges 3 Na^+ for each Ca^{++}) derives its energy for Ca^{++} transport from the steep electrochemical gradient for Na^+ across the cell. Another transcellular pathway involves sequestration and extrusion of Ca^{++} via membrane vesicles, which are formed at an increased frequency in enterocytes during Ca^{++} absorption. Calbindins are thought to promote Ca^{++} transport by the vesicular pathway, which releases Ca^{++} across the basolateral membrane by exocytosis (Figure 7-8).

Paracellular transport is an important mechanism for Ca^{++} transport in the human small bowel. The passage of Ca^{++} through this pathway is a passive process that depends on concentration and electric gradients across the enterocytes. The rate of permeation of Ca^{++} in the paracellular pathway is influenced by the concentration of proteins that comprise the tight junctions, such as claudins and occludins. This route is the more important mechanism for Ca^{++} absorption when Ca^{++} intake (and luminal concentration) is high, while the transcellular mechanism dominates when Ca^{++} intake is low.

The intestine, along with kidney, bone, and parathyroid glands play critical roles in the regulation of extracellular fluid Ca^{++} concentration. Whole body Ca^{++} homeostasis is achieved by the controlled movement of calcium within and among these tissues, and is regulated by hormones (e.g., parathyroid hormone, PTH) and by vitamin D and its metabolites. Through the actions of these regulatory factors, the small intestine is able to adjust the level of Ca^{++} absorption (by the mechanisms described above) in response to alterations in extracellular Ca^{++} concentration (Fig. 7-8). For example, when dietary calcium intake is low and there is a consequent fall in plasma Ca^{++} concentration, the parathyroid gland senses the altered Ca^{++} level and secretes PTH. The PTH exerts an indirect effect on intestinal Ca^{++} absorption by regulating the metabolic conversion of vitamin D_3, which gains access to extracellular fluid from the diet and/or from exposure of skin to ultraviolet light, to 1,25 dihydroxyvitamin D3 [$1,25(OH)_2D_3$]. This product of vitamin D metabolism in the kidney acts as a hormone to exert a powerful influence on Ca^{++} absorption in the small intestine. $1,25(OH)_2D_3$ engagement with the vitamin D receptor (VDR) in enterocytes, results in enhanced synthesis of messenger RNA that encodes proteins that contribute to both the transcellular and paracellular pathways for Ca^{++} absorption (Fig. 7-8). Consequently, in response to $1,25(OH)_2D_3$ stimulation, the expression and function of Ca^{++} channels on the brush border membrane (TRPV6, $Ca_v1.3$), calbindin, the Ca^{++} export pumps (Ca^{++}-ATPase, Na+-Ca++ exchanger), and claudins are increased, leading to increased intestinal calcium absorption and the restoration of extracellular Ca^{++} concentration to its normal value.

Magnesium

Approximately 300 mg of magnesium is ingested each day by the adult human. Of this amount, between 10% and 65% is absorbed in the small intestine, depending on the dietary Mg^{++} load. The characteristics of intestinal Mg^{++} absorption closely resemble those described above for Ca^{++}, including the involvement of all segments of the small bowel, both passive (paracellular) and active (transcellular) pathways of absorption, and the engagement of identical (e.g., TRPV6 channels on the brush border membrane) or similar (Na^+-Mg^{++} exchanger on the basolateral membrane) transport proteins. At low luminal concentrations, Mg^{++} is primarily absorbed via the active transcellular route, while passive paracellular transport increases with rising intraluminal concentration.

Iron

Iron is an essential trace element that plays a critical role in a number of physiological processes, including aerobic respiration and erythropoiesis. Since the body has no effective means of excreting iron, the regulated absorption of dietary iron in the small intestine is essential for iron homeostasis. The average daily intake of iron in humans is 10 to 20 mg, of which approximately 10% (1–2 mg/day) is absorbed by the small intestine. This rate of iron uptake is balanced with daily losses via blood (menstruation), the gut (sloughing of senescent enterocytes), skin (exfoliation), urine, and bile. In the carnivorous diet, iron is ingested mainly in the form of heme, that is, sequestered in the protoporphyrin ring of myoglobin and hemoglobin. Nonheme iron, which largely exists in the ferric (Fe^{+++}) form and is highly insoluble, is also abundant in meat, and it represents the dominant form of iron in plants. The bioavailability of nonheme iron in the GI tract is significantly enhanced as a result of the changes in the luminal environment that accompanies the ingestion of a meal. Gastric acid and proteolytic enzymes derived from the stomach and pancreas facilitate the release of iron from ingested food. HCl and dietary ascorbic acid (and to a lesser extent citric acid, amino acids, sugars, and alcohol) also promote the reduction ($Fe^{+++} \rightarrow Fe^{++}$) and solubilization of iron in the intestine. Conversely, some dietary factors commonly found in plants, such as phytate, oxalate, tannins, and polyphenols, form complexes with nonheme iron to reduce its solubility and impede absorption by the intestine.

The major sites of iron absorption are the duodenum and proximal jejunum. Iron enters the intestinal epithelium either as heme or an inorganic ion, with the ferrous form more rapidly absorbed than the ferric, and heme-iron more readily absorbed than inorganic iron. Heme-iron is transported as heme into the enterocyte by a process involving heme carrier protein-1 (HCP1) and endocytosis (Figure 7-9). Once within the enterocyte, ferrous iron is liberated

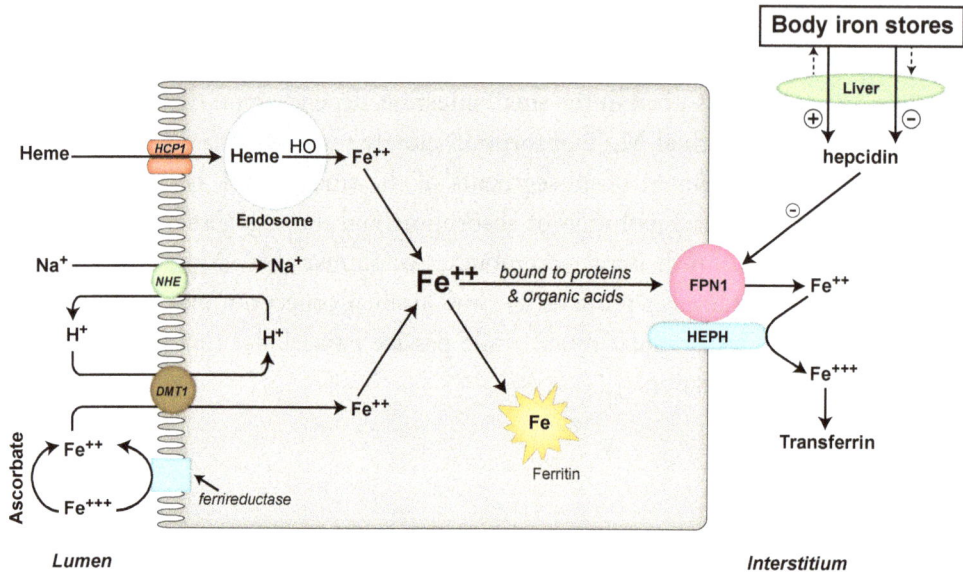

FIGURE 7-9: Iron absorption by enterocytes in the small intestine. Iron is absorbed both as intact heme and as Fe^{++}. Heme enters the enterocyte by a heme carrier protein-1 (HPC1) and by endocytosis. Once within the enterocyte, ferrous iron is liberated from heme by the enzyme heme oxygenase (HO). Luminal Fe^{+++} is converted to Fe^{++} either by dietary ascorbate or by a ferrireductase localized in the brush border membrane. Subsequently, Fe^{++} enters the enterocyte via the divalent metal-ion transporter 1 (DMT1), which is coupled to proton (H^+) extrusion from the cell. Within the cytosol, Fe^{++} is either bound by ferritin for storage or transported across the basolateral membrane by ferroportin 1 (FPN1). During its extrusion from the cell by FPN1, Fe^{++} is oxidized to Fe^{+++} by a functionally coupled ferroxidase, hephaestin (HEPH). This oxidation of Fe^{++} is required for iron binding to transferrin. The liver-derived peptide, hepcidin (HEPC) regulates iron absorption in the intestine (by modulating FPN1 expression) in accordance with the status of whole body iron levels.

from most of the absorbed heme by the enzyme heme oxygenase (HO). Nonheme iron enters the enterocyte via a specific carrier protein, the divalent metal-ion transporter 1 (DMT1), which has an affinity for ferrous (but not ferric) iron and is coupled to proton (H^+) extrusion from the cell. Since most dietary iron is in the ferric form, it must be reduced before engaging with DMT1. In the small bowel, Fe^{+++} is converted to Fe^{++} either by dietary ascorbate or by a ferrireductase (such as cytochrome B) localized in the brush border membrane (BBM). The protons that produce the driving force for Fe^{++} uptake by DMT1 are provided by sodium-hydrogen exchangers (NHE) that also reside in the BBM.

The ferrous iron that enters the enterocyte from both heme and nonheme sources is likely chelated by intracellular proteins, amino acids and low molecular weight organic acids (e.g., citrate). The fate of this iron is determined by the status of body iron stores. If the absorbed iron is not immediately required by the body, it is bound by ferritin for storage, and most will be lost from the body when the senescent enterocytes are sloughed from the villus tips at the end of their life cycle. When the body iron is deficient, Fe^{++} in the enterocyte is rapidly transferred to the basolateral membrane where it is exported into interstitial fluid by the iron transport protein, ferroportin 1 (FPN1). During its extrusion from the cell by FPN1, Fe^{++} is oxidized to Fe^{+++} by a functionally coupled ferroxidase, hephaestin (HEPH). This oxidation of Fe^{++} is required for iron binding to transferrin in interstitial fluid and the subsequent distribution of absorbed iron throughout the body. A small portion of heme that is absorbed by enterocytes is exported across the basolateral membrane into interstitial fluid as intact heme, where it binds with hemopexin for distribution throughout the body.

Exquisite regulatory mechanisms have evolved that enable intestinal absorption of dietary iron to precisely control overall iron balance in the body. Hepcidin (HEPC), a liver-derived peptide that regulates the expression of ferroportin 1, accounts for this link between intestinal iron absorption and body iron homeostasis (Fig. 7-9). HEPC production and release by the liver is responsive to molecules, such as transferrin receptor 2 and hemojuvelin, that "sense" body iron levels. When iron stores are replete, HEPC synthesis is increased, while low iron stores lead to suppression of HEPC production. HEPC blocks intestinal iron absorption by binding to ferroportin 1 (FPN1) and causing its internalization and degradation. Hence, high body iron stores promote the expression/release of HEPC, leading to diminished export of iron from enterocytes and increased storage as ferritin, which is lost when the enterocyte in sloughed into the lumen. During iron deficiency (iron stores depleted), HEPC production is very low, FPN1 is fully functional, less iron is stored as ferritin, and the gut plays a major role in restoring whole body iron homeostasis. Additional mechanisms are activated to promote intestinal iron absorption when iron stores are depleted. For example, low iron levels are known to elicit transcription-dependent increases in the expression of proteins that are engaged in the uptake of iron by enterocytes, including the brush border membrane ferrireductase and the iron transporter, DMT1. The transcripts for these iron uptake proteins contain an iron-responsive element (IRE) that interacts with (and is under the control of) cytosolic iron-sensing proteins, which allows for feedback regulation of the uptake and accumulation of iron in enterocytes.

SECRETION OF WATER AND ELECTROLYTES

Net water and electrolyte transport across the mucosa of the small intestine represents the difference between lumen-to-blood and blood-to-lumen fluxes. While the difference between the unidirectional fluxes normally favors net water absorption, it is estimated that 1.5 to 2 liters of isotonic fluid are secreted by the small intestine each day. The physiologic role of intestinal secretion remains unclear; however, its purpose may be to maintain the fluidity of chyme and/or to wash away (or dilute) potentially injurious substances near the epithelial surface. It is now well recognized that excess stimulation of the normal secretory processes can upset the normal balance between the absorptive and secretory water fluxes so that net fluid and electrolyte secretion and, ultimately, diarrhea result. The water and electrolyte loss from patients with prolonged and severe diarrhea (e.g., cholera-induced secretion) can result in dehydration of the body tissues and a severe electrolyte imbalance that can have rapidly fatal consequences (e.g., metabolic acidosis due to excess bicarbonate loss). Diarrheal disorders such as cholera are a leading cause of death worldwide, particularly in underdeveloped countries.

Secretion of water and electrolytes in the small intestine can be induced by a variety of chemical and physical stimuli (Table 7-2). For many stimuli, the underlying cause of the net water secretion is enhanced transport of chloride ions by the epithelial cells lining the crypts of Lieberkuhn. Several bacterial toxins, neurotransmitters (substance P), and inflammatory mediators (e.g., histamine) that induce net water secretion in the small bowel do so by stimulating active chloride secretion. This secretory process is often elicited by a change in the intracellular concentration of either cyclic AMP or calcium. The transporters that are engaged in chloride secretion in the small intestine are similar to those discussed more extensively in Chapter 8 (Figure 8-3, panel B) for the colon. Briefly, chloride ions enter crypt epithelial cells via an electroneutral transporter that moves 2 Cl^- along with a single Na^+ and K^+ ion across the basolateral membrane into the cell. The energy required for this Na^+-K^+-$2Cl^-$ co-transporter is derived from the Na^+ gradient created by the Na^+-K^+ pump. The accumulating intracellular Cl^- then moves out of the cell and is secreted into the crypt lumen via the cystic fibrosis conductance regulator (CFTR) channel located at the apical membrane. Secretagogues such as cholera enterotoxin bind to receptors on the apical membrane of crypt epithelium and enhance adenyl cyclase activity, thereby increasing intracellular cAMP concentration, which activates protein kinase A (PKA). The PKA then activates pre-existing CFTR and promotes the insertion of more Cl^- channels into the apical membrane, thereby resulting in enhanced Cl^- secretion. Other secretagogues, such as acetylcholine, substance P, and histamine elicit Cl^- secretion in intestinal crypts by causing a rise in intracellular Ca^{++}, which stimulates a different population of chloride channels on the apical membrane. These calcium-activated chloride channels produce a shorter lived and less intense secretory response compared cAMP-mediated, CFTR dependent secretion. The plasma to lumen Cl^- movement that results from either cAMP- or Ca^{++} mediated secretion creates a potential difference (lumen negative) across crypt epithelium

TABLE 7-2: Agents and Conditions that Can Elicit Net Fluid Secretion in the Small Intestine	
Paracrines, neurotransmitters, inflammatory mediators:	
Histamine	Acetylcholine
Secretin	Serotonin
Vasoactive intestinal polypeptide	Substance P
Prostaglandins	Guanylin
Bacterial toxins:	
Vibrio cholera	Staphylococcus aureus
Escherichia coli	Clostridium perfringens
Bacillus cereus	Pseudomonas aeruginosa
Shigella dysenteriae	Klebsiella pneumonia
Laxatives, irritants:	
Ricinoleic acid (castor oil)	Long-chain fatty acids
Dihydroxy bile acids	Magnesium salts
Physical factors:	
Increased luminal osmolality	Luminal distension
Hypoproteinemia	

that drives the movement of Na^+ through paracellular pathways into the lumen. The osmotic pressure gradient generated by NaCl accumulation in the crypt lumen draws water to maintain isotonicity.

Another mechanism by which cholera enterotoxin produces net secretion of water and electrolytes is by inhibition of sodium absorption via the sodium hydrogen exchanger (NHE) on the apical membrane of villus epithelial cells. The inhibitory influence of cholera toxin on

NHE-dependent sodium absorption also appears to be mediated by a rise in cell cAMP concentration. Because cholera toxin (and cAMP) does not inhibit glucose-stimulated sodium absorption, ingestion of glucose-electrolyte solutions has proven to be an effective therapeutic means of reducing the fluid loss in patients with cholera.

Increased luminal osmolality is the primary mechanism responsible for net fluid secretion in the proximal small bowel of normal individuals. As discussed above, when hypertonic chyme enters the duodenal lumen from the stomach, the blood-to-lumen water flux increases to keep the luminal contents isotonic, and net fluid secretion results. Osmotic equilibration in the small intestine is also the basis of action of some laxatives. Magnesium sulfate is a poorly absorbed salt, and therefore it must retain its osmotic equivalent of water in the intestinal lumen either by preventing water absorption or by inducing water secretion (passive). A clinical example of luminal osmolality as a mechanism for net fluid secretion is the diarrhea associated with lactase deficiency. In this condition, an increased blood-to-lumen water flux results from the increased luminal osmotic pressure arising from lactose retention in the lumen.

Net water and electrolyte secretion can also be induced by an increase in the hydrostatic pressure gradient across the mucosal membrane. This type of secretion is commonly referred to as "secretory filtration" because it denotes the filtration of interstitial fluid into the bowel lumen. Secretory filtration does not occur across the mucosal membrane of the normal villus for two reasons: a low interstitial hydrostatic pressure (0 to 2 mm Hg) and a high resistance to water movement (low hydraulic conductance). Conditions (e.g., portal hypertension, hypoproteinemia) that greatly enhance capillary fluid filtration and lead to interstitial fluid accumulation produce a rise in interstitial hydrostatic pressure. When interstitial hydrostatic pressure increases to 4 to 6 mm Hg, the mucosal membrane ruptures at the villus tips, producing a one thousand-fold increase in mucosal hydraulic conductance. The decreased mucosal resistance to water flow, coupled to the rise in interstitial hydrostatic pressure, can lead to secretory filtration. Secretory filtration can also occur following bowel obstruction, when luminal distension compresses the venous drainage and greatly enhances intestinal capillary filtration rate.

CARBOHYDRATE DIGESTIONS AND ABSORPTION

Dietary Carbohydrates

Carbohydrates are an inexpensive source of calories worldwide. The principal dietary carbohydrates are starches, sucrose, and lactose. Adult humans in the Western world consume 100 to 800 gm of carbohydrates per day. Starch accounts for 45% to 60% of the digestible carbo-

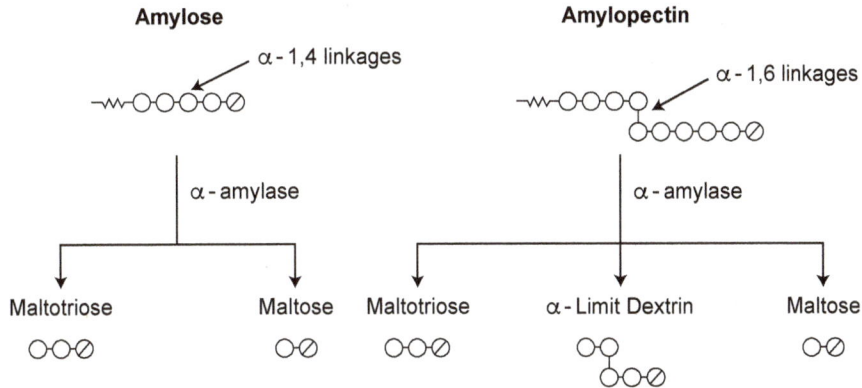

FIGURE 7-10: The major polysaccharides of starch (amylose, amylopectin) and their hydrolytic products after exposure to α-amylase. Maltose and maltotriose are products of α-amylase hydrolysis of both amylose and amylopectin. α-limit dextrins are also formed during amylopectin hydrolysis due the presence of α 1,6-linkages. Reducing glucose units are indicated by ⊘.

hydrates ingested, whereas disaccharides (sucrose and lactose) and monosaccharides (glucose and fructose) account for 30% to 40% and 5% to 10%, respectively. Indigestible carbohydrates, which are the main constituents of dietary fiber, normally account for a relatively small fraction of the Western diet.

Starch is a glucose-containing polysaccharide primarily found in plants with a molecular weight of 100,000 to more than 1 million. The two major polysaccharides of starch are amylose and amylopectin (Figure 7-10). Amylose is a straight chain of α 1-4 linked glucose molecules and it accounts for approximately 20% of dietary starch. Amylopectin, accounting for 80% of dietary starch, differs from amylose in that there are α 1-6 linkages approximately every 25 glucose molecules along the straight (α 1,4) chain. Glycogen, a high molecular weight polysaccharide found in animal tissues, is similar to amylopectin; however, the α 1-6 branch points occur with greater frequency, that is, every 12 glucose residues.

Digestion

The initial step in the digestion of starch involves the hydrolysis of α 1,4-linkages by salivary and pancreatic α-amylases (Fig. 7-10). Salivary and pancreatic a-amylase exhibit optimal activity at near neutral pH and salivary amylase is largely inactivated by gastric acid. Consequently, in healthy adults, little starch hydrolysis ($\leq$5%) occurs under the influence of salivary α-amylase because of the short exposure time before gastric acid inactivates the enzyme. However, when food is chewed for an extended period of time, over half of the starch can be digested to the disaccharide stage by salivary α-amylase. In humans, pancreatic amylase is considered the primary enzyme that mediates the initial digestion of carbohydrates.

Intraluminal digestion of carbohydrates in the duodenum is extremely rapid (completed within 30 min after chyme mixes with pancreatic juice) because of the tremendous amount of amylase secreted by the pancreas. The final oligosaccharide products of luminal digestion are formed before a starch meal reaches the jejunum. Some pancreatic α-amylases is adsorbed to the brush border membrane of the mucosa where it hydrolyzes starch exposed to the surface of epithelial cells.

As an endoenzyme, α-amylase preferentially cleaves the interior α 1,4-linkages and will not cleave terminal α 1,4-linkages. α-amylase is incapable of cleaving the β 1,4-glucose linkages in cellulose, and it cannot break the α 1,6-branching links of amylopectin and glycogen. The final hydrolytic products of α-amylase action on amylose are maltose (glucose-glucose, α 1,4-linkage) and maltotriose (glucose-glucose-glucose, α 1,4-linkages) (Fig. 7-10). Maltose (disaccharide) and maltotriose (trisaccharide) are also products of amylase action on amylopectin and glycogen. Branched oligosaccharides (with an average of 6 to 8 glucose units) containing both α 1,6- and α 1,4-linkages, which are known as α-limit dextrins, are also products of amylolytic digestion of amylopectin and glycogen (Fig. 7-10). α-limit dextrins are formed owing to the fact that the α 1,6-linkages in amylopectin and glycogen are resistant to hydrolysis by a-amylase. Glucose is not a product of starch digestion by α-amylase.

The products of luminal starch digestion and the other dietary sugars (sucrose and lactose) must be hydrolyzed to their monosaccharide constituents, because there are no mucosal transport mechanisms for saccharides having two or more hexose units. The enzymes responsible for the complete hydrolysis of oligosaccharides are glycoproteins, which are an integral part of the brush border membrane and are positioned in it so that the active hydrolytic site is on the luminal surface of the membrane. Thus, their oligosaccharide substrates are hydrolyzed when they contact the brush border membrane. The activity of the mucosal membrane oligosaccharidases parallels the absorptive capacity of the small bowel for monosaccharides, that is, it begins to rise in the duodenum, reaches a maximum in the mid-jejunum and falls off to low levels in the ileum.

The major brush border membrane oligosaccharidases are glucoamylase (also called maltase), sucrase-isomaltase and lactase. The substrates and products of each of the enzymes are summarized in Table 7-3. Glucoamylase, the most active of the oligosaccharidases, hydrolyzes maltose and maltotriose to glucose. Sucrase-isomaltase represents two enzymes that are bound

TABLE 7-3: Intestinal Oligosaccharidases Expressed on the Brush Border that are Responsible for the Mucosal Phase of Carbohydrate Digestion

ENZYME	SUBSTRATES	BONDS CLEAVED	PRODUCTS
Glucoamylase	Maltose, maltotriose	α 1–4 linkage	Glucose, maltose
Sucrase-isomaltase	Sucrose, maltose α-limit dextrins	α 1–4 linkage α 1–6 linkage	Glucose, fructose Glucose, maltose
Lactase	Lactose	β 1-4 linkage	Glucose, galactose

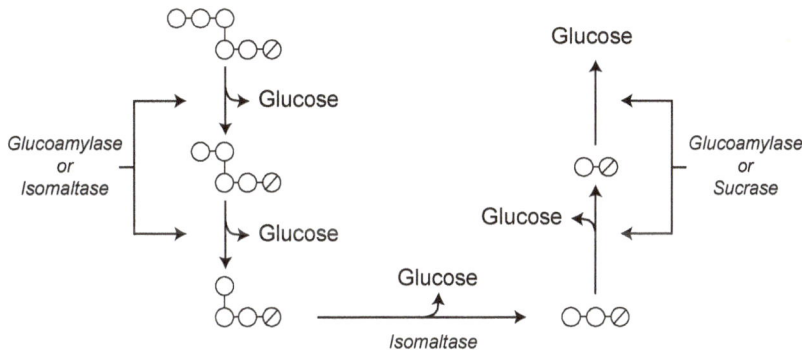

FIGURE 7-11: Sequential hydrolysis of α-limit dextrins by the brush border oligosaccharidases glucoamylase and isomaltase. Reducing glucose units are indicated by ⊘. [Modified from Gray GM. Assimilation of dietary carbohydrates. *Viewpoints Dig. Dis.* 12:3, 1980.]

together in the brush border membrane. The sucrase component of this enzyme complex hydrolyzes the α 1–4 linkage in sucrose to produce glucose and fructose, but it is also very active against maltose and maltotriose. The essential physiologic function of the isomaltase moiety (also called α-dextrinase or debranching enzyme) is hydrolysis of α 1,6-linkages in α-limit dextrins; however, the enzyme is also capable of removing α 1-4-linked glucose residues from the non-reducing end of α-dextrins. Lactase, the least active of the disaccharidases, catalyzes the hydrolysis of lactose to glucose and galactose. As illustrated in Figure 7-11, the α-limit dextrins are hydrolyzed by the complimentary action of glucoamylase and isomaltase, each of which acts at discrete regions on the oligosaccharide.

The activity of membrane-bound oligosaccharidases in the small bowel is so high that the absorption of hexoses, rather than hydrolysis of oligosaccharides, is the rate-limiting step in the overall assimilation of most carbohydrates. An exception to this is lactose, whose rate of hydrolysis by lactase is slower than the rate of absorption of its monosaccharide constituents (maltose is hydrolyzed as rapidly as sucrose, but lactose is cleaved only half as quickly.) The activity of the brush border oligosaccharidases is influenced to some extent by the carbohydrate composition of the diet. Changing from a sucrose-deficient diet to a high-sucrose diet leads to a doubling of sucrase and maltase activities within 2 to 5 days. Oligosaccharidase activities are not increased by diets rich in either lactose, maltose, or both.

Transport of Monosaccharides

The principal final products of surface hydrolysis of oligosaccharides are glucose, galactose, and fructose. Once the monosaccharides are released by surface digestion, they diffuse a small distance to transport sites (carrier proteins) located very near the brush border oligosaccharidase. Since monosaccharides are too large to cross the mucosal membrane by simple diffusion, specific carrier-mediated processes exist to ensure adequate absorption of glucose, galactose, and fructose. It is estimated that the human small intestine can absorb an amount of hexoses equivalent to 22 pounds of sucrose daily.

The aldohexoses, glucose and galactose, have a nearly identical structure and therefore share a single transporter (protein carrier in the brush border membrane) for entry into the enterocyte (Figure 7-12). The sodium/glucose cotransporter (SGLT-1) mediates the uphill transport of glucose and galactose from the gut lumen into the cell. This active process is powered by the electrochemical gradient for Na^+ across the brush border, which is maintained by the extrusion of Na^+ across the basolateral membrane via the Na-K pump. SGLT-1 transports two sodium ions per monosaccharide molecule into the cell. A separate, Na^+-independent, transport system mediates the absorption of fructose in the small intestine. Fructose uptake by enterocytes occurs by facilitated diffusion and is mediated by the carrier protein GLUT5. The exit of glucose, galactose, and fructose from the absorptive cell is mediated by a single transporter (GLUT2) located on the basolateral membrane and also occurs by facilitated diffusion (Fig. 7-12).

The active transport of either glucose or galactose is inhibited by the presence of the other aldohexose; that is, the absorption of glucose depresses the rate at which galactose is absorbed and vice versa, indicating competition for a site on the carrier. Inasmuch as glucose has a greater affinity for the carrier site, its absorption depresses galactose transport more than galactose absorption depresses the transport of glucose. The structural specificity of the aldohexose

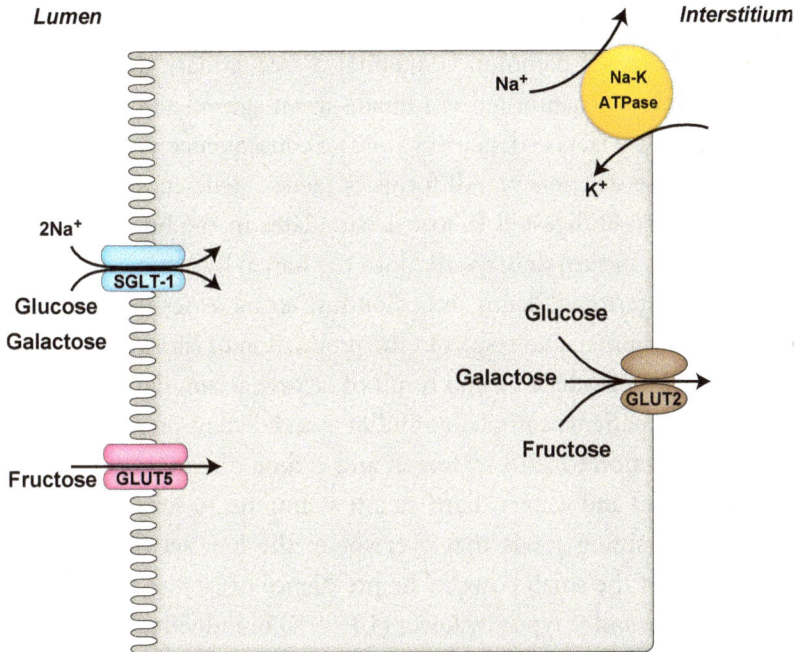

FIGURE 7-12: Transport mechanisms involved in the absorption of glucose, galactose, and fructose. Glucose and galactose are actively transported into the enterocyte via a Na^+-dependent mechanism that is mediated by SGLT-1. This process is powered by the electrochemical gradient for Na^+ across the brush border, which is maintained by the extrusion of Na^+ across the basolateral membrane via the Na-K pump. Fructose enters the cell via facilitated diffusion using the carrier protein GLUT5. Glucose, galactose and fructose exit the cell via facilitated diffusion using GLUT2.

carrier molecule is also demonstrated by the fact that modifications or substitutions at certain positions in the glucose or galactose molecule prevent their active transport. Fructose is also less well absorbed than glucose. The rate of fructose absorption in man is normally one-sixth to one-third of the rate of glucose absorption.

Carbohydrate Intolerance

Although pancreatic amylase-related maldigestion of starch is virtually nonexistent in the adult human (because the enzyme is secreted in great excess), deficiencies of membrane-bound disaccharidases are a frequent cause of carbohydrate malabsorption. The enzyme deficiency may be genetic or acquired, that is, resulting from a disease of the small intestine such as celiac sprue. Lactase deficiency is the most common congenital disaccharidase deficiency. However, this condition is more commonly manifested in humans as an age-related decline in lactase activity after childhood (primary lactase deficiency) or as a consequence of disease-related mucosal damage (secondary lactase deficiency). All forms of lactase deficiency are presented clinically as lactose intolerance. The undigested lactose accumulates in the bowel lumen and increases luminal osmolality, which in turn draws water into the lumen by osmosis. The fermentation of lactose by bacteria in the terminal ileum and colon further increases in the number of osmotically active solutes in the lumen, and results in the production of short chain fatty acids, hydrogen, carbon dioxide and methane. The end result of lactose accumulation (and metabolism) is an osmotic diarrhea. The patients complain of fullness and distention of the abdomen followed by increased flatus production (due to hydrogen and carbon dioxide gases produced by bacterial degradation of lactose) and watery diarrhea a few minutes to several hours after ingesting a volume of lactose containing foods that overwhelm the low activity of lactase present in the mucosal membrane of the small bowel. The prevalence of lactose intolerance in Northern America is highly variable and is typically lower (15–30%) in individuals of northern European ancestry and higher (79–90%) in African and Asian populations. The symptoms of lactase deficiency can be eliminated by excluding or significantly reducing dairy products from the diet, or by supplementation with exogenous lactase.

Dietary Fiber

The term dietary fiber is generally used to describe the carbohydrate components of food that are not digested by the endogenous secretions of the human digestive tract. The major indigestible carbohydrates are cellulose, hemicelluloses, gums, and pectins. The richest natural food sources of dietary fiber include grains, fruit, vegetables, nuts and seeds. In the Western diet, much of the dietary fiber is removed during food processing. There is a growing body of evidence that dietary fiber is beneficial to human health. For example, higher dietary fiber intake is associated with reduced risk for cardiovascular disease, type-2 diabetes, and cancer (particularly colorectal cancer). While the mechanism(s) that underlie these health benefits of increased fiber intake remain poorly understood, there is clear evidence that dietary fiber improves digestive health. The positive influence of fiber on digestive function has been attributed to two characteristics, that is, to its ability to bind water and to its vulnerability to fermentation by colonic bacteria. A high fiber diet increases the weight of the stool by retaining water and significantly accelerates its transit through the bowel. This bulking effect (increased fecal biomass) of dietary fiber is the basis of pharmacologic action of certain laxatives, such as psyllium (Metamucil) and methylcellulose (Citrucel). The fermentation of dietary fiber and the physiologic benefits of this process in the colon are discussed in Chapter 8.

PROTEIN DIGESTION AND ABSORPTION

Dietary Requirements

Unlike carbohydrate, protein is essential for growth of the young and the maintenance of health in adults. The protein content of the average American diet is 70 to 100 grams per day. The minimum daily protein intake required to achieve nitrogen balance ranges between 35 and 50 grams per day. Growing children and pregnant and lactating women require a proportionately greater protein intake relative to body weight to maintain nitrogen balance; that is, young children require 4 grams per kg body weight per day, compared to 0.6 grams per kg body weight per day for adults. The proteins of the body are comprised of 20 different amino acids, only twelve of which can be synthesized in the body; the remaining eight (essential amino acids) must be provided in the diet. The dietary proteins are derived almost entirely from meats, fish, and vegetables. In general, meat and other animal proteins (e.g., milk, eggs) contain all of the essential amino acids, while vegetable proteins are deficient in one or more amino acids (e.g., zein, the protein of corn is deficient in lysine and tryptophan). Therefore, even if the daily intake of protein exceeds 35 to 50 grams, normal body growth is not likely to occur if the protein is derived from a single vegetable. Finally, food processing can have a significant influence on the digestibility of proteins. For example, heating associated with the preparation or storage of food, can result in the formation of inter- and intramolecular covalent bonds that renders protein resistant to enzymatic digestion.

Protein that is available for absorption in the small intestine is derived not only from food but also from digestive secretions (enzymes, mucoproteins), desquamated cells, and plasma proteins. Approximately 30 grams of protein per day enter the intestinal lumen via the digestive secretions (saliva, gastric juice, bile, pancreatic secretions), while epithelial shedding (desquamated cells) can produce as much as 50 grams per day. Although the intestinal mucosa is relatively impermeable to macromolecules, a significant quantity of plasma proteins enters the gut lumen each day. It is estimated that approximately 15% of the normal· turnover of albumin and "γ"-globulin can be accounted for by enteric protein loss, that is, 3 grams per day. Thus, the proteins of endogenous origin nearly double the total amount of protein presented to the intestine for digestion and absorption. Both exogenous (food) and endogenous (secretions, desquamation) sources of proteins entering the gut lumen are processed identically by pancreatic proteases. Despite the large amounts of protein presented daily to the small intestine, less than 10% escapes absorption and is lost in the stool.

Digestion

The complete digestion of proteins to amino acids requires a larger number of enzymes than is needed to reduce starch or fat to their constituent components. The first enzyme involved in the hydrolysis of protein is pepsin, which is secreted as pepsinogen from the chief cells of the gastric mucosa and is optimally active under acidic conditions (see Chapter 4). When the stomach contents reach the duodenum, it mixes with pancreatic juice, the alkalinity of which inactivates pepsin. The extent of gastric protein digestion is determined by the physical state of the ingested protein (type of protein and extent of mastication), the activity of pepsin in gastric juice, and the length of time that it stays in the stomach. Ingested protein is usually transported to the duodenum as a mixture of native protein, protein fragments, large polypeptides, and only a minute quantity of free amino acids. Therefore, the gastric phase of protein digestion is generally considered to be of limited importance, relative to intestinal digestion, in the overall assimilation of protein, accounting for 10–15% of dietary protein hydrolysis. The nonessential nature of pepsin-mediated protein digestion is demonstrated by the fact that patients with either achlorhydria (failure of the stomach to secrete acid and pepsinogen) or a total gastrectomy exhibit no impairment of protein assimilation; i.e., they remain in nitrogen balance on normal protein diets.

Luminal Digestion

When ingested protein and the products of pepsin-mediated digestion reach the duodenum, they are mixed with the proteolytic enzymes secreted by the pancreas. Secretion of pancreatic enzymes is enhanced when food is present in the duodenal lumen, a response that is elicited by cholecystokinin-dependent mechanisms (see Fig. 5-7 in Chapter 5). The proteolytic enzymes of the pancreas are secreted as inactive precursors (proenzymes), which are subsequently activated in the duodenum (see Fig. 5-4 in Chapter 5). Intraluminal activation initially involves the conversion of trypsinogen (inactive precursor) to trypsin (active protease), a reaction catalyzed by the brush border enzyme, enteropeptidase. The presence of trypsin in the lumen further enhances its own formation (from trypsinogen), and it rapidly activates the other pancreatic proteases.

   The proteolytic enzymes produced by the pancreas function as either endopeptidases (trypsin, elastase, chymotrypsin) or exopeptidases (carboxypeptidases A and B). Endopeptidases act on specific peptide bonds at the interior of protein and polypeptide molecules. Trypsin hydrolyzes only the peptide bonds formed by amino acids with a positively charged side chain (e.g., arginine, lysine); chymotrypsin attacks only the peptide bonds formed by aromatic amino acids (e.g., tyrosine and phenylalanine); and elastase targets bonds formed by aliphatic amino acids (e.g., alanine, and leucine). In contrast, the exopeptidases hydrolyze, in sequence, the carboxyl terminal peptide bond in the peptide chain. Intraluminal digestion of dietary protein occurs by sequential or essentially simultaneous action of the endopeptidases and exopeptidases. The endopeptidases yield oligopeptides that are ideal substrates for the exopeptidases, which in turn remove an amino acid from the carboxyl terminus of the peptide. The combined and coordinated actions of the endopeptidases and exopeptidases result in a mixture of small peptides of two to six amino acid units and neutral and basic amino acids (particularly tyrosine, arginine, and methionine). Small peptides account for 70% of the final products of intraluminal protein digestion, while free amino acids account for the remainder (Figure 7-13). Inasmuch as the pancreas secretes a tremendous overabundance of proteolytic enzymes in response to a meal, the hydrolysis and absorption of small peptides (and amino acids) by the enterocytes, rather than intraluminal digestion, are the rate-limiting processes in overall protein assimilation.

Cellular Digestion of Peptides

The oligopeptides produced by pancreatic enzyme digestion are further hydrolyzed to free amino acids either at the surface of the brush border membrane or within the cytosol of the enterocyte (Fig. 7-13). The site (brush border or cytosolic) of oligopeptide hydrolysis is largely determined by the number of amino acid residues in the peptide substrate, with the larger oligopeptides (3-8 amino acids) digested by brush border peptidases while di- and tripeptides are hydrolyzed by cytosolic peptidases. The peptide-hydrolyzing brush border enzymes largely consist of multiple exopeptidases, endopeptidases, and dipeptidases. A large number of brush border peptidases are needed to ensure complete hydrolysis of peptides because each peptidase recognizes only a limited number of peptide bonds and the oligopeptides generated within the gut lumen contain 20 different amino acids. Specific carrier mechanisms (discussed below) also exist at the brush border membrane that allow dipeptides and tripeptides to enter the cytoplasm of the enterocyte where they are hydrolyzed by specific peptidases (dipeptidase, amino-tripeptidase) to free amino acids. Oligopeptides with more than three amino acid residues either are not transported across the brush border membrane (e.g., hexapeptide) or are very poorly absorbed (e.g., tetrapeptide). Most of the di- and tripeptides that are transported into the cytosol of the enterocyte then undergo hydrolysis to amino acids. Consequently, the

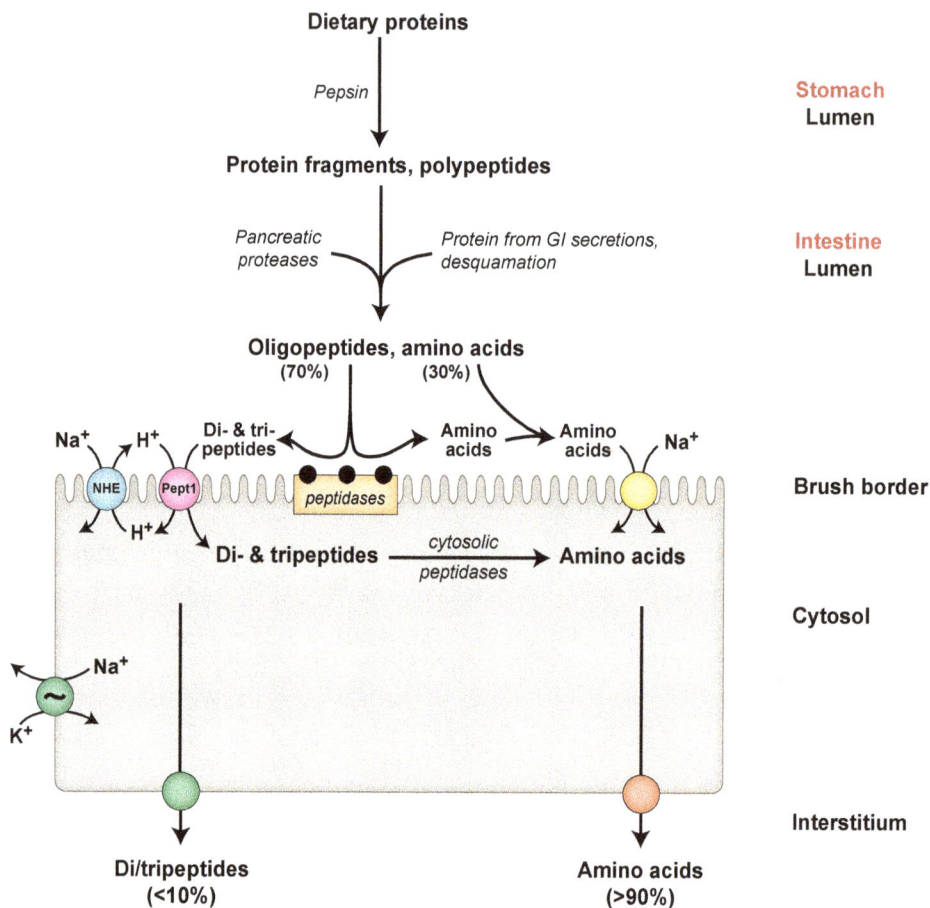

FIGURE 7-13: Summary of the different steps involved in the assimilation of ingested proteins by the gastrointestinal tract. Protein digestion begins in the stomach via the actions of pepsin. The large protein fragments and polypeptides generated in the stomach undergo further hydrolysis by pancreatic proteases in the intestinal lumen. The oligopeptides produced during luminal digestion are further hydrolyzed to free amino acids either at the surface of the brush border membrane or within the cytosol of the enterocyte. Di- and tripeptide absorption by enterocytes is mediated by a H^+/oligopeptide co-transporter (Pept1) that is driven by the hydrogen ion gradient that exists across the apical brush border membrane due to the Na^+-H^+ exchanger (NHE). Amino acid uptake by enterocytes is mediated by a diverse group of transport proteins that are typically Na^+-dependent. Transport proteins, located on the basolateral membrane, move absorbed amino acids and di/tripeptides out of the enterocyte.

end product of cellular peptide digestion is almost entirely free amino acids, which appear in portal blood after a protein meal. A small fraction (<10%) of the di- and tripeptides appear to be resistant to cytosolic hydrolysis and also gain access to portal blood.

Peptide and Amino Acid Transport Mechanisms

Peptides

Absorption of peptides is now considered to be the predominant mechanism for intestinal assimilation of dietary protein. It is well recognized that amino acids are absorbed at a much higher rate if introduced into the gut lumen as dipeptides and tripeptides rather than as free amino acids. Peptide absorption by enterocytes is mediated by a H^+/oligopeptide co-transporter (Pept1) that is driven by the hydrogen ion gradient that exists across the apical brush border membrane (Fig. 7-13). An acidic microenvironment is maintained on the luminal aspect of the brush border membrane by a Na^+-H^+ exchanger (NHE) that pumps H^+ into the intestinal lumen in exchange for Na^+. NHE uses the steep transmembrane Na^+ gradient, created by the Na^+-K^+ ATPase on the basolateral membrane, to drive H^+ into the gut lumen. The resulting inwardly directed proton gradient enables Pept1 to move peptides into the enterocyte even against a concentration gradient. This peptide-proton cotransport mechanism accounts for the absorption of a variety of di- and tripeptides (and possibly tetrapeptides), exhibiting broad substrate specificity but strict stereospecificity (only L-amino acids). Pept1 also accounts for the ability of the intestine to absorb some dipeptide-like antibiotics (e.g., cephalosporins) and angiotensin converting enzyme (ACE) inhibitors.

Amino Acids

In comparison to peptide absorption, the transport of free amino acids at the brush border membrane is relatively slow and appears to be the rate-limiting step in the assimilation of that fraction of dietary protein that is hydrolyzed to amino acids within the intestinal lumen and at the brush border membrane. Intestinal transport of amino acids exhibits specificity for L-stereoisomers, which can be transported against very steep concentration gradients, whereas D-isomers cannot. The uptake of amino acids by enterocytes is mediated by a diverse group of transport proteins that exhibit variable (and sometimes overlapping) specificity and are typically Na^+-dependent (Fig. 7-13). Seven distinct amino acid transport systems are present on the apical membrane (Table 7-4). The predominant transport system (B) is largely responsible for the Na^+-dependent transport of neutral amino acids into the enterocyte. This transport system functions similar to the glucose transporter SGLT1, which also relies on a steep

TABLE 7-4: Transport Systems for Amino Acids in the Brush Border Membrane of the Small Intestine		
TRANSPORT SYSTEM	**SUBSTRATES**	**DEPENDENCE ON NA$^+$**
B^0	Neutral amino acids	Yes
$B^{0,+}$	Neutral & cationic amino acids	Yes
$b^{0,+}$	Neutral & cationic amino acids Cystine	No
IMINO	Imino acids	Yes
β	Taurine, β-alanine	Yes
X^-_{AG}	Anionic amino acids	Yes
ASC	Neutral amino acids	Yes
N	Glutamine, asparagine, histidine	Yes
PAT	Small neutral amino acids	No

[From V. Ganapathy. Protein digestion & absorption. In: Physiology of the gastrointestinal tract. (LR Johnson, ed), Chapter 59, 1595–1617, 2012.]

transmembrane sodium gradient created by the basolateral Na-K pump to drive its co-transported organic solute into the cell. Other transport systems are also present on the apical membrane to move cationic (basic), anionic (acidic) imino, and β-amino acids into the cell. The amino acids that accumulate in enterocytes via these transport systems, as well as the amino acids generated intracellularly by cytosolic peptide hydrolysis, are either used for intracellular protein synthesis (~10%) or transported across the basal membrane to enter the interstitium and portal blood. There are a number of transport proteins for amino acids located on the basolateral membrane. Most of these transporters move amino acids out of the cell and their function is typically Na$^+$-independent, while the remaining basolateral membrane transporters are Na$^+$-dependent pathways that supply enterocytes with amino acids for cellular nutrition during the interdigestive period. Although functionally distinct from the apical membrane amino acid transporters, the basolateral transport systems also exhibit variable and sometimes overlapping specificity for different types of amino acids, such as neutral vs cationic amino acids.

Absorption of Native Proteins

Normal adults absorb only minute amounts of undigested protein, because the intestinal mucosa is normally too impermeable for diffusion of macromolecules through the tight junctions. Very low amounts of intact protein enter the enterocytes by endocytosis however these molecules are rapidly degraded by lysosomes. When the mucosal barrier is disrupted, large quantities of intact proteins can gain access to the lamina propria of intestinal villi in adult humans. A more physiologically important route for the movement of intact protein across the gut mucosa exists in regions that overlie the Peyer's patches. At these sites, epithelial cells are replaced by "M" cells, which are specialized for the uptake of proteins (antigens). Lysosomal degradation of endocytosed protein is limited in these cells. Instead, the ingested proteins are packaged in vesicles and released intact across the basolateral membrane into the lamina propria for processing by lymphocytes and the initiation of an immune response. In the newborn, passive immunity is conferred to the suckling neonate by absorption of immunoglobulins contained in mother's milk. This mechanism for uptake of intact protein by gut epithelial cells is rendered inactive by the sixth month via a process that is hormonally mediated.

Nucleoprotein Digestion and Absorption

Nucleoproteins are normal constituents of ingested animal and plant products that are also hydrolyzed and absorbed in the small intestine. Intraluminal hydrolysis of dietary nucleoproteins is catalyzed by the spectrum of proteolytic enzymes secreted by the pancreas. The endopeptidases and exopeptidases sequentially hydrolyze the histone and protamine components of nucleoproteins to oligopeptides and amino acids, which are ultimately absorbed by the enterocytes as described above. The nucleic acid components (DNA and RNA) are hydrolyzed by the pancreatic enzymes, deoxyribonuclease and ribonuclease, to yield polynucleotides. There are specific phosphodiesterases in the brush border membrane that catalyze the surface hydrolysis of polynucleotides to purine and pyrimidine nucleotides. The nucleotides, for which a transport mechanism does not exist, then undergo surface hydrolysis (catalyzed by nucleotidases) to form nucleosides, which is the major vehicle for entry of purines and pyrimidines into enterocytes. There are specific transport mechanisms in the brush border membrane that allow entry of purine and pyrimidine nucleosides into the enterocyte. Although more than 90% of dietary and endogenous nucleosides and bases are absorbed in the small intestine, the capacity of the nucleoside transport pathways is small in comparison to the absorptive capacities for carbohydrates and amino acids; however, the dietary load for nucleosides is also much smaller.

WATER-SOLUBLE VITAMINS (TABLE 7-5)

TABLE 7-5: Absorption of Water-Soluble Vitamins in the Small Intestine

VITAMIN	MECHANISM OF TRANSPORT	MAJOR PHYSIOLOGIC FUNCTIONS	DAILY ADULT REQUIREMENT	DEFICIENCY
Ascorbic acid (Vitamin C)	Na$^+$-dependent transporter (*SVCT*)	Cofactor for variety of critical metabolic reactions	Male: 90 mg Female: 75 mg	Scurvy
Cobalamin (Vitamin B$_{12}$)	Receptor-mediated endocytosis	DNA synthesis, formation of erythrocytes	2.0 µg	Pernicious anemia
Folate	Proton-coupled transporter (*PCFT*)	Nucleic acid biosynthesis	Male: 200 µg Female: 180 µg	Megalo-blastic anemia
Thiamin (Vitamin B$_1$)	Active Na$^+$-dependent	Carbohydrate & energy metabolism	Male: 1.5 mg Female: 1.1 mg	Beriberi
Niacin (Vitamin B$_3$)	Proton-coupled transporter	NAD & NADH redox reactions	Male: 19 mg Female: 15 mg	Pellagra
Riboflavin (Vitamin B$_2$) neuropathy	Carrier-mediated Na$^+$-independent	Co-enzymes FAD & FMN	Male: 1.7 mg Female: 1.3 mg	Peripheral
Biotin (Vitamin B$_7$)	Carrier-mediated Na$^+$-dependent	Gluconeogenesis, fatty acid production	30 µg	Neurologic changes
Pyridoxine (Vitamin B$_6$)	Proton-coupled transporter	Heme synthesis	Male: 2.0 mg Female: 1.6 mg	Stomatitis glossitis

Vitamins are organic substances that are necessary for normal metabolic functions, yet they cannot be manufactured by the body. Therefore, these essential compounds must be derived from the diet. Although the gut microbiota can synthesize some water-soluble vitamins (e.g., cobalamin, folic acid, and biotin), the quantitative importance of this source remains unclear. The water soluble vitamins generally function as essential cofactors for enzymatic reactions, and they range in size from 176 MW (ascorbic acid) to 1300 MW (cobalamin). A variety of protein carriers and transport mechanisms are involved in the absorption of water-soluble vitamins. All are largely absorbed in the small intestine, but there is growing evidence for the absorption of some water-soluble vitamins in the colon (see Chapter 8). When a carrier molecule is involved in vitamin transport, it is located at the brush border membrane, and vitamin transport into the cell is often coupled to Na^+, H^+, or some other ion. Certain vitamins (e.g., folates and thiamine) are metabolized either while crossing the brush border membrane or within the cytosol. The mechanism(s) by which water soluble vitamins leave the enterocyte across the basolateral membrane are less well understood; however, it appears that the molecular size and degree of ionization are major factors influencing the mode and rate of exit from the cell. Although all of the water-soluble vitamins have been extensively studied, folic acid and cobalamin (vitamin B_{12}) have received the most attention from gastrointestinal physiologists and gastroenterologists because blood disorders commonly result from malabsorption of these vitamins.

Folates

Folates are found in a wide variety of foods. Excellent sources of folate include dark green leafy vegetables, dried beans and peas, orange juice, strawberries, and liver. The average American diet provides approximately 700 µg of folates per day, and the recommended daily allowance (RDA) is 200 µg for males and 180 µg for females. Seventy to 80% of the dietary folates are polyglutamate conjugates (pteroylpolyglutamates), linked by a γ- rather than the more common α-peptide bond. The remaining dietary folates exist in free form. Gastric and pancreatic enzymes are incapable of cleaving the γ-peptide bond. However, the brush border membrane of enterocytes in the proximal small bowel contains an enzyme (folate hydrolase) which functions as a glutamate carboxypeptidase that sequentially hydrolyzes the polyglutamyl conjugates of folate (Figure 7-14). Surface hydrolysis of the polyglutamate-folate conjugates ($PteGlu_7$) ultimately yields a monoglutamyl product ($PteGlu_1$) that gains entry into the enterocyte by two specific folate transporters located on the apical membrane, that is, the proton-coupled folate transporter (PCFT) and the reduced folate carrier (RFC) (Fig. 7-14). PCFT is a folate-H^+ cotransporter that is fueled by the hydrogen ion gradient created by NHE across the apical brush border membrane, as described above for Pept1. RFC is an anion exchanger that capitalizes on

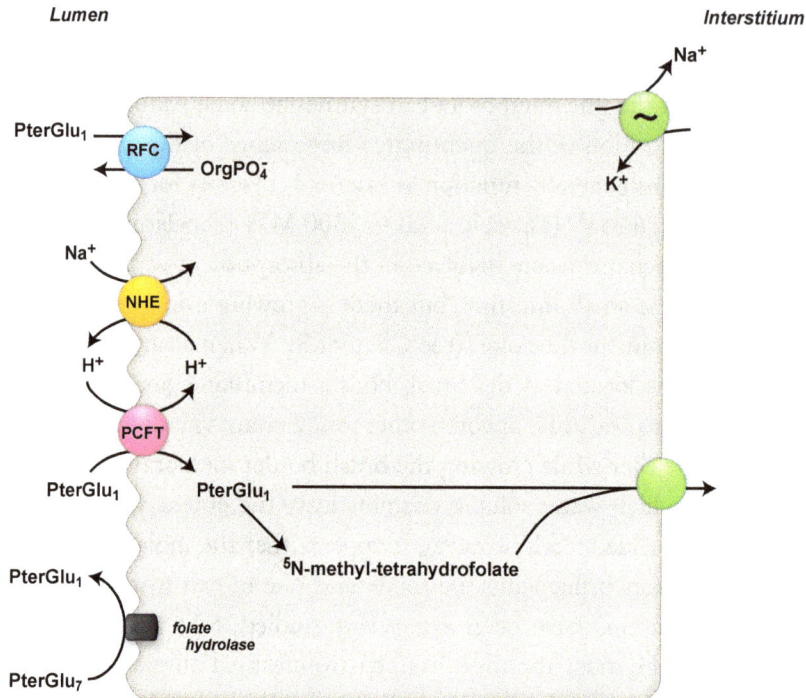

FIGURE 7-14: Digestion and absorption of dietary pteroylpolyglutamates (PteGlu$_7$) in the small intestine. Surface hydrolysis of the PteGlu$_7$ by enterocytes yields a monoglutamyl product (PteGlu$_1$), which gains entry into the enterocyte by two specific folate transporters located on the apical membrane, that is, the proton-coupled folate transporter (PCFT) and the reduced folate carrier (RFC). PCFT is fueled by the H$^+$ gradient created by NHE across the apical brush border membrane. RFC relies on the transmembrane organic phosphate concentration gradient to achieve an uphill transport of PteGlu$_1$ into cells. Once within the cell, PteGlu$_1$ is methylated and reduced to form ^{5}N-methyl-tetrahydrofolate. The methylated and non-methylated forms of folic acid then exit the cell at the basolateral membrane.

the transmembrane organic phosphate concentration gradient to achieve an uphill transport of PteGlu$_1$ into cells. Organic phosphates are highly concentrated within cells and the resulting gradient across the apical cell membrane provides the driving force for RFC-mediated transport of folate into cells. PCFT-mediated folate transport is considered the dominant mechanism in the proximal half of the small intestine where the surface pH is acidic, while RFC is believed to operate in the distal small intestine, where the enterocyte surface pH is neutral.

Once within the cell, a large proportion of the folic acid is methylated and reduced to the tetrahydro form (^{5}N-methyl-tetrahydrofolate). The methylated and non-methylated forms of folic acid then exit the cell at the basolateral membrane (via an undefined carrier) and transported to the liver in the portal vein. Tetrahydrofolate is a critical cofactor for chemical reactions related to the synthesis of thymines and purines. Hence, folate deficiency compromises DNA synthesis and cell division, and its effects are most notable in bone marrow where the turnover of cells is rapid. Dietary folate deficiency is quite common worldwide and deficiency of this vitamin is linked to various metabolic and blood disorders. In the US, Canada, and some European countries, mandatory fortification (140 µg/g) of all grain products has been instituted to reduce the occurrence of dietary folate deficiency. Folate malabsorption is a common accompaniment to gastrointestinal disease (e.g., celiac disease) and in patients on a variety of drugs (e.g., sulfasalazine, an agent used to treat ileitis and colitis).

Cobalamin

Bacteria are the dominant source of cobalamin (vitamin B_{12}) in nature, and in conventional diets B_{12} is exclusively derived from animal food sources, such as meat, liver, fish, eggs and dairy products. The average Western diet contains 5 to 10 µg per day, and the daily adult requirement is 2 µg. Cobalamin is bound to protein (primarily enzymes) in food sources, from which it is liberated by pepsin in the presence of acid (Figure 7-15). Once cobalamin is released from proteins in the stomach, it binds to haptocorrin (HC), which is produced by the salivary glands to protect B_{12} against acid degradation. Upon entering the duodenum, B_{12} is released from the B_{12}-HC complex due to the alkaline environment created by pancreatic HCO_3- and the action of pancreatic proteases. In this environment, B_{12} is free to bind to intrinsic factor (IF), a glycoprotein synthesized by gastric parietal cells with a high binding affinity for B_{12} (one molecule of IF binds two B_{12} molecules). The intrinsic factor-cobalamin (IF-B_{12}) complex is delivered to the ileum where it attaches to the multi-ligand receptor, cubam, that is expressed on the brush border membrane (Fig. 7-15). Free cobalamin does not bind to the ileal receptor. Following its attachment to cubam, the IF-B_{12} complex enters the enterocyte by receptor-mediated endocytosis. Lysosomal action leads to dissociation of the IF-B_{12} complex, which allows B_{12} to enter the cytosol, where it can exist in the free form or can bind to the transport protein, transcobalamin II (TCII). Cobalamin exits the cell by one of two mechanisms: 1) the TCII-bound B_{12} leaves the cell via exocytosis, and 2) free B_{12} exits across the basolateral membrane, facilitated by the multidrug resistance protein 1 (MDR1), an ATP-dependent efflux pump with broad substrate specificity. The free B_{12} that enters interstitial fluid by the latter pathway is rapidly bound to TCII, which transports the vitamin, via the portal vein, the to liver for storage.

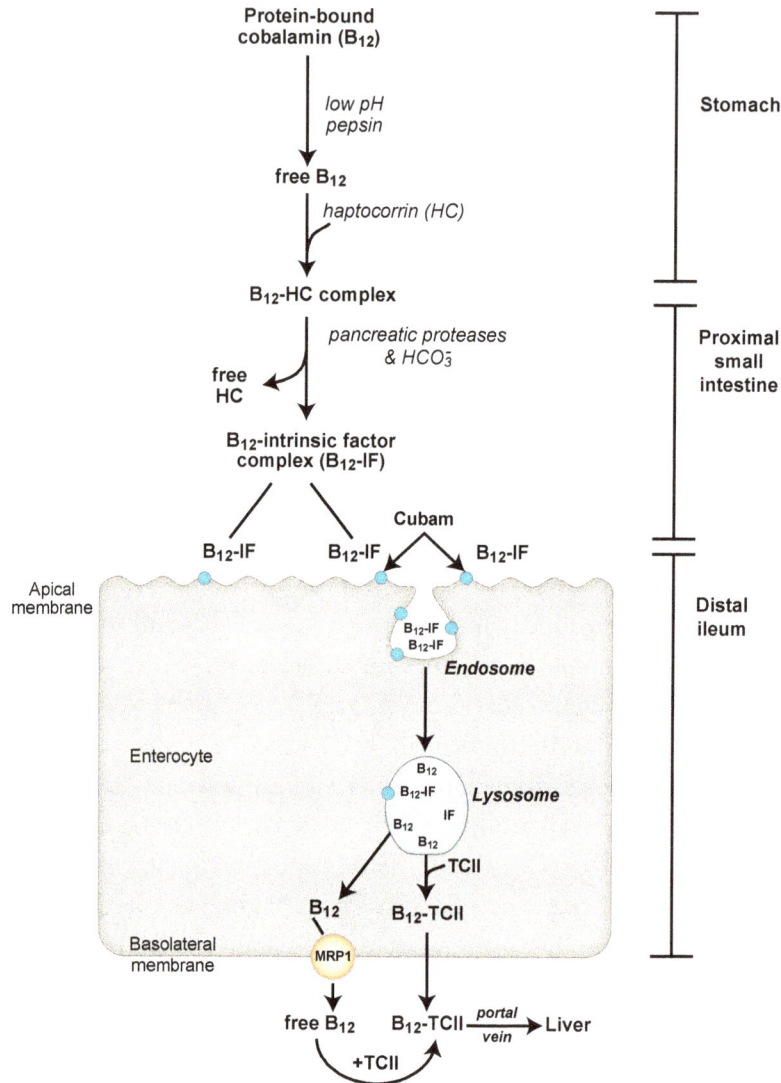

FIGURE 7-15: Sequence of events involved in the assimilation of ingested vitamin B_{12} (cobalamin) by the gastrointestinal tract. Pepsin in the stomach releases B_{12} from ingested proteins, allowing it to bind to haptocorrin (HC). The B_{12}-HC complex is dissociated in the proximal small intestine due to the actions of pancreatic proteases and HCO_3^-. The released B_{12} can then bind to intrinsic factor (IF). The B_{12}-IF complex is delivered to the ileum, where it is internalized by endocytosis in enterocytes. Lysosomes dissociate the IF-B_{12} complex, allowing B_{12} to enter the cytosol, where it can exist in the free form or can bind to the transport protein, transcobalamin II (TCII). Cobalamin exits the cell either bound to TCII or as free B_{12} (via the MDR1 pump). Any free B12 leaving the cell binds to TCII in interstitial fluid.

Approximately 2 mg of cobalamin is normally stored in the liver, with another 2–3 mg stored in other tissues. With a total body store of ~5 mg and a daily requirement of 2 µg per day, a normal individual would require several years to become deficient in B_{12} if intake or absorption suddenly ceases. Nonetheless, cobalamin malabsorption can result from any disorder that chronically depresses intrinsic factor secretion (pernicious anemia), interferes with attachment of the cobalamin-intrinsic factor complex to the ileal receptors (severe celiac disease), or impairs uptake into blood (transcobalamin deficiency). Vitamin B_{12} is essential for the metabolism of carbohydrates, fats and proteins, the formation and regeneration of red blood cells, and maintenance of the central nervous system. Consequently, cobalamin deficiency typically results in ineffective erythropoiesis and megaloblastic anemia and can lead to irreversible neurologic and psychiatric abnormalities.

LIPID DIGESTION AND ABSORPTION

Dietary Lipids

Fats account for 30–50% of total caloric intake in the Western diet, which averages between 120 and 150 grams consumed per day. An additional 30–40 grams of fat enters the bowel each day from biliary secretions (largely lecithin and cholesterol), dequamated cells and dead bacteria. Newborn infants consume 3–5 times more lipid than adults, relative to body weight. The primary dietary lipids are triglycerides (triacylglycerol), cholesterol esters, phospholipids, and the fat-soluble vitamins. Triglycerides (TG) account for over 90% of dietary lipid. Most of the ingested TG contain long-chain fatty acids (palmitic, stearic, oleic, and linoleic acids) with a smaller proportion (about 10%) containing short- and medium-chain fatty acids (e.g., butyric acid). Dietary TG is derived from animal body and milk fats and from vegetable oils. Lipids, in general, are the most concentrated source of energy and calories in the diet, yielding more than twice as many calories as carbohydrates (9 calories per gram for lipids and 4 calories per gram for carbohydrates). Fats also make food more palatable and represent the primary nutrient responsible for postprandial satiety.

Triglyceride Digestion

The transformation of ingested fat into an emulsion of fine oil droplets and water is an early and important step in the digestion of lipids. This emulsification process begins with chewing and continues in the stomach due to antral peristalsis. The mechanically-induced emulsion is stabilized by the actions of emulsifying agents, such as ingested proteins and polysaccharides, that prevent the small dispersed lipid droplets from coalescing. Because a low pH inhibits emulsification, this process proceeds slowly in the stomach and is greatly accelerated in the proximal small intestine. In the alkaline environment of the duodenum, fats are emulsified by lecithin, bile salts, fatty acids, monoglycerides, lecithin, and proteins. These emulsifying agents act as detergents that physically disperse the fat into minute droplets. Prior to emulsification, fat globules have an average diameter of 1000 Å, which is reduced to 50 Å by emulsification. Emulsification facilitates TG digestion by increasing the total surface area available for the action of lipases, which act only at the surface of the lipid droplets.

Hydrolysis of lipids begins in the stomach through the actions of lingual (salivary) and gastric lipase. The cellular sources and characteristics of these enzymes are addressed in the preceding chapters that address the salivary glands (Chapter 2) and stomach (Chapter 4). Lingual and gastric lipases (often referred to as acidic lipases) exhibit optimal activity in an acidic (pH between 2 and 6) environment, are rapidly inactivated by pancreatic proteases, and do not require either bile salts or colipase to hydrolyze triglycerides. The actions of lingual and gastric lipases in the stomach are estimated to account for 10–15% of total fat digestion in healthy adult humans. However, in the setting of chronic pancreatitis, the chief cells of the stomach exhibit a compensatory increase in gastric lipase secretion that allows for the digestion of as much as 30% of ingested fat. Both lingual and gastric lipases also play a more important role in fat digestion in newborn infants, in which pancreatic lipase activity is not fully established. Breast-fed infants also benefit from the presence of a bile salt-stimulated milk lipase, normally found in human milk, which digests fats. Cow's milk does not possess this lipase. Human milk lipase is stable both in the acidic environment of the stomach and alkaline pH in the duodenum and jejunum. Unlike pancreatic lipase, whose hydrolytic actions are limited to TG, milk lipase hydrolyzes monoglycerides, diglycerides, cholesterol esters as well as TG. Bile salts not only activate milk lipase, but also protect it against the proteolytic action of pancreatic enzymes.

The presence of fat (particularly long-chain fatty acids) in the duodenum stimulates the release of cholecystokinin, which, in turn, causes the pancreas to secrete digestive enzymes into the bowel lumen (via the mechanisms depicted in Fig. 5-7 of Chapter 5). Among these are the lipolytic enzyme lipase and its cofactor, colipase, both of which play a major role in the hydrolysis of triglycerides. Pancreatic lipase (also known as glycerol ester hydrolase), which is secreted in an active form (rather than a proenzyme), is highly specific for TG, acts only at the oil-water interface, but is inactive in the presence of bile salts. Colipase must also be present for lipase to initiate lipid digestion. Unlike lipase, colipase is secreted by the pancreas in the form of a precursor, that is, procolipase. Procolipase is converted to colipase by trypsin. Colipase recognizes and binds to TG at the oil-water interface, a process requiring the presence of bile salts. Colipase also binds to lipase at a 1:1 molar ratio, thereby allowing TG hydrolysis to proceed. Colipase acts as an "anchor" for lipase attachment at the oil-water interface; without colipase, bile salts would fully occupy the oil-water interface and not allow the enzyme to adhere to the surface of fat droplets (Figure 7-16). Other physiologic advantages of the lipase-colipase interaction are that (1) the optimum pH for lipase action, which is normally 8, is reduced to 6 (near duodenal pH) when colipase binds to lipase and (2) the inactivation of lipase by bile salts is prevented when colipase displaces bile salts from the surface of oil droplets.

Triglyceride is acted upon by pancreatic lipase to yield two fatty acids (hydrolyzed at the 1 and 3 positions) and a 2-monoglyceride (Figure 7-17). The lipase sequentially attacks the two outer bonds of the triglyceride molecule, producing, at first, a diglyceride and one fatty acid, then two fatty acids and a monoglyceride. The monoglyceride can be converted into glycerol and a fatty acid by pancreatic lipase. However, the latter reaction requires shifting of the ester bond to the 1 position. This isomerization proceeds at a slower rate than the absorption of 2-monoglycerides by the enterocyte. Gastric lipase differs from its pancreatic counterpart by preferentially hydrolyzing the fatty acid at the 1 position, yielding a diglyceride and a fatty acid as its sole products.

The rate of triglyceride hydrolysis by lipase is extremely rapid, owing to the large excess of enzyme secreted by the pancreas in response to a meal. It has been estimated (based on the amount of lipase in intestinal contents) that humans have the capacity to digest 140 grams of fat per minute. Since daily fat ingestion rarely exceeds 150 grams, there is over a thousand-fold excess of lipase in the intestinal lumen. Up to 80% of the total fatty acids present in triglyceride are liberated by the time the fat has reached the middle of the duodenum.

Phospholipid Digestion

Dietary phospholipids (e.g., lecithin, a mixture of phospholipids) are hydrolyzed in the intestinal lumen by the pancreatic enzyme phospholipase A_2. The phospholipase is secreted in pancreatic juice as an inactive proenzyme (prophospholipase A_2), which is activated in the lumen by trypsin. Phospholipase A_2 hydrolyzes lecithin at the fatty acid ester bond at position 2 to yield lysolecithin and a fatty acid (Fig. 7-17). The enzyme cannot cleave the fatty acyl ester at position 1 of the phospholipid. Other phospholipids not contained in lecithin (e.g., phosphatidylserine) are also substrates for phospholipase A_2. The activity of phospholipase A_2 is dependent on luminal calcium, requires bile salts for optimal activity, and has a pH optimum of 7.5.

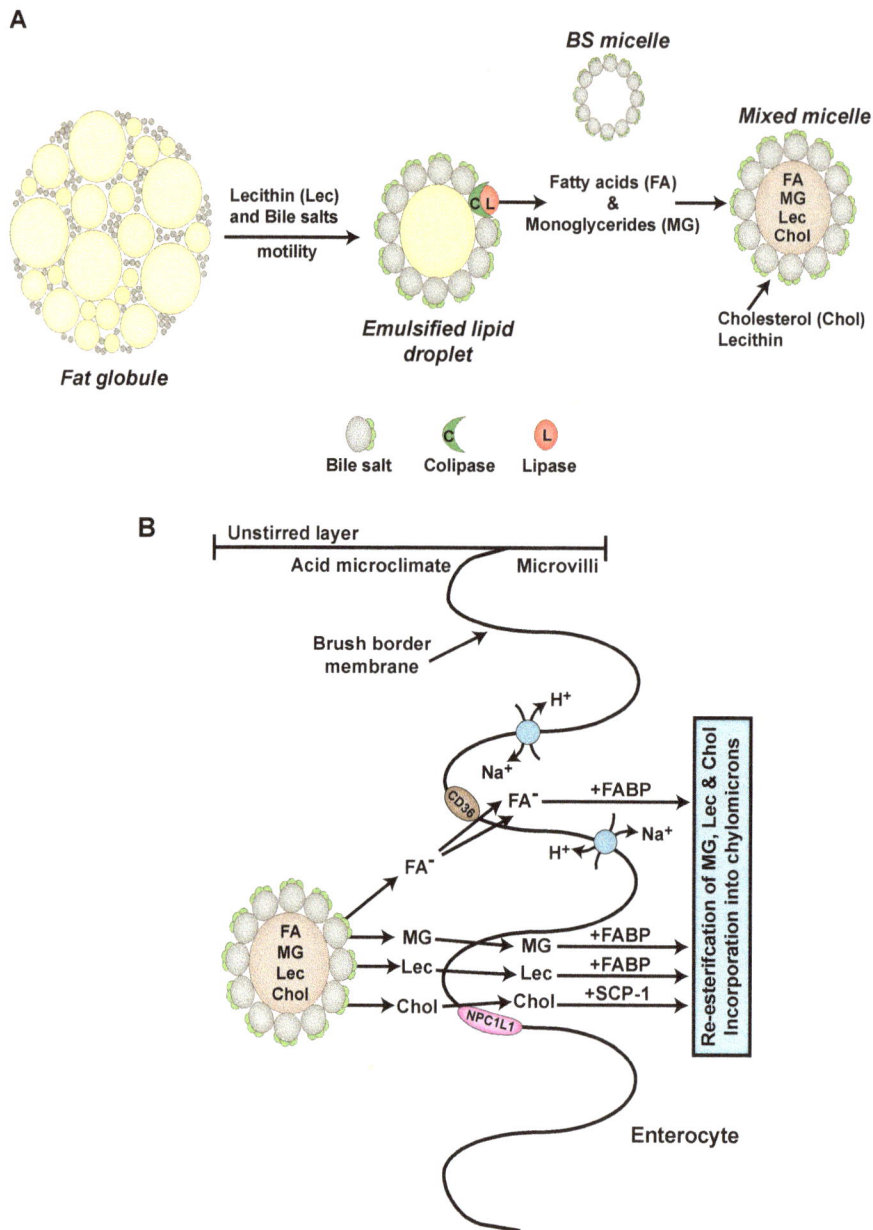

FIGURE 7-16: Emulsification, solubilization, and absorption of ingested fats in the small intestine. A. Fats enter the gut lumen as large fat globules, which are transformed into smaller lipid droplets via the agitation (mechanical forces) provided by peristalsis and the emulsifying actions of lecithin and bile salts. Emulsification facilitates triglyceride digestion by increasing the total surface area available for the action of pancreatic lipase, which is anchored to the surface of the lipid droplets by colipase. The products of lipase action (fatty acids,

monoglycerides) as well as de-esterified cholesterol and lecithin (phospholipids) enter (and are solubilized within) the hydrophobic domain of simple bile salt micelles to produce mixed micelles. B. The mixed micelles ferry the lipid digestion products to the brush border membrane of enterocytes, allowing their access to the cell interior by either simple diffusion or with the aid of protein carriers such as CD36 (fatty acids) or NPC1L1 (cholesterol). The acidic microenvironment created on the luminal aspect of the brush border membrane by the Na^+-H^+ exchanger leads to the protonation of fatty acids (FA●), which enhances their passive diffusion into the cell. Once within the cell, the products of lipid digestion bind to carrier proteins (FABP, SCP-1), which solubilizes the lipids and transfers them to the endoplasmic reticulum for re-esterification and incorporation into chylomicrons.

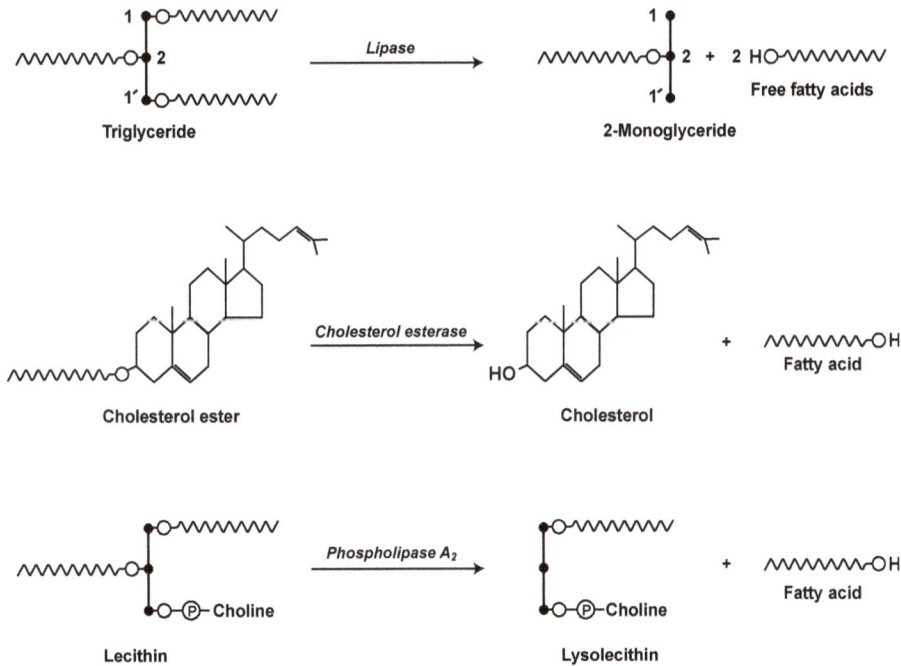

FIGURE 7-17: Mechanisms of hydrolysis of the three major dietary lipids by pancreatic enzymes in the small intestine. Pancreatic lipase, also called glycerol ester hydrolase, mediates the hydrolysis of triglyceride to yield two fatty acids (hydrolyzed at the 1 and 3 positions) and a 2-monoglyceride. The enzyme cholesterol esterase, also called cholesterol ester hydrolase, hydrolyzes cholesterol esters to yield cholesterol and a free fatty acid. Phospholipase A_2 (PLA$_2$) hydrolyzes lecithin to release the fatty acid with an ester bond at position 2 to yield lysolecithin (lysophosphatidylcholine). PLA$_2$ cannot cleave the fatty acyl ester at position 1 of phospholipids.

Cholesterol Digestion

Cholesterol is found in the human diet principally in the form of esters. Intestinal hydrolysis of cholesterol esters to cholesterol and free fatty acids is catalyzed by the enzyme cholesterol esterase, also called cholesterol ester hydrolase (Fig. 7-17). This enzyme is present in pancreatic juice and acts primarily on unsaturated fatty acids. Intraluminal hydrolysis of cholesterol esters is of importance, since cholesterol is absorbed as free cholesterol rather than in the ester form. Optimal cholesterol esterase activity is achieved in the presence of certain bile salts such as taurocholate and taurochenodeoxycholate. Cholesterol esterase polymerizes in the presence of these trihydroxy bile acids, which greatly enhances enzyme activity. The enzyme also functions as a nonspecific esterase that cleaves fatty acid linkages in a variety of lipid substrates, including triglycerides, 2-monoglycerides, and fatty acyl esters of vitamins A, D, and E.

Solubilization of Lipid Digestion Products (Fig. 7-16)

Before the products of lipid digestion can be absorbed by the intestine, they must be solubilized in the aqueous phase of the lumen contents. This "solubilization" is achieved by incorporating the lipolytic products into bile salt micelles. At the concentrations found in bile and the intestinal lumen, bile salts form cylindrical aggregates (simple micelles) in which the hydrophilic portion of the molecule faces the aqueous phase of the solution while the hydrophobic portion faces inward, away from the aqueous phase (see Fig. 6-5 in Chapter 6). Lipids such as cholesterol, lysolecithin, monoglycerides, and free fatty acids (particularly long-chain) enter the lipid-soluble interior of the bile salt micelle, forming larger mixed micelles. Micellar solubilization increases the effective concentration of long chain fatty acids in the aqueous phase near the brush border membrane by a factor of 1,000,000, when compared to the free monomer dissolved in water. Although the micelle (70 to 400 Å, diameter) is much larger than a fatty acid molecule, it profoundly increases (>100,000 times) the rate of delivery of fatty acids to the enterocyte surface owing to the ability of the particle to traverse the unstirred water layer adjacent to the brush border and diffuse among the microvilli. Micellar solubilization of monoglycerides and fatty acids also removes them from the oil-water interface at which pancreatic lipase acts, thereby enhancing digestion of the remaining triglyceride.

Cellular Uptake of Lipolytic Products

Mixed micelles formed in the intestinal lumen must interact with three diffusion barriers before the lipid contents can enter the enterocyte: (1) the unstirred water layer overlying the enterocyte, (2) the mucous coat covering the brush border membrane, and (3) the lipid bilayer membrane that makes up the brush border. Both the unstirred water layer and mucous coat significantly reduce the movement of micelles (and long-chain fatty acids) relative to free diffusion in water. Although micelles per se do not permeate the brush border membrane, their lipid contents (monoglycerides, fatty acids, cholesterol, lysolecithin) very rapidly enter the cell. Once the lipid contents of the micelle enter the intestinal cell, the emptied bile salt micelle picks up another load of lipid hydrolysis products for delivery to the enterocyte.

It has long been assumed that the hydrophobicity of the products of lipid hydrolysis allows for their partitioning into the lipid bilayer of the cell membrane and subsequent entry into the cell by passive non-ionic diffusion. With such a mechanism, the rate of exchange of lipolytic products between the micelles and lipid membrane of the cell would be determined by the solubility of the lipolytic product in the cell membrane and its concentration gradient across the membrane. In this regard, the acidic microenvironment created on the luminal aspect of the brush border membrane by the Na^+-H^+ exchanger (NHE) appears to enhance the entry of fatty acids into the cell because protonated fatty acids are more lipid soluble. While passive diffusion of lipid hydrolytic products remains a viable explanation for their entry into enterocytes, there is mounting evidence for the involvement of carrier-mediated transport systems in the transfer of lipid hydrolytic products across the brush border membrane. Transport proteins have been identified that appear to facilitate the uptake of long-chain fatty acids (e.g., CD36) and cholesterol (e.g., NPC1L1 transporter) by enterocytes. However, the relative importance of the facilitated versus passive diffusion mechanisms to overall absorption of long chain fatty acids and cholesterol remains unclear (Fig. 7-16).

While long-chain fatty acids and cholesterol must be incorporated into bile salt micelles before absorption, solubilization in micelles is not a requirement for the absorption of short- (C2 to C4) and medium- (C6 to C10) chain fatty acids. Short- and medium-chain fatty acids and the triglycerides of those acids are sufficiently water soluble that they can reach the enterocyte by simple diffusion; that is, passage through the unstirred water layer and mucous coat is not the rate-limiting step in their absorption. This is demonstrated by the fact that approximately 30% of an oral dose of medium-chain triglyceride is absorbed intact. Nonetheless, a proportion of the medium-chain fatty acids in the intestinal lumen is incorporated into micelles and is transported to the enterocyte membrane in the same manner as long-chain fatty acids.

Intracellular Metabolism of Lipid Digestion Products
Monoglycerides and Fatty Acids
After monoglycerides and fatty acids cross the brush border membrane and enter the cell, they bind to specific fatty acid binding proteins (FABP) that solubilize and transfer the lipids to

the smooth endoplasmic reticulum (SER) for resynthesis into triglycerides (Figure 7-18). The endoplasmic reticulum, which is engorged with lipid following a fatty meal, contains all of the enzymes and cofactors necessary for re-esterification of fatty acids and monoglycerides into triglycerides. Two metabolic pathways are involved in the synthesis of triglycerides in enterocytes: the monoglyceride acylation pathway and the α-glycerophosphate (phosphatidic acid) path-

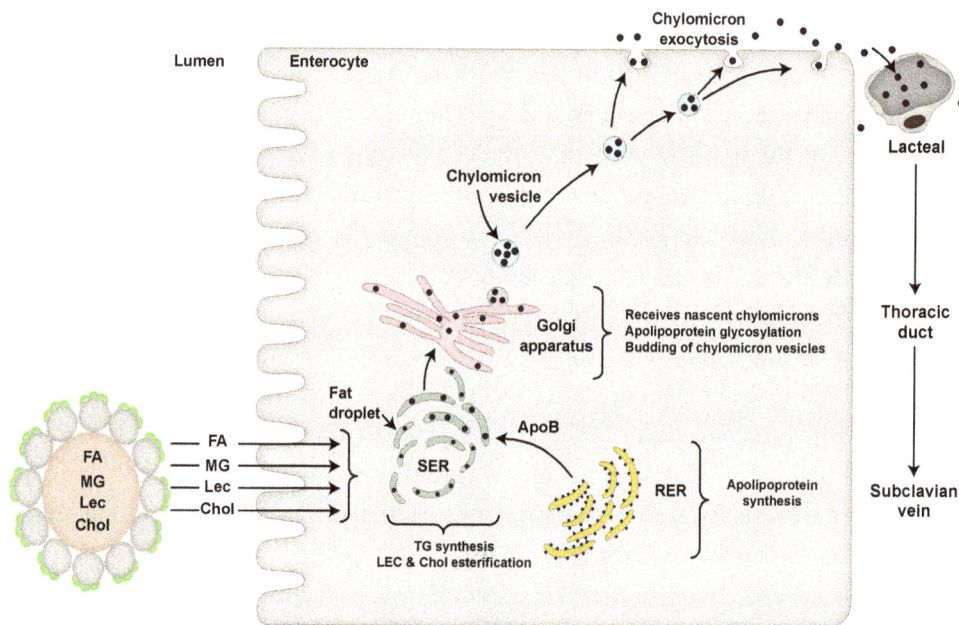

FIGURE 7-18: Cellular processes involved in the production of chylomicrons from lipids absorbed by enterocytes. Once fatty acids (FA), monoglycerides (MG), lecithin (Lec) and cholesterol (Chol) enter the enterocyte from gut lumen, they are transferred to the smooth endoplasmic reticulum (SER), where droplets of resynthesized triglycerides and newly esterified cholesterol accumulate to yield lipid droplets. The surface of the lipid droplet is coated with newly synthesized phospholipid and surrounded by apolipoprotein B (apoB), which is synthesized in the rough endoplasmic reticulum and transported to the SER. Vesicles in the SER carry the lipoprotein particles to the Golgi apparatus, where the surface apoB molecules are glycosylated and additional apoproteins (e.g., apoA1) are added to the particle surface. Vesicles containing the fully processed chylomicrons are formed via budding from the Golgi membrane and subsequently released into the cytosolic compartment. The vesicles then move toward, and fuse with, the basolateral membrane to release their contents into the sub-epithelial interstitial space. The chylomicrons then enter the lamina propria and accumulate in the terminal lacteals, which transfer the lipoprotein particles to the blood stream via the thoracic duct, and subclavian vein.

way. Both pathways utilize absorbed fatty acids with chain lengths greater than twelve carbons; smaller fatty acids are not re-esterified into triglyceride but diffuse directly into portal blood. The monoglyceride acylation pathway is the more important pathway for triglyceride resynthesis during fat absorption. This involves the activation of absorbed fatty acids by the enzyme fatty acid CoA ligase and subsequent attachment of the activated fatty acid to monoglyceride, a reaction catalyzed by the enzyme monoglyceride acyltransferase. The third fatty acid is added by diglyceride acyltransferase to complete the resynthesis of triglyceride. In addition to the monoglyceride acylation pathway, the enterocyte can produce triglyceride by esterification of absorbed fatty acid with glycerol phosphate derived from intracellular metabolism of glucose. The phosphatidic acid (glycerol phosphate-fatty acid ester) is dephosphorylated and then esterified with an additional fatty acid to produce a triglyceride. The phosphatidic acid pathway contributes to enterocyte triglyceride formation during fasting.

Cholesterol

Cholesterol is almost insoluble in water and therefore enters the intestinal cell less readily than long-chain fatty acids and monoglycerides. Only half of the luminal cholesterol enters the mucosal cell. Once within the cell, dietary cholesterol binds to a sterol carrier protein (e.g., SCP-1), which transfers the cholesterol to the SER, where it is re-esterified mainly with oleic acid, a reaction catalyzed by cholesterol esterase (Fig. 7-17). Some absorbed cholesterol is not esterified in enterocytes and the relative amount of cholesterol that is esterified appears to depend on the amount consumed in the diet, with more free cholesterol released from the enterocyte as cholesterol consumption decreases.

Phospholipids

The products of pancreatic enzyme digestion of lecithin are fatty acids and lysolecithin; the latter is less efficiently absorbed by the enterocytes. Once inside the absorptive cell (Fig. 7-16), lysolecithin binds to FABP, which ensures its transfer to the SER, where the lysolecithin is acylated by microsomal enzymes to form esters (e.g., lecithin). The rate of re-esterification of phospholipids is related to its need in the assembly and transport of chylomicrons. Other phospholipids (e.g., sphingolipids) are degraded in the enterocyte. Phospholipid is essential for normal structure and function of membranes such as the brush border membrane. A large fraction of absorbed phospholipid exits the cell in chylomicrons; however, some of the dietary phospholipid is incorporated into the cellular membranes of the enterocyte.

Chylomicron Formation and Transport (Fig. 7-18)

As fat absorption proceeds, droplets of resynthesized triglycerides fill the smooth endoplasmic reticulum (SER), with newly esterified cholesterol added to the lipid droplet. The surface of the lipid droplet (called a pre-chylomicron or nascent chylomicron) is covered with a monolayer of newly synthesized phospholipid and surrounded by apolipoprotein B (apoB), which is synthesized in the rough endoplasmic reticulum and transported to the SER. [ApoB, as well as the microsomal triglyceride transfer protein (MTTP), which promotes lipoprotein synthesis by shuttling lipid within the endoplasmic reticulum to the apoB molecule, are considered key regulators of lipoprotein synthesis.] Vesicles in the SER (pre-chylomicron transport vesicles) that contain the nascent chylomicrons carry the lipoprotein particles to the Golgi apparatus, where the surface apoB molecules are glycosylated and additional apoproteins (e.g., apoA1) are added to the particle surface. Vesicles containing the fully processed chylomicrons are formed via budding from the Golgi membrane and subsequently released into the cytosolic compartment. The vesicles then move toward, and fuse with, the basolateral membrane to release their contents into the sub-epithelial interstitial space. The entire process of chylomicron formation and release, from cell uptake of fat to release into the lateral intercellular spaces, requires only 12 to 15 minutes. The chylomicrons must then traverse the lamina propria, and finally accumulate in the lymph of the terminal lacteals, to which chylomicrons impart a milky appearance. The chylomicrons ultimately gain access to the blood stream via the thoracic duct, which empties into the subclavian vein. The particles are too large to cross the endothelial cell fenestrae in blood capillaries, leaving lymphatic vessels as the sole route for chylomicron transfer to the blood stream.

Chylomicrons, the largest of the lipoprotein particles that appear in blood, have a diameter ranging between 750 and 6000 Å. The variable size of chylomicrons appears to be related to the rates of fatty acid absorption and triglyceride resynthesis. Their chemical composition is approximately 90% triglyceride, 7% phospholipid, 2% cholesterol, and 1% protein. The inner core of the chylomicron contains almost all the triglyceride and most of the cholesterol. Phospholipid covers 80% to 90%, while protein covers 10% to 20% of the chylomicron surface.

Enterocytes also form smaller lipoprotein particles (300 to 800 Å diameter) called very low density lipoproteins (VLDL), which appear in intestinal lymph during fasting. The mechanism of VLDL production is independent of chylomicron synthesis, although ApoB is needed for the production of both particles. Both chylomicrons and VLDL are triglyceride rich, but VLDL contains more phospholipid, cholesterol and protein than chylomicrons. The liver is the major source of VLDL appearing in blood, however, the intestine accounts for as much as 40% of VLDL in blood during fasting.

The partition of absorbed fatty acids between portal venous blood and intestinal lymph is largely determined by the chain length of the fatty acid. Long-chain fatty acids (>14C) are transported mainly as resynthesized triglyceride in the chylomicrons and VLDL of lymph. The lipoproteins (VLDL, chylomicrons) cannot gain access to the blood circulation via intestinal capillaries because the fenestral openings in the capillary endothelium (500 Å diameter) are too small to allow permeation by the lipid particles. As much as 20% of the long-chain fatty acids absorbed by the small bowel enter the portal vein bound to plasma proteins such as albumin. Fatty acids of intermediate chain length (10 to 14C) are removed from the intestine both in chylomicrons and by diffusion into blood. Because short-chain fatty acids (<8C) are water-soluble, they readily diffuse from the enterocyte into blood capillaries, which is their primary route of transport out of the intestine.

Absorption of Fat-Soluble Vitamins

The fat-soluble vitamins (A, D, E, and K) are absorbed primarily in the proximal small intestine by mechanism(s) that require bile salts. These vitamins are all relatively nonpolar molecules and must be incorporated into micelles for efficient absorption. After entry into the mucosal cell, the fat-soluble vitamins become associated with chylomicrons, VLDL, or both, from which they gain access to the blood circulation via lymph.

Vitamin A (retinol) found in animal-based foods (liver, dairy products, eggs) exists as reti-nyl esters, such as retinyl palmitate. These esters are hydrolyzed by pancreatic lipase and cho-lesterol esterase in the duodenal lumen. In fruits and vegetables, retinol exists as provitamin A carotenoids (mainly β-carotene and α-carotene), which are metabolized into retinol after absorption by enterocytes. The β-carotene molecule is split by an intracellular enzyme (dioxy-genase) to yield two molecules of retinaldehyde, which is subsequently reduced to retinol. Reti-nol is required for the formation of visual (retinol) pigments, and it is necessary for the normal growth and proliferation of epithelial cells. Enterocytes take up retinol using specific transport proteins, such as Stimulated by Retinoic Acid 6 (STRA6). Since vitamin A is fat soluble, retinol esters and retinol have to be re-esterified (usually with palmitic acid) by the enterocyte before incorporation into chylomicrons with other lipids. Vitamin A deficiency, a public health problem in most developing countries, occurs mainly in children with poor nutrition and/or fat malabsorption.

Vitamin D is produced endogenously as a consequence of skin exposure to ultraviolet light; however, significant quantities of both vitamins D_2 (ergocalciferol) and D_3 (cholecalcif-erol) are absorbed in the small intestine. Vitamin D_3 occurs naturally in high concentrations in fish liver oils, and vitamins D_2 and D_3 are added to fortify certain foods (e.g., cereals). Vitamin D absorption in the proximal intestine takes place via a passive process that may involve transport proteins in the brush border membrane, similar to vitamin A absorption. Most of the vitamin D absorbed appears in intestinal lymph incorporated in chylomicrons. The most biologically active form of vitamin D, $1,25(OH)_2D_3$, plays an important role in the regulation of plasma calcium levels; a major effect of this vitamin D metabolite is stimulation of intestinal calcium absorption (see Fig. 7-8).

Vitamin E actually represents a group of lipid-soluble derivatives with varying biologic potency; the most active of these compounds is α-tocopherol. The richest sources of vitamin E in the diet are seeds and vegetable oils. A major physiologic function of α-tocopherol is to prevent the oxidation of unsaturated fatty acids in cellular membranes. Intestinal absorption of vitamin E requires both pancreatic juice and bile. Pancreatic enzymes facilitate vitamin E absorption by hydrolyzing esters of the vitamin and by providing lipolytic products for enhanced solubilization of the vitamin. A variety of lipid transport proteins (e.g., NPC1L1, CD36) have been implicated in the transfer of vitamin E across the brush border membrane. Once in the enterocyte, vitamin E is incorporated into chylomicrons and appears in intestinal lymph.

Vitamin K represents a group of compounds that are required for the activation of clotting factors such as prothrombin and factors II, VII, IX and X. Dietary vitamin K is generally categorized into two forms: phylloquinone (vitamin K_1) and menaquinones (vitamin K_2). Vitamin K_1 accounts for approximately 90% of dietary vitamin K, and is enriched in green leafy vegetables, beans, and certain plant oils. This form of vitamin K is largely absorbed in the proximal small intestine. Vitamin K_2 is produced by bacteria in the distal small intestine, either by de novo synthesis or via metabolic conversion of vitamin K_1 to K_2. Vitamin K_1 is absorbed by enterocytes via a NPC1L1-dependent pathway (similar to cholesterol) while vitamin K_2 enters the enterocytes in the distal intestine via a passive mechanism. Once within the mucosal cell, both forms of vitamin K are incorporated into chylomicrons and then transferred to the lymphatics.

MOTILITY

Coordinated contractions of the external muscle layers of the small bowel optimize the processes of digestion and absorption by (1) mixing the chyme with digestive secretions, (2) repeatedly exposing the lumen contents to the digestive and absorptive surface of the mucosa, and (3) moving the chyme down the intestine. The inner circular muscle layer is considered to play a more important role in the mixing and propulsion of chyme. Intestinal motility represents a coordinated series of contractions that serve to assure the slow passage of chyme through the bowel so that the residue of one meal leaves the ileum as another meal is being ingested. A sphincter at the ileocecal junction regulates the passage of chyme into the colon and, more importantly, serves to prevent colonic reflux of bacteria into the small intestine.

Electrical and Contractile Activity of Smooth Muscle

As in the stomach, contractions of the small intestine are ultimately governed by the basal electrical rhythm (slow waves) of the smooth muscle cells. Intestinal slow waves are generated by the interstitial cells of Cajal (ICC) located in the myenteric region (ICC-MY) and relayed to the smooth muscle cells via gap junctions. The smooth muscle slow waves per se do not evoke muscular contractions; instead, they govern the rate at which spike activity and muscle contractions can occur. Spike potentials, which occur only during the depolarization phase of the slow waves, lead to contraction of the muscle. The strength of contraction is directly proportional to the frequency of the spike potentials. The relationships between slow waves, spike potentials, and muscle contractions are illustrated in Fig. 1-7 of Chapter 1.

As evident in the stomach, where the slow waves propagate aborally from the pacemaker region in the upper corpus to the pylorus, intestinal slow waves generated in the proximal duodenum propagate aborally to the terminal ileum. However, the characteristics of intestinal slow waves (frequency, velocity, and amplitude) are in stark contrast to those of gastric slow waves. While gastric slow waves have a constant frequency dictated by the gastric BER (3 cycles/min), intestinal slow waves display a decreasing frequency gradient from proximal duodenum (12 cycles/min) to terminal ileum (8 cycles/min). In the stomach, both the velocity and amplitude of the slow waves increase as they progress aborally (see Fig. 4-9 of Chapter 4). By contrast, the velocity and amplitude of slow waves decrease along the small intestine; the steepest decrements occurring in the upper small intestine.

Although technically difficult, there have been comprehensive assessments of the propagation of slow waves along the entire small intestine. It is generally agreed that a pacemaker region (where the slow waves exhibit the highest frequency) is located in the proximal duodenum close to the pylorus. However, the mechanisms by which the slow wave frequency decreases from 12 cycles/min in the duodenum to 8 cycles/ min in the terminal ileum is still a subject of debate. The crux of the issue lies in why the measured frequency decrease is stepwise, rather than linear, particularly in the duodenum and jejunum (Figure 7-19).

The most widely accepted explanation for the existence of slow wave frequency plateaus is based on coupled oscillators, analogous to the conduction system of the heart. In the heart

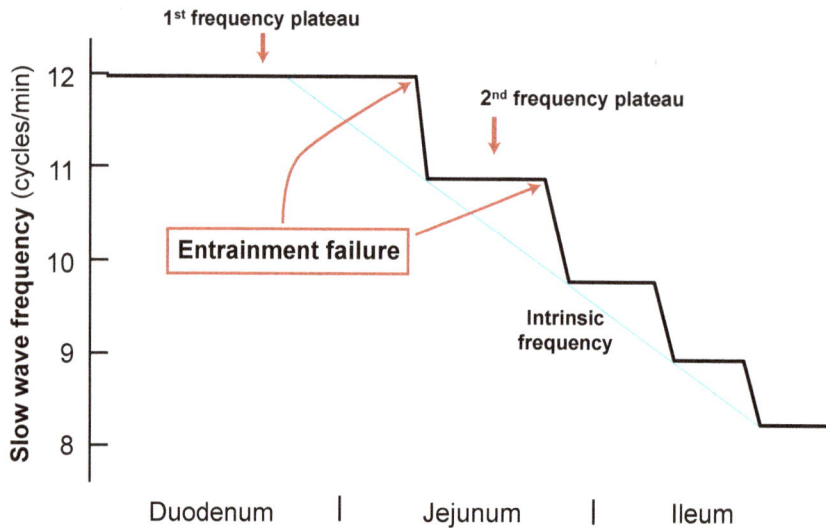

FIGURE 7-19: The frequency of slow waves decreases along the small intestine; the decrement is not linear, but occurs in a stepwise fashion. Slow wave frequency in the upper duodenum is 12 cycles/min and remains at this rate as far downstream as the mid jejunum, generating a "frequency plateau." Several more progressively lower frequency plateaus develop in the remainder of the small bowel, culminating in 8 cycles/min at the terminal ileum. The generation of frequency plateaus is attributed to the presence of coupled oscillators (interstitial cells of Cajal) of declining intrinsic frequencies along the small intestine (blue line). The highest frequency oscillator in the upper duodenum assumes the role of pacemaker and entrains downstream oscillators of lower frequencies to follow its higher frequency, thereby creating a frequency plateau. When this entrainment fails (e.g., mid jejunum), the downstream lower frequency oscillator takes over the role of dominant pacemaker and generates another plateau of lower frequency. This sequence of events continues until a slow wave frequency of 8 cycles/min is reached in the terminal ileum.

the, the sinoatrial (SA) node serves as a dominant pacemaker firing action potentials at a rate of 60–100/min. The SA node entrains the lower frequency oscillators of the AV node (40–60/min) and Purkinje system (20–40/min) to its frequency, thereby setting heart rate at 60–100 beats/min. In the small intestine, the myenteric region of the muscle layers is populated by coupled oscillators (ICC-MY) that exhibit a gradually lower intrinsic frequency from duodenum to terminal ileum. The ICC-MY of the upper duodenum have the highest frequency and serve as pacemakers which can entrain adjacent downstream oscillators of lower intrinsic frequency. Accordingly, the dominant ICC-MY pacemaker would impose its frequency on downstream ICC via cell-to-cell electrical coupling, thereby generating a frequency plateau. The slow wave frequency would remain synchronized along the bowel until a region is reached downstream where the intrinsic

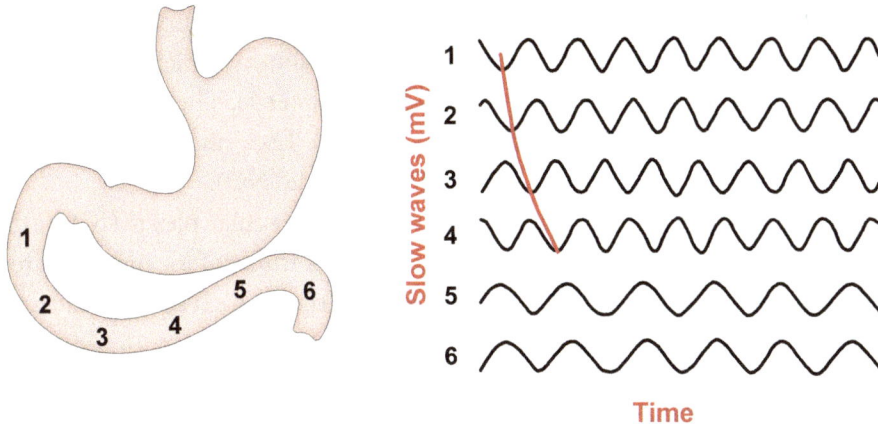

FIGURE 7-20: Slow waves recorded from multiple site along the duodenum and jejunum. Sites 1–4 correspond to sites along the duodenum and upper jejunum and the red line indicates the "phase lag" or aboral progression of a slow wave along a frequency plateau (1st frequency plateau in Fig. 7-19). Sites 5 & 6 show the decrease in slow wave frequency noted downstream to entrainment failure (2nd frequency plateau in Fig. 7-19).

frequency of the ICCs is too low to be entrained by the upstream higher order oscillator. At this point, entrainment fails and the "freed" lower order oscillator would assume the role of dominant pacemaker and create another lower frequency plateau by a similar entrainment mechanism.

The aboral gradient in the frequency plateaus ensures that chyme moves from a region of higher potential contractile frequency to a region of lower potential frequency. However, the question arises as to how chyme moves along a given frequency plateau, where the frequency of slow waves remains constant. The answer lies in the observation that the slow waves along a frequency plateau do not occur simultaneously at all loci, that is, there is an aborally oriented phase lag. This phase lag is created by the time required for the propagation of the pacemaker frequency (Figure 7-20).

Neural Regulation of Motility

The mixing and propulsive patterns of small intestinal motility are regulated by intrinsic and extrinsic neuronal input. The contractile activity of the circular and longitudinal muscle layers is under the influence of the enteric nervous system (primarily the myenteric plexus). The enteric nervous system (ENS) operates autonomously; independently of any extrinsic innervation. For example, the ENS contains all of the necessary components to elicit a reflex in response to luminal contents, that is, sensory neurons, interneurons, and motor neurons. The motor neurons of the ENS are generally characterized as excitatory or inhibitory, which elicit smooth muscle contraction or relaxation, respectively (see Table 1-3 of Chapter 1). The primary excitatory neurotransmitters are acetylcholine (ACh) and substance P (SP). The primary inhibitory neurotransmitters are vasoactive intestinal peptide (VIP) and nitric oxide (NO).

Recently, a population of ICC located in the deep muscular plexus (ICC-DMP) have been implicated in transduction of neurotransmitter signals from enteric motor neurons to circular smooth muscle. These ICC-DMP are analogous to the gastric ICC-IM and are found in close proximity to nerve varicosities.

The extrinsic nerves serve to simply modulate the patterns established by the enteric nervous system. This form of regulation is achieved via parasympathetic and sympathetic inputs to the myenteric plexus. Viewed simply, the vagus nerve contains two populations of preganglionic parasympathetic fibers; some synapsing with excitatory neurons, others synapsing with inhibitory neurons. Although vagal stimulation generally elicits contractions of the intestine, suppression of ongoing spontaneous contractions has also been observed. The sympathetic supply to the intestinal musculature is derived from preganglionic axons leaving the spinal cord (at T-9 to T-10) and synapsing in the celiac and superior mesenteric ganglia (see Fig. 1-6 of Chapter 1). The postganglionic fibers issuing from the celiac ganglia innervate the upper duodenum, whereas those leaving the superior mesenteric ganglia innervate the remainder of the small bowel. The adrenergic fibers from both ganglia synapse primarily on neurons within the myenteric plexus, although a few do terminate on smooth muscle cells. Sympathetic activation consistently results in inhibition of intestinal contractions at non-sphincteric sites, either by preventing the release of acetylcholine from intramural neurons or by activating adrenergic (β_2) receptors on the smooth muscle cells. At the ileocecal junction, sympathetic activation causes smooth muscle contraction.

The involvement of extrinsic autonomic nerves in the motility patterns observed in the fed or fasted states is minimal. However, the extrinsic nerves are important in modifying motor activity via reflex arcs originating either in the gastrointestinal tract or elsewhere. Importantly, most of the fibers in both the parasympathetic and sympathetic nerves are afferent. They receive chemical and mechanical sensory input from intestines and relay information through spinal or CNS reflexes back to the myenteric plexus.

Functional Patterns of Intestinal Muscle Contractions

There several different patterns of contractions that allow the small intestine to facilitate various digestive and non-digestive functions. In the postprandial state, coordinated contractions/

relaxations of adjacent segments of the intestine serve to mix chyme with digestive secretions as well as propel it slowly aborally for a few centimeters. Between meals, a coordinated complex of similar (but more powerful) contractions migrates down the small intestine and propels any undigested/unabsorbed material, including bacteria, into the colon. In addition, the presence of toxic materials in the lumen can induce even more powerful contractions that propagate either aborally (e.g., diarrhea) or orally (e.g., emesis). The latter contractions serve to rapidly clear the intestine of noxious material. All of these functional contractions are based on a basic integrated neuronal pattern of firing within the myenteric plexus ENS that regulates contraction and relaxation of circular smooth muscle. The role of longitudinal smooth muscle in these motor patterns is not entirely clear.

The myenteric plexus contains both excitatory and inhibitory neurons. Excitatory neurons, which are predominantly cholinergic (release ACh), contract smooth muscle, while inhibitory neurotransmitters are predominantly nitrergic (release NO) relax smooth muscle. Under basal conditions, the ENS exerts an inhibitory effect on intestinal motility, presumably due to tonic activity of inhibitory neurons and quiescence of excitatory neurons (Figure 7-21, A). This is exemplified by the increase in intestinal tone when the neural activity within the myenteric plexus is compromised by pharmacologic blockade (local anesthetics). In general, two stereotypical contractile events are the basis for all of the functional motility patterns that occur in the intestine: stationary contractions and peristaltic waves.

A common contractile event in the small intestine is a *stationary contraction* in which one segment contracts while adjacent segments on either side remain quiescent. These types of contractions can be simple isolated contractions or, more frequently, cycles of contractions and relaxations in the same segment, that are referred to as segmentation. The basic ENS firing pattern involves removal of the tonic inhibitory input, while simultaneously activating excitatory input, to a narrow band of circular muscle (Figure 7-21 B). These local contractions that does not spread within the syncytium to adjacent segments, since the latter are still under the influence of the inhibitory neurons. This contractile pattern cycles between adjacent segments during segmentation. Specifically, a small segment of bowel (<2 cm in length) contracts while adjacent segments remain quiescent. Subsequently, the contracted segment relaxes while the previously quiescent adjacent segments contract. This segmentation pattern results in propulsion of material both proximally and distally from the site of contraction and serves to mix luminal contents.

Another motor pattern consists of a wave of contraction coupled to relaxation that serves to propel luminal contents. This propulsive contractile wave is referred to as *peristalsis*. As with isolated contractions, peristalsis primarily involves circular muscle contractile activity that is

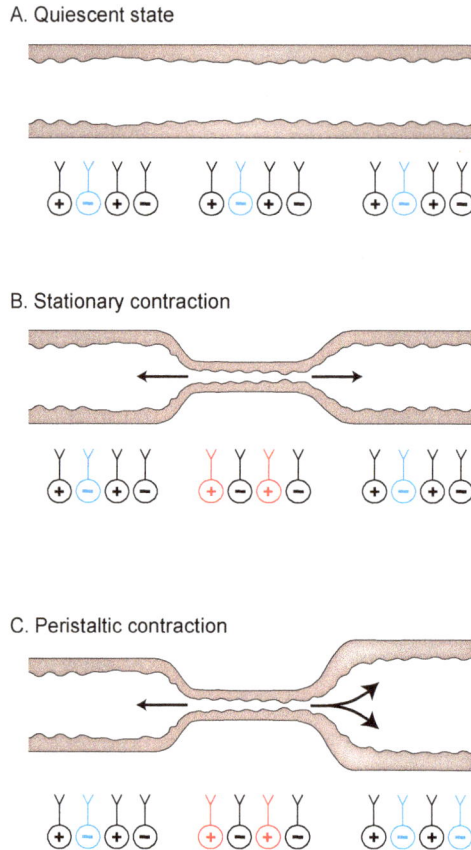

FIGURE 7-21: Motility responses of an intestinal segment (≈1 cm) to the relative activity of excitatory (ACh, SP) and inhibitory (VIP, NO) motor neurons of the ENS. Inactive inhibitory (−) or excitatory (+) neurons are depicted in black, active inhibitory (−) neurons are depicted in blue, while active excitatory neurons are depicted in red. A. In the basal quiescent state, some of the inhibitory neurons are active, maintaining the intestine in a relatively relaxed state. B. During a stationary contraction (mid-portion of segment), inhibitory neurons are inactivated (black) and excitatory neurons are activated (red). Adjacent sections to either side of the contracted unit remain under the influence of active inhibitory neurons (blue). Thus, if any chyme is present in the lumen, it will be forced in both directions. C. During a peristaltic contraction, a contraction (mid-portion of segment) is a result of inactivation of inhibitory neurons (black) and activation of excitatory neurons (red). Depending on the specific input from neurons of the ENS, one of the adjacent units remains relaxed (one blue neuron; left side), while the other one dilates further due to enhanced activity of inhibitory motor neurons (two blue neurons; right side). Thus, if any chyme is present in the lumen, it will preferentially move to the more relaxed region of the segment. The neuromuscular units depicted are repeated along the intestine (much like "beads on a string") allowing for the coordination of the two major functional motility patterns: segmentation and peristaltic waves.

coordinated by a network of excitatory and inhibitory nerves of the myenteric plexus (Figure 7-21 C). The ENS suppresses tonic inhibitory input and enhances excitatory input to a narrow band of circular muscle. However, one of the adjacent segments, rather than simply remaining quiescent, will relaxes further due to an enhancement of inhibitory input from the ENS. Compression of the lumen in the contracting segment tends to propel contents into the adjacent segment that is relaxed and thus, offers less resistance to flow. In general, activation of the excitatory and inhibitory neurons is coordinated such that the peristaltic wave propels intestinal contents in the aboral direction. However, the polarity can be reversed with propulsion occurring orally (retropulsion) in certain situations, that is, emesis.

Interdigestive Pattern

As first addressed in Chapter 4, the migrating motility complex (MMC) is a major contractile motor pattern of the gastrointestinal tract between meals. Although most of the contractile activity originates in the gastric antrum, approximately 30% of the MMC are of duodenal origin. Irrespective of its origin, the MMC sweeps down the length of the small intestine with most of waves of contraction terminating at the ileum; none of the complexes enter the colon (Figure 7-22). The MMC occur approximately every 90 to 120 minutes; the speed of migration is coordinated so that as one complex terminates in the ileum, another is beginning in the upper gut. In addition, there is a gradient in the velocity of migration along the small intestine, with the velocity greater in the duodenum than in the ileum. This gradient in velocity of the motor complex is attributed to the gradient of slow wave frequency along the gut. As discussed in Chapter 4, the MMC consists of three phases: phase I is a quiescent period, followed by phase II characterized by sporadic contractile activity, and finally, phase III consisting of very strong peristaltic contractions (Fig. 7-22). Phase III contractions are considered the activity front and serve to propel undigested material, including bacteria, along the small intestine and into the colon. The importance of the MMC in clearance of bacteria from the small intestine is evidenced by the susceptibility to small intestinal bacterial overgrowth (SIBO) in individuals with MMC derangements (loss or reduced cycling).

The mechanisms that regulate intestinal MMC are not clear. The recurrent nature of the MMC suggests an ultradian rhythm (cycle repeated throughout a 24-hour day) with potential endocrine/neurocrine regulatory mechanisms. Both motilin and the vagus have been implicated in the initiation of gastric phase III contractions of the MMCs. However, the initiation and aboral progression of the intestinal phase III contractions are not affected by blockade of

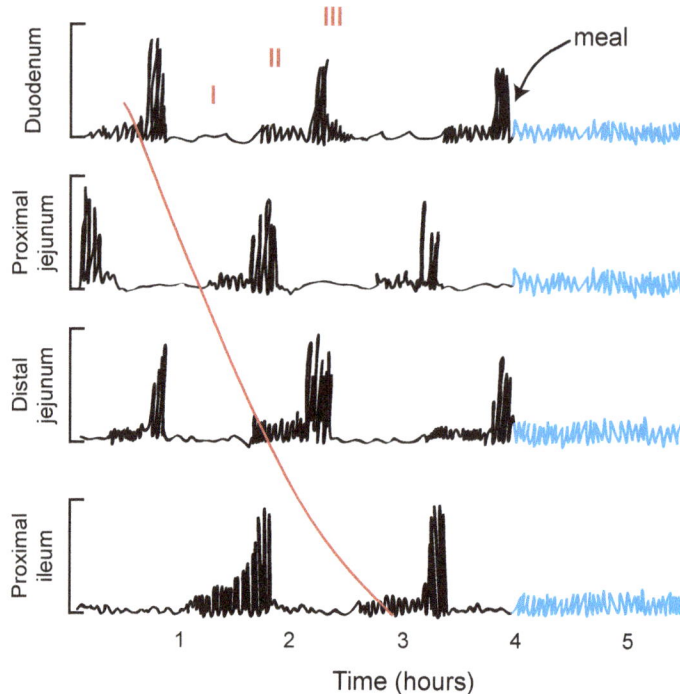

FIGURE 7-22: The migrating motility complex (MMC) in the human small intestine travels down the length of the intestine to the ileum; very few reach the terminal ileum. The MMC consists of three phases (identified in red): phase I, no contractions, phase II, irregular contractions of moderate amplitude, and Phase III, maximum number of high amplitude contractions. The MMC is interrupted by ingestion of a meal and replaced with a digestive pattern of contractions of moderate amplitude.

motilin. It is generally accepted that the intestinal MMC requires an intact ENS; the vagus has a modulatory role. Vagal blockade or intestinal transplantation diminishes intestinal phase II contractions without affecting phase III activity. The importance of neural influences on the MMC is exemplified by the proposed use of the MMC as a marker for the integrity of neuro-muscular function.

It is difficult to ascribe a role for luminal factors in initiating the MMC, since MMC occur only in the fasted state. However, other motor and secretory events also cycle in association with the MMC, such as gall bladder contraction and relaxation of the sphincter of Oddi, as well as pancreatic secretion. There is some evidence that periodic changes in duodenal pH or bile content during fasting may play a role in regulating the MMC of duodenal origin. Nonetheless, a role for luminal factors in the intestinal MMC has not gained wide-spread acceptance.

Digestive Pattern

Feeding interrupts the migrating motility complex and initiates a different pattern of intestinal contractions (Fig. 7-22). Unlike the discrete phases of the MMC, the fed pattern is character-ized by random motor activity (groups of 1 to 3 sequential contractions) with much shorter intervals of inactivity (5 to 40 seconds). Further, the strength of the contractions is similar to those of phase II contractions, that is, much weaker than the phase III contractions. The num-ber of contractions per unit time is dictated by the slow waves and their intensity depends on the physical and chemical composition of the food ingested. A larger number of contractions are induced by ingestion of solid food than when an equicaloric amount of liquid is consumed. The postprandial contraction pattern is dependent on the nutrient composition of the con-sumed meal; caloric value is relatively less important. While all of the three major foodstuffs (lipids, carbohydrates, and proteins) interrupt the MMC and initiate a postprandial contraction pattern, lipids have the greatest impact (see Fig. 24). Lipids stimulate the largest number of bursts and interrupt the MMC for the longest duration. Long chain triglycerides are more ef-fective than medium chain triglycerides. Hydrolysis of triglycerides is critical, since long chain fatty acids are effective while their parent triglycerides are not. In general, the functional motil-ity patterns that occur during the fed state have been classified into two types: segmental and peristaltic.

Segmentation

Rhythmic segmental contractions represent the most frequent type of motor activity occurring after meals. Segmental contractions are initiated by introduction of liquid nutrients (especially fatty acids) into the intestinal lumen. Segmentation consists of cyclic contractions and relax-ations of the circular muscle layer at any given site along the gut (Figure 7-23). Initially, a small segment of bowel <2 cm in length) contracts while adjacent segments are relaxed. Subsequently, the contracted segment relaxes while the previously relaxed adjacent segments contract. At any given site along the small bowel, the rate of segmental contractions is either equal to or some multiple of the intrinsic slow wave frequency. Segmental contractions can occur as independent single events or in clusters. Although the basal electrical rhythm determines the maximal number

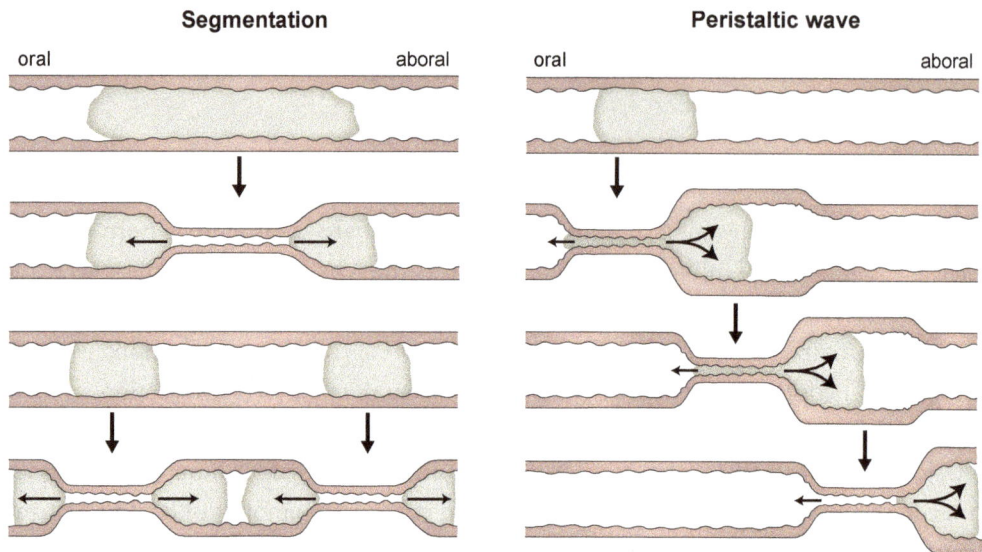

FIGURE 7-23: Effects of intestinal segmentation and peristaltic waves on the movement of chyme. Segmental contractions (left) displace chyme to and fro randomly, thereby mixing the chyme with digestive secretions. Peristaltic waves (right) are propagating contractions consisting of a contractile wave preceded by a relaxation wave that tend to move chyme aborally. The role of the ENS in these two types of functional contractions is depicted in Fig. 7-21.

of segmental contractions, neural input from the myenteric plexus plays an important role in coordinating the contraction-relaxation cycles of adjacent segments (Fig. 7-21A).

Segmental contractions of the small intestine serve to mix the chyme by continuously displacing it to and fro within the lumen. They also tend to move chyme in an aboral direction. The underlying basis for aboral transport is the relationship between slow waves and segmentation. Because of the proximal-to-distal phase lag of the slow waves on a frequency plateau (Fig. 7-20), individual contractions tend to follow contractions in more proximal segments; the net result is an aboral propulsion of chyme along a given frequency plateau. The aborally declining frequency gradient along the entire small bowel also provides a mechanism for downward movement of chyme. Contractions of the bowel wall not only generate a pressure for propulsion of chyme but also impose a resistance to the movement of chyme. It is generally held, however, that intestinal contractions contribute more to an increase in resistance than propulsion of chyme. Since segmental contractions are less frequent aborally than orally, the average resistance is lower in the ileum than the duodenum. Thus, chyme gradually moves toward the ileum.

Peristaltic Wave

Peristaltic contractions occur much less frequently than segmental contractions. These contractile waves move chyme aborally for a few centimeters and therefore involve coordinated sequential contractions and relaxation cycles of the circular muscle (Fig. 7-23). The peristaltic wave consists of a wave of contraction coupled to a downstream wave of relaxation. The polarity of the peristaltic wave is programmed by the neuronal circuitry of the myenteric plexus. Placement of a bolus of material in the lumen of the small intestine generally results in contraction oral to the bolus and relaxation aboral to the bolus. However, retroperistalsis, or a reversal of the polarity of the peristaltic wave, can occur and chyme can be moved in the oral direction. Retroperistalsis is rare and usually occurs under abnormal circumstances such as chemical irritation or distension of the bowel.

Initiation of Postprandial Motility Pattern

Predictably, luminal factors have been implicated in the initiation and control of the postprandial contractile activity. Non-nutrient (cellulose) meals induce peristaltic contractions, while meals containing hydrolytic products of proteins, carbohydrates and lipids induce primarily segmental contractions. Cellulose elicits primarily peristaltic contractions at a maximum frequency and propels chyme at a rapid rate along the intestine. Of the three classes of macronutrients, long chain fatty acids elicit the most pronounced contractile response (Figure 7-24). As illustrated, the contractions produced by an oleic acid meal are stronger and occur in clusters. The mechanism underlying this response to the oleic acid is unclear, but may be related to the mucosal irritation induced by this long chain fatty acid. In general, some of the contractions induced by nutrients, whether isolated or in clusters, tend to migrate short distances aborally.

The enteric nervous system plays a critical role in initiating and maintaining the postprandial segmental and propulsive contractions (Fig. 7-21). Activation of the ENS can result from stimulation of mucosal EEC by chemical/physical stimuli present in the lumen. Various EEC contain receptors for hydrolysates of protein, carbohydrates, and lipids (Table 1-1, Chapter 1). Enterochromaffin cells (EC) and I cells have been implicated in the segmental contractions elicited by intraluminal fatty acids. Both EC and I cells have fatty acid receptors and, when stimulated, release their products (serotonin and cholecystokinin, respectively) into

Motor patterns	Motility index (#/5 min)	Propagative (%)	Transit rate (cm/sec)
mN Cellulose meal	5.6	90	1.9
Casein meal	3.8	45	1.0
Glucose meal	3.1	30	0.8
Oleic acid meal	5.2	35	1.0

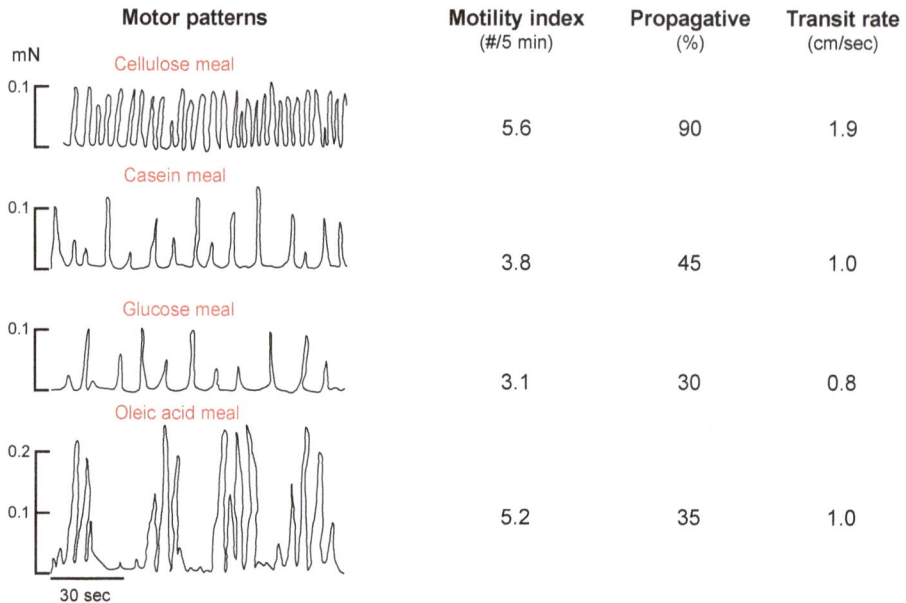

FIGURE 7-24: Motor patterns in the canine jejunum induced by different meals. The amplitude/force of the contractions are given in Newtons (N). Cellulose, a non-nutrient, viscous meal, served as a control to which were added casein hydrolysate, glucose, or oleic acid. The cellulose meal elicited high frequency, propulsive contractions, which propelled chyme along the segment for short distances. The contractions induced by the nutrient containing meals were primarily segmental. The oleic acid meal produced the strongest contractions which generally occurred in clusters [From Schemann M and Ehrlein H-J. *Gastroenterology* 1986, 90: 991–1000.]

the interstitium. Sensory neurons contain both serotonin (5-HT) and cholecystokinin (CCK) receptors that can induce an intrinsically programmed ENS response for segmentation when activated. The EC, which represent 70% of the total population of intestinal EEC, are also responsive to mucosal deformation. The peristaltic waves elicited by intraluminal instillation of non-nutrient material, such as cellulose or small pellets, or by mucosal stroking (villus deformation) can be inhibited by serotonin antagonism. Severe distension of the lumen can also induce peristaltic activity; a response independent of serotonin release by EC. It has been proposed that distention-induced peristalsis is due to the activation of tension sensors in the muscularis externa, which subsequently impact on the ENS neural circuitry.

Perfusion of isolated innervated jejunum or ileum with nutrient solutions induces a shift from the MMC to a postprandial motility pattern in the duodenum, not exposed to nutrients. This suggests that extrinsic neurohumoral factors play a role. The demonstration of a cephalic phase in the interruption of the MMC has prompted the proposal that extrinsic nerves may be involved; specifically, the vagus. However, while vagotomy can modulate the fed pattern of motility (e.g., reduced number and duration of contractions), neither vagotomy nor splanchnicectomy (dissection of the splanchnic nerves) prevents the induction of the fed pattern of motility. Although several potential humoral factors have been proposed (e.g., CCK, gastrin), currently there is no consensus regarding the candidate humoral factor(s) that mediates the transition from the interdigestive to digestive motility pattern in the small intestine.

Villus Contractions

The villi of the intestinal mucosa contain smooth muscle fibers that enable them to exhibit contractile activity. Villus contractile activity is independent of the contractile activity of the muscularis externa discussed above. Individual villi contract at irregular intervals; the predominant movement is a piston-like retraction and extension. The frequency and duration of villus contractions is highest in the duodenum and decreases progressively toward the ileum, where villi seldom contract. The frequency of villus contractions increases after meals. Of the major macronutrients, oleic acid and amino acids, but not glucose, increase villus contraction frequency; long chain fatty acids are the most potent stimulus. Intestinal lymph flow is directly related to villus contraction frequency, indicating that villus shortening facilitates lipid absorption by emptying the central lacteal.

Villus contractions do not appear to be under extrinsic neurohumoral control. Neither vagotomy nor atropine significantly alters basal villus contraction frequency. In a similar vein, neither ablation of the sympathetic nerves or adrenergic blockade affects villus contractions. Intraduodenal instillation of predigested food does not affect villus contractility in an isolated (empty) jejunal segment. Finally, removal of chyme from the intestine of a fed animal, results in a villus contractile frequency similar to that of fasted animals. Collectively, the available information indicates that villus contractions are regulated by as yet unidentified local factors, rather than extrinsic neurohumoral mechanisms.

Transit Time of Chyme

The transit of chyme through the small bowel after a meal is slow enough to allow for adequate digestion and absorption of nutrients. The rate of aboral transit is not uniform but is dependent on the relative frequency of segmental and peristaltic contractions. At any given time, chyme may be either moving to and fro as a result of segmental contractions or moving rapidly for a few centimeters by peristaltic waves. In general, however, chyme moves aborally at a rate of 1.5 cm per minute, and the velocity of transit is more rapid in the duodenum than in the ileum. The average meal takes a few hours to traverse the small intestine, contrasting with the transit of chyme in the colon, which can be on the order of days.

The Ileocecal Junction

Sphincter Function

As chyme moves gradually down the intestine most of the nutrients, electrolytes, and water are absorbed, leaving only a small volume of residue to pass into the colon (approximately 1.5–2.0 liters per day). The flow of chyme between the small intestine and colon is regulated by a 4-cm segment of the terminal ileum, called the ileocecal junction (Figure 7-25). This structure has a thickened muscular (circular) coat that can generate an intraluminal pressure of about 15 mm Hg at rest; adjacent ileal and cecal pressures are approximately 6–7 mm Hg. The ileal papilla, a small protrusion of the ileal oriface into the cecum, has been occasionally referred to as a valve. However, it is not a true valve and the possibility of the junction acting as a mechanical barrier

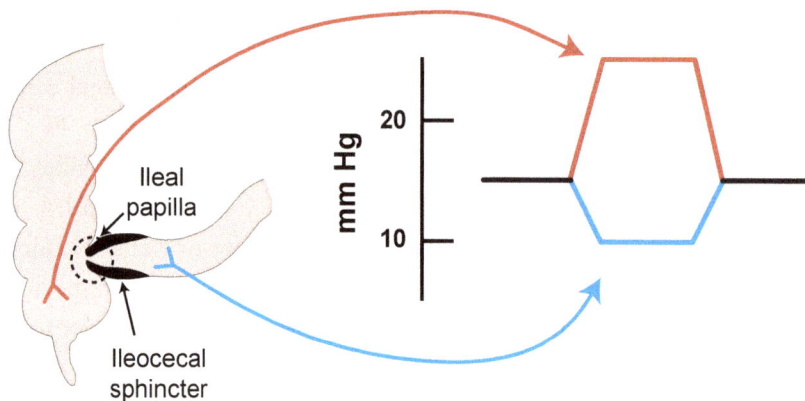

FIGURE 7-25: The ileocecal sphincter of humans generates a pressure of approximately 15 mm Hg while the pressures in the ileum and cecum are approximately 6 mm Hg. Depicted are the responses of the ileocecal sphincter to distension (≈ 14 mm Hg) of either the ileum (6.0 ml saline) or cecum (55.5 ml saline). Distension of the ileum reduces sphincter pressure (blue arrow), resulting in ileal-cecal pressure gradient favoring movement of chyme into the cecum. Distension of the cecum increases sphincter pressure (red arrow), reinforcing the pressure gradient preventing reflux of cecal contents into the ileum [Data from Shafik A et al. *Med Sci Monit*, 2002; 8 (9): CR629–635.]

to the flow of chyme is not generally accepted. Current consensus is that the ileocecal junction is an area of high pressure functioning as a sphincter.

The ability of the sphincter to generate tone is attributed to the ENS, since total autonomic denervation of the junction does not affect existing intraluminal pressure and isolated segments of the junction exhibit tone in vitro. Reflex arcs in the myenteric plexus are also responsible for the response of the ileocecal sphincter to distention of either the ileum or cecum (Fig. 7-25). In general, distension of the ileum will relax the sphincter, while distension of the cecum will contract it. Since local anesthesia (xylocaine) can abolish these responses, it is apparent that both excitatory and inhibitory neural reflexes within the ENS regulates ileocecal sphincter tone.

With respect to extrinsic neural regulation, stimulation of the sympathetic nerves constricts the sphincter; a response blocked by α-adrenergic-, but not β-adrenergic-antagonists. Surprisingly, stimulation of the vagus also constricts the sphincter; a response blocked by atropine. Pharmacologic studies support these findings that α-adrenergic and cholinergic agents constrict the sphincter, while β-adrenergic agents relax the sphincter.

Transjunctional Flow

A major function of the ileocecal junction is to regulate the flow of chyme between the ileum and colon. The movement of chyme across the ileocecal junction is pulsatile. As chyme progressively accumulates in and distends the distal ileum, reflex relaxation of the sphincter occurs, and chyme passes into the colon. During fasting, ileocecal flow is erratic. There is some transjunctional movement of chyme after passage of an MMC, but since very few of the phase III contractions reach the terminal ileum, some residual chyme is usually present in the ileum. After meals, strong propagating ileal contractions, which upon arrival at the junction, relax the sphincter and drive ileal chyme into the cecum. The ileal contractions have been attributed to a long reflex elicited by the presence of food in the stomach, that is, a "gastroileal reflex." The specific mediators of this reflex are unkown.

If colonic contents are not moved analward, but accumulate and distend the cecum, reflux back into the ileum is prevented by reflex constriction of the sphincter. In health, there is relatively little colo-ileal reflux of chyme across a patent ileocecal sphincter. However, an incompetent sphincter can lead to small intestinal bacterial overgrowth (SIBO) and its sequelae. The endogenous bacteria that reside in the small intestine compete for nutrients with the host (e.g., fermentation of carbohydrates) and inhibit nutrient absorption (e.g., fat malabsorption due to deconjugation of bile salts). The symptoms of SIBO range from mild abdominal discomfort to steatorrhea and malnutrition.

Defensive Motor Patterns

Power Propulsion

Occasionally, intense contractile waves are generated in the mid-small intestine and propagate rapidly either aborally or orally for long distances. The aborally migrating contractions are referred to as "giant migrating contractions (GMC)" whereas the orally migrating ones are termed "retrograde peristaltic contractions (RPC)." Both GMCs and RPCs are greater in amplitude than phase III contractions of the MMC. The duration of a contractile wave is longer than several slow wave cycles, thus RPCs and GMCs are not regulated by slow waves. These contractions generally occur in response to severe mucosal irritation, representing an effort by the intestine to clear/remove the noxious material.

Giant Migrating Contractions (GMCs)

GMCs can occasionally occur in health, but are more frequently a consequence of pathologic events, such as intraluminal irritants or bacterial infection. GMCs are single contractions that originate in the jejunum or ileum and propagate aborally at a rate of approximately 2 cm/sec. Like MMC, GMCs predominately occur during fasting. However, unlike MMC, which terminate proximal to the ileocecal junction (IJ), roughly half of the GMCs pass through the IJ and continue on into the colon. The intestinal GMCs are not to be confused with those that occur in the colon after meals and are responsible for the "mass movement" of chyme. GMCs of small intestinal origin can be associated with abdominal cramps and diarrhea. It has been proposed that the GMCs serve to rapidly clear the lower small intestine of noxious material and facilitate their elimination via the rectum.

The ENS is assumed to play an important role in coordinating the GMCs. However, since intestinal GMCs can also be induced by ischemia and ionizing radiation, a potential role for extrinsic factors, neural or humoral, also exists. However, little information is available regarding extrinsic regulation of intestinal GMCs.

Retrograde Peristaltic Contractions (RPCs)

RPCs are intense contractions that originate in the mid-small intestine and propagate very rapidly (up to 10 cm/sec) toward the stomach. It is believed that RPCs serve to transport irritants from the upper small intestine to the stomach from which the material can be expelled by vomiting. Since emetic agents can induce RPCs, they are considered a prelude to vomiting. However, RPCs can occur in the absence of retching or vomiting and vice versa. In addition to the critical role of the ENS, several lines of evidence support a role for extrinsic nerves in regulation of RPCs. Retrograde contractions do not occur in autotransplanted (denervated) small bowel, and vagotomy abolishes RPCs in the intact intestine. The involvement of the CNS in some RPCs is evidenced by their induction via activation of the vestibular system (motion sickness) and chemoreceptive trigger zone (e.g., opioids, chemotherapeutics).

Inhibition of Propulsion

Abnormal intraluminal stimuli can also elicit quiescence or inhibition of motor activity. In general, these inhibitory reflexes have a distal-to-proximal orientation, that is, stimuli in distal segments of the small intestine inhibit the motor activity of upstream segments. They serve as inhibitory feedback mechanisms to prevent the transit of chyme to the affected site. Two prominent examples of such reflexes are the "intestino-intestinal reflex" elicited by distension of the gut and the "ileal brake" initiated by lipids. Although technically, both are intestinointestinal reflexes, this moniker generally refers to the distension-induced reflex.

Intestino-Intestinal Reflex

The intestino-intestinal reflex is initiated by distension of a portion of the small intestine and results in the cessation of contractile activity in more proximal segments. In humans, distension of the distal jejunum inhibits motility of the proximal jejunum, while distension of the proximal jejunum does not affect motility of the more distal segment. Since the discomfort induced by distension of proximal and distal segments is similar, the retrograde relaxation induced by distension is not likely due to a CNS-mediated response to the distension stimulus. Sympathetic nerves appear to be a major component of this reflex arc, since bilateral section of the splanchnic nerves, but not vagotomy, abolishes the reflex. The proposed function of this reflex is to prevent the continued accumulation of chyme at a site already distended with material.

The intestino-intestinal reflex should not be confused with "ileus" or "adynamic ileus" which is a general cessation of peristaltic activity of the small bowel in the absence of mechanical obstruction. Ileus is generally a result of surgical procedures (laparotomy, manipulation or injury of the gut), local inflammation, or certain pharmaceuticals (e.g., opiates).

Ileal Brake

The ileal brake is the reflex inhibition of proximal motility and transit of chyme by perfusion of the distal small intestine with nutrients. Of the major macronutrients, hydrolyzed long chain triglycerides are the most potent stimulant for this response; isotonic amino acids or glucose solutions are relatively ineffective. Both the ENS and extrinsic nerves have been implicated in the ileal brake. Local anesthesia of the ileum inhibits the ileal brake response, indicating that afferent sensory nerves are involved. The EEC of the ileal mucosa that appear to play a major role in responding to luminal lipids and signaling to the sensory neurons are the L cells (Table 1-1 in Chapter 1). Peptide YY (PYY) is released from ileal L cells in response to fatty acids and immunoneutralization of PYY abolishes the ileal brake. The efferent limb of the reflex appears to involve noradrenergic nerves which activate β1 adrenergic receptors; analogous to the sympathetic efferent limb of the intestino-intestinal reflex.

The jejunum can also act as a "brake," inhibiting proximal intestinal transit when perfused with long chain fatty acids. A comparison of the jejunal brake to the ileal brake indicates that, of the two, the ileal brake has a more potent inhibitory effect. The ileal brake may actually be a component of an overall "intestinal brake," since a "duodenal" and "colon" brake have also been described. The "duodenal brake," by which lipids in the duodenum inhibit gastric emptying, was discussed in Chapter 4.

INTESTINAL CIRCULATION

Blood flow to the small intestine accounts for 10% to 15% of resting cardiac output in the adult human. Blood is supplied to the small bowel almost entirely by the superior mesenteric artery; the celiac artery is a major contributor of blood to the duodenum, while the inferior mesenteric artery provides a small amount of blood to the terminal ileum.

The mucosa, submucosa, and external muscle layers of the small intestine have discrete microvascular networks that are arranged as series and parallel-coupled circuits. Under resting conditions, approximately 75% of total intestinal blood flow is distributed to the mucosa, 5 percent to the submucosa, and 20% to longitudinal and circular muscle layers. The villi receive approximately 60% of the blood flow to the mucosal layer, with the crypts receiving the remainder. The relative distribution of blood flow to the different regions of the intestinal wall presumably reflects the varying metabolic demands of cells within each region. The metabolically active villus and crypt epithelium receive the largest proportion of flow, while the muscle layer, with its lower demand for oxygen, receives a somewhat lower fraction of total blood flow. The dependence of blood flow distribution on tissue demand is demonstrated by observations that blood flow is preferentially distributed to the mucosa during periods of enhanced nutrient absorption or electrolyte secretion, while increased intestinal motility is associated with blood flow redistribution to the muscle layers.

Blood flow to the small intestine increases by 30% to 130% after a meal. The hyperemia is confined to that segment of bowel exposed to chyme, and the magnitude of the hyperemia is related to the type of food that is ingested. The hydrolytic products of food digestion are primarily responsible for the hyperemia in the duodenum and jejunum. Long-chain fatty acids (solubilized in bile) and glucose are the major luminal stimuli for the hyperemia. Although bile does not produce a hyperemia in the proximal intestine, it greatly increases blood flow in the terminal ileum. Bile salts, which are actively absorbed in the terminal ileum, are largely responsible for the bile-induced ileal hyperemia.

A major function of intestinal capillaries and lymphatics is removal of absorbed nutrients and water from the mucosal interstitium. Both the capillaries and lacteals of the intestine are highly permeable to absorbed nutrients such as glucose and amino acids; however, the capillaries remove virtually all of the absorbed water-soluble nutrients from the mucosal interstitium. The much greater capacity of capillaries to "clear" nutrients from the interstitium results from the much higher (1,000-times) flow of blood in these vessels compared to the slow flow of lymph in the lacteals. However, the lymphatics play a more important role in removing large molecules (proteins) and particles (chylomicrons) from the mucosal interstitium. The lacteals are highly permeable to even the largest chylomicron (6000 Å diameter), while the fenestrated capillaries of the intestinal mucosa are relatively impermeable to proteins as small as albumin (75 Å diameter). Thus, the immunoglobulins (IgA), enzymes, and chylomicrons that are liberated into the mucosal interstitium gain access to the blood stream exclusively by way of lymph.

The process by which absorbed fluid is driven into the mucosal capillaries and lymphatics can be explained in terms of the hydrostatic and oncotic forces described in Starling equation of transcapillary fluid exchange (Figure 7-26). In the nonabsorbing small intestine, the small imbalance in the hydrostatic and oncotic forces acting across the capillary wall that favors net movement of fluid from blood to interstitium, that is, net filtration. Although small (~1.0 mm Hg), this imbalance of forces is sufficient to cause net fluid filtration (Jv) owing to the relatively high hydraulic conductivity of capillaries in the intestine. This low rate of capillary fluid filtration is balanced by an equal outflow of fluid via the lacteals, which ensures a constant interstitial fluid volume. When net water absorption is stimulated (Fig. 7-26), mucosal interstitial volume rises, owing to accumulation of protein-free absorbed fluid in the sub-epithelial spaces. The interstitial volume expansion produces an increase in interstitial hydrostatic pressure (P_t) and a reduction in interstitial oncotic pressure (π_t), due to dilution of interstitial proteins. Associated with the changes in interstitial forces is a doubling of capillary hydraulic conductivity due to perfusion of a larger number of capillaries. The absorption-induced changes in capillary and interstitial forces modify the balance of pressures across the capillaries to produce a net

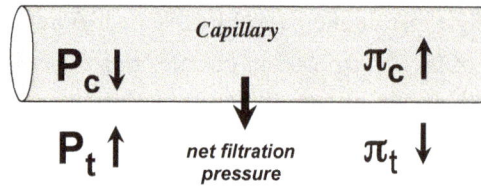

$$J_v = K_f \left[(P_c - P_t) - (\pi_c - \pi_t) \right]$$

Epithelial status							Capillary status
Resting	+0.10	=	0.10	[(12 - 1) -	(28 - 18)]		*Low filtration*
Absorption	-0.60	=	0.20	[(12 - 3) -	(28 - 16)]		*Absorbing*
Secretion	+0.75	=	0.15	[(13 - 0) -	(28 - 20)]		*Hyperfiltration*
	ml/min		ml/min/ mmHg	mmHg mmHg	mmHg mmHg		

FIGURE 7-26: Alterations of hydrostatic (capillary, P_c, and interstitial, P_t) and oncotic (capillary, π_c, and interstitial, π_t) forces enable capillaries to absorb or filter water (J_v) in accordance with the transport function of intestinal epithelial cells. In the resting state (absence of net water transport), capillaries filter at a low rate. During periods of net water absorption, water enters the mucosal interstitium and capillaries assume an absorptive phenotype due to the rise in P_t and K_f (a measure of the number of capillaries open to perfusion), and a reduction in π_t. When there is net fluid secretion (e.g., cholera toxin stimulation), interstitial volume falls, leading to a fall in P_t and an increase in π_t. These changes in interstitial forces are accompanied by increases in P_c and K_f. Collectively, the alterations in P_t, π_t, P_c and K_f result capillary hyper-filtration, which provides the fluid needed for the secretory pump. [Modified from Kvietys PR and Granger DN, The splanchnic circulation, Chapter 10. In: Gastrointestinal Anatomy and Physiology: The Essentials. Reinus JF and Simon D, eds, Wiley-Blackwell, 2014.]

absorptive force. This force, coupled to the elevated capillary hydraulic conductivity, drives approximately 80% of the absorbed fluid into the mucosal capillaries. Intestinal lymph flow also increases during absorption because of the increased lymphatic filling resulting from the rise in interstitial hydrostatic pressure. The enhanced lymph flow removes the remaining 20% of absorbed fluid from the mucosal interstitium. Thus, as a consequence of the reactions initiated by absorbed fluid entering the interstitium, two important changes occur: (1) filtering capillaries are converted to absorbing capillaries, and (2) the rate of lymph formation is increased.

Consequently, absorbed fluid is removed from the mucosal interstitium via two routes, the capillaries and lacteals.

Significant changes in capillary and interstitial forces also occur when the intestine is stimulated (e.g., by cholera toxin) to act as a net secretory organ (Fig. 7-19). In this instance, capillary filtration rate must rise to ensure that sufficient fluid is made available for the epithelial cells to perform their secretory function. As fluid leaves the interstitial compartment, crosses the epithelial lining and enters the bowel lumen, interstitial volume decreases, which result in a decline in P_t and a rise in π_t (because interstitial proteins are more concentrated following water removal). These changes enhance capillary fluid filtration and diminish the drainage of interstitial fluid via the lymphatics (due to the fall in P_t). Arteriolar dilation (and the resultant increase in capillary pressure, P_c) as well as capillary recruitment also facilitates the delivery of fluid from blood to the secretory pump. The net result is the delivery of more fluid from the blood stream to meet the greater need of the secretory epithelial cells for water.

SMALL INTESTINAL PAIN

As with other hollow viscera, pain from the small intestine arises from stretching or distension. The nerves transmitting this pain are located in the muscle layer. Pain arising from the small intestine is felt in the central abdomen around the umbilicus. Pain from the distal ileum can be experienced in the right iliac fossa.

PATHOPHYSIOLOGY AND CLINICAL CORRELATIONS

Deranged small intestinal function is demonstrated by the condition of celiac disease (gluten-sensitive enteropathy), a disorder that affects the small intestinal mucosa in a diffuse continuous fashion. The disease is due to the immunologic effect of gluten, a protein constituent of wheat or barley flour. In susceptible individuals, gluten elicits an immunological response resulting in a marked alteration in the architecture of the small intestinal mucosa. The histologic lesion in the full-blown condition is termed "total villous atrophy" and is most severe in the proximal small intestine, but may extend to involve the terminal ileum. The villi are absent; thus the mucosal surface is quite flat with the crypts opening onto the surface. The epithelium of the crypts is dividing very actively, but the enterocytes do not survive long enough to form villi. Since the acquisition of many important digestive enzymes such as disaccharidases and peptidases by the enterocyte occurs during its progress from crypt to villus tip, the surface epithelium in this disease is functionally immature. The clinical picture, a generalized malabsorption, is explicable both by the loss of absorptive surface area and a deficiency of mucosal enzymes involved in nutrient digestion and transport.

Patients often present with diarrhea and moderately increased stool fat. Abdominal cramps may occur; bacterial production of gas from unabsorbed nutrients may contribute to this. Various mechanisms are responsible for the diarrhea: malabsorption of nutrients causes osmotic effects with increased stool water; unabsorbed fatty acids are converted by enteric bacteria to hydroxy-fatty acids, which are powerful secretagogues in the colonic mucosa; and there is evidence that the abnormal small intestinal mucosa actively secretes water and electrolytes in this condition. If there is extensive involvement of the terminal ileum, malabsorption of bile acids occurs and a greater amount than the normal daily load of bile acids (approximately 300 mg) enters the colon. The bile acids and their bacterial derivatives such as deoxycholic acid have a secretagogue action on the colonic mucosa similar to hydroxy-fatty acids.

Although the primary physiologic defect in celiac disease is malabsorption, there is some evidence that cholecystokinin release from the diseased mucosa of the upper small intestine may be defective; thus the pancreatic enzyme response to a meal is diminished and a component of maldigestion may result. The degree of steatorrhea in celiac disease is, however, seldom as great as that in severe pancreatic insufficiency, where lipolysis may be greatly impaired.

Since most foodstuffs are absorbed in the upper small intestine, a wide spectrum of nutritional deficiencies may be seen in celiac disease. Weight loss is common, and in severe cases hypoproteinemic edema may occur. Anemia due to malabsorption of iron or folic acid or both is very common and indeed may be the presenting and only feature of the disease. Vitamin BI2 deficiency is seen only when the disease has extended to the terminal ileum. A variety of vitamin deficiencies may occur: night blindness (vitamin A), osteomalacia with pathologic fractures (vitamin D), and a bleeding tendency due to hypoprothrombinemia (vitamin K). Oral lesions may be seen, some of which are the result of various B vitamin deficiencies. Bone disease may not be due simply to vitamin D deficiency. Calcium may be malabsorbed by the diseased mucosal cells, and unabsorbed fatty acids may sequester calcium in the form of insoluble soaps. It should be stressed that since there is a spectrum of severity of mucosal damage, this entire galaxy of deficiencies is seldom encountered in most cases where only one or two deficiencies may occur.

As increasing interest in gluten free diets has evolved in the last several years, attention has focused on testing for celiac disease. Diagnosis is made by detecting serology for tissue *transglutaminase* with confirmation by biopsies of the duodenum identifying villous atrophy. Biopsies are targeted in the duodenum due to ease of tissue acquisition with upper GI endoscopy (esophagogastroduodenoscopy, EGD). Additional imaging can also be completed with small bowel capsule endoscopy (small camera in a plastic pill that records video images) that searches for the appearance of a "scalloped" (flat) mucosa. Tissue transglutaminase is an autoantigen that identifies individuals with celiac disease. An immunoglobulin IgA level should also be obtained in these patients, since 2% of patients with celiac disease will have an IgA deficiency.

Primary treatment of celiac disease is the absolute avoidance of gluten containing foods and personal products. Gluten is a common ingredient/component of many foods and personal grooming products and recent labeling has assisted patients in identifying these products. Many restaurants and food outlets have created menus and gluten free products for patients with celiac disease. The immunologic response of gluten is also seen with barley and oats, primarily due to contamination during foodstuffs production. No reaction is seen with exposure to rice, potatoes, almonds, or quinoa, thus providing alternative sources for flours and other food products. Upon adopting a gluten free diet, it may take several weeks before a reduction in symptoms of gas, bloating, and diarrhea will occur. Continued exposure to gluten over several years places a patient at risk for development of a T-cell lymphoma.

Tropical sprue is a syndrome characterized by either acute or chronic diarrhea, weight loss, and malabsorption of nutrients. Classically, the syndrome is found in visitors or residents in the tropics. No causative factor has been determined, but a microbial infection is considered to be the inciting event. Small intestinal villi are destroyed, which can lead to the development of significant malabsorption of nutrients, particularly folate, vitamin B-12, and iron. Following an initial bacterial infection, bacterial overgrowth may occur and lead to additional mucosal injury from slowed intestinal transit. Patients can present with fat malabsorption as defined by stool diagnostic tests. Treatment with broad-spectrum antibiotics and nutritional supplementation leads to recovery in most patients.

REFERENCES

Bharucha AE and Hasler WL. Motility of the small intestine and colon. In: Yamada's Textbook of Gastroenterology, 6th Ed. 2016, Edited by DK Podolsky, M Camilleri, G Fitz, AN Kalloo, F Shanahan, and TC Wang. Publishers John Wiley & Sons, Chapt 21, pp 367–385.

Bishu S, Quigley EMM. Nutrient digestion, absorption, and sensing. In: Yamada's Textbook of Gastroenterology, 6th edition, (DK Podolsky et al., editors) Chapter 29, pp 538–555, 2016, John Wiley & Sons.

Deloose E, Janssen P, Depoortere I, and Tack J. The migrating motor complex: control mechanisms and its role in health and disease. *Nature Reviews: Gastroenterology & Hepatology*, 9: 271–285, 2012.

Diaz de Barboza, G, Guizzardi S, Tolosi de Talamoni N. Molecular aspects of intestinal calcium absorption. *W. J. Gastroenterol.* 21:7142–7154. doi: 10.3748/wjg.v21.i23.7142.

Ehrlein HJ and Schemann M. *Gastrointestinal motility*, Technische Universität München, Munich, 2005.

Furness JB, Rivera LR, Cho H-J, Bravo DM, and Callaghan B. The gut as a sensory organ. *Nature Reviews: Gastroenterology & Hepatology*, 10: 729–740, 2013.

Gulec S, Anderson GJ, Collins JF. Mechanistic and regulatory aspects of intestinal iron absorption. *Am J Physiol: GI & Liver Physiol.* 307:G397–G409, 2014.

Jaladanki RN, Wang JY. *Regulation of Gastrointestinal Mucosal Growth*, 2nd Ed. In: Colloquium Series on Integrated Systems Physiology: From Molecule to Function to Disease (Granger DN, Granger JP, eds), Vol. 68. Morgan and Claypool Publishers, 2016. doi: 10.4199/C00145ED2V01Y201610ISP068

Pleuvry BJ. Physiology and pharmacology of nausea and vomiting. *Anesth Intensive Care Med*, 16 (9): 462–466.

Rao MC, Sarathy J, Ao M. *Intestinal Water and Electrolyte Transport in Health and Disease*. In: Colloquium Series on Integrated Systems Physiology: From Molecule to Function to Disease (Granger DN, Granger JP, eds), Vol. 31. Morgan and Claypool Publishers, 2012. doi: 10.4199/C00049ED1V01Y201112ISP031

Rubio-Tapia, A., Hill, I., Kelly, C., Calderwood, A., Murray, J. Diagnosis and Management of Celiac Disease. *Am J Gastroenterol* 2013; 108:656–676.

Said HM. Intestinal absorption of water-soluble vitamins in health and disease. *Biochem J* 437: 357–372, 2011.

Sanders KM, Kito Y, Hwang SJ, Ward SM. Regulation of smooth muscle function by interstitial cells. *Physiology* 31: 316–326, 2016.

Wood JD. Enteric nervous system: brain-in-the-gut. In: Physiology of the Gastrointestinal Tract, 6th Edition, 2018, Volume I, Said HM (editor), Chapt. 15, pp 361–372.

CHAPTER 8

The Large Intestine

INTRODUCTION

Although the large intestine is generally perceived as a simple storage depot for fecal material, it is now recognized as a digestive and absorptive organ of significant import. One of the most important functions of the large intestine is to absorb most of the 1.5 liters of water and electrolytes that escape absorption in the small bowel each day. In addition, the large number of bacteria in the colon transforms food residues and dietary fiber into substances of caloric (e.g., short chain fatty acids) and physiologic (e.g., vitamins) value that are subsequently absorbed. The mixing and propulsive activities of the colonic smooth muscle are designed to facilitate absorption by continuously re-exposing "chyme" to the absorptive cells. Once "chyme" has been sufficiently dehydrated and the residual material reaches the rectum, there is an urge to defecate. The process of defecation is complex and involves both voluntary and involuntary reflexes. The voluntary component of the defecation reflex allows for discretionary fecal elimination.

ANATOMY

The large intestine extends from the ileocecal junction to the anus. It can be anatomically subdivided into the cecum (and its appendix), ascending colon, transverse colon, descending colon, sigmoid colon, rectum, and anus (Figure 8-1). The length of the large bowel, from ileocecal valve to rectum, is approximately one meter, which is much shorter than the small intestine. The "large" intestine derives its name from the characteristically widened external diameter, relative to the

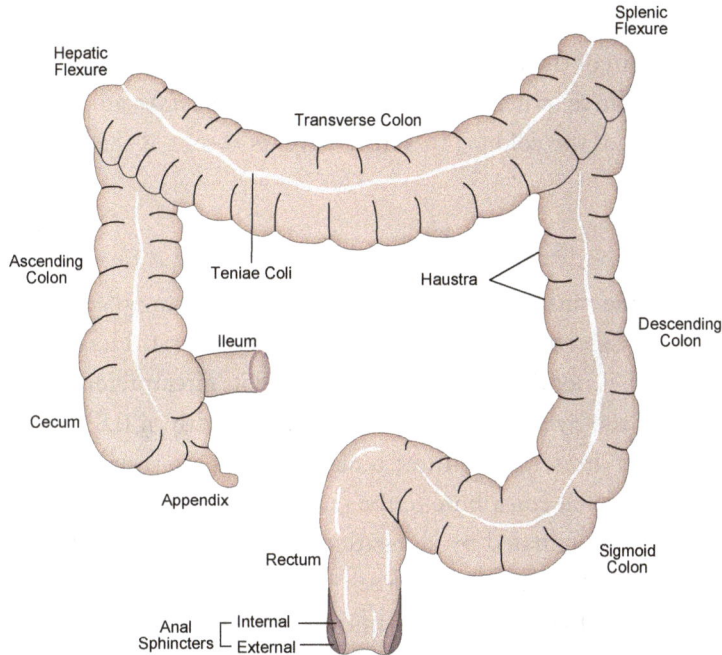

FIGURE 8-1: Anatomical features of the large intestine. The large intestine extends from the ileocecal junction to the anus. It can be anatomically subdivided into the cecum (and its appendix), ascending colon, transverse colon, descending colon, sigmoid colon, rectum, and anus. Bands of longitudinal smooth muscle (called teniae coli) that are situated equidistantly around the circumference of the colon contribute to the sacculated (haustral) appearance of the colon. The haustra (recesses) are more prevalent in the proximal portions of the large intestine.

small bowel. Although the proximal large intestine is twice as wide as the small bowel, the more distal portions of the colon have a caliber similar to that of the small bowel. The cecum is the widest portion of the colon, with a diameter of about 8.5 cm, whereas its diameter narrowed to 2.0 to 2.5 cm at the level of the sigmoid colon. As a result of the reduced diameter of the sigmoid colon, neoplastic lesions are more likely to cause obstruction in this segment. The most striking feature of the large intestine is the presence of haustra (recesses), which give it a sacculated appearance. Haustra are more prevalent in the proximal portions of the large intestine.

The wall of the colon, like that of the small intestine, is divided into four layers: serosa, muscularis externa, submucosa, and mucosa. However, there are several anatomic specializations in the colon that distinguish this portion of the gut from the small intestine.

Serosa

The squamous epithelium of the colonic serosa is studded by elongated protrusions of the peritoneum containing fat. These structures, called *appendices epiploicae*, are larger and more prevalent in the distal portion of the colon. The serosal layer is absent at the level of the rectum.

Muscularis Externa

As in the small bowel, the muscle coat of the colon consists of an outer longitudinal layer and an inner circular layer, collectively called the muscularis propria. However, the longitudinal muscle layer of the colon is not distributed evenly around its circumference, but is concentrated into three bands called the *teniae coli*. The bands, with a width of 0.8 cm, are separated by a thin sheet of longitudinal muscle. The three teniae coli that are situated equidistantly around the circumference of the colon contribute to the sacculated (haustral) appearance of this organ. Haustra formation is also facilitated by elongation of the circular muscle such that the inner muscle layer appears to bulge outward between the teniae coli. The bulge is interrupted at various sites by constricted bands of circular muscle, the plicae circulares. The size and shape of the haustrations are not fixed but depend on the contractile activity of the circular and longitudinal muscle layers. In the distal portion of the sigmoid colon, the teniae coli broaden and converge so that the longitudinal muscle is one uniform and continuous sheet covering the rectum.

Nerve plexuses lie in the connective tissue bordering the muscle layers. The myenteric plexus is located between the circular and longitudinal muscles. The neurons of the myenteric plexus are concentrated beneath the teniae coli and are more sparsely distributed between them. The submucous plexus lies beneath the circular muscle layer and is more prominent in the rectum than elsewhere along the colon. Interstitial cells of Cajal (ICC) are found within the muscularis propria and are closely associated with smooth mucle fibers in the submucosa. The ICC form neuroeffector junctions with nerve fibers in these regions and serve as pacemakers for the slow waves that result in colonic smooth muscle contractions.

Mucosa

The mucosa of the colon is characterized by a relatively smooth surface. The anatomic modifications of the colonic mucosa, which increase its surface area, are not as distinctive as those of the small intestine. The colonic mucosal surface is thrown into irregular folds (plicae semilunares) that are less prominent than the plicae circulares of the small bowel mucosa. The mucosa of the colon is devoid of villi, and the microvilli of the colonic absorptive cells are less abundant than their counterparts in the small intestine. Nonetheless, the plicae semilunares and micro-villi increase the surface area of the colonic mucosa by as much as 10 to 15 times that predicted for a simple cylinder.

The colonic mucosa is partitioned into three layers: the muscularis mucosae, the lamina propria, and the epithelium. The muscularis mucosa is a layer of smooth muscle approximately 8 to 12 cells thick that separates the submucosa from the lamina propria of the mucosa. The lamina propria provides structural support for the epithelium; within its connective tissue lie the blood vessels, lymphatics and nerves.

The lamina propria of the large intestine contains numerous T-lymphocytes, macro-phages, plasma cells, and occasional lymphoid nodules that protrude into the submucosa. The plasma cells secrete immunoglobulins (primarily IgA), and the lymphoid tissue can mount an immune response. The immune function of the colon is particularly important because of its extensive bacterial flora. The lamina propria of the fetal colon is devoid of immune cells. Only after birth, when the colon is colonized by bacteria, do the lymphoid and plasma cells appear.

Epithelial Cells

Although the mucosal surface of the colon is devoid of villi, there are numerous crypts (about 0.7 mm deep) extending from the flat absorptive surface to the muscularis mucosa. The cells in the crypts and on the mucosal surface are somewhat similar to those found in the small intes-tine. The goblet and enteroendocrine cells (EEC) of the colonic mucosa are morphologically indistinguishable from their counterparts in the small bowel (see Fig. 7-4 in Chapter 7). The goblet cells, with their mucus-laden granules, are more prominent in the colon than in the small intestine. The large number of goblet cells in the colon minimizes the frictional interaction be-tween the semisolid feces and the mucosal surface. There are fewer EEC in the colon, and they differ from those in the small intestine only in regard to the type of peptides secreted.

The epithelia in the lower half of the colonic crypts consist primarily of undifferentiated cells that ultimately give rise to the goblet, enteroendocrine, and absorptive cells. The undiffer-entiated cells migrate up the crypt as they proliferate and mature. Upon reaching the mucosal surface, they degenerate and eventually slough off into the lumen. The kinetics of epithelial cell proliferation, migration, and extrusion are slightly slower in the colon than in the small intes-tine, requiring approximately four to seven days for a complete turnover.

MUCOSAL GROWTH AND ADAPTATION

The proliferative activity of the colonic mucosa is largely dependent on lumen contents. Food deprivation results in a reduction of mucosal surface area in the colon, primarily as a result of decreased cell proliferation in the crypts. Reinstitution of feeding with a bulk-free diet does not alleviate the mucosal atrophy, because the nutrients are almost completely absorbed by the small intestine. If, however, bulk is added to the diet, the mucosal surface area, as well as cell proliferation kinetics, is returned to normal. Extensive surgical resection of the small bowel, which is associated with enhanced delivery of unabsorbed nutrients to the colon, increases colonic weight, length, and circumference. Feeding non-metabolizable bulk to experimental animals results in hypertrophy of the proximal colon and an increase in its solute transport capacity. Thus, in contrast to the small intestine, luminal nutrients are of secondary importance in maintaining cell turnover rate and growth in the colon. The presence of bacteria in the lumen is also important in maintaining a normally functioning mucosa, for in their absence there is a decrease in cell proliferation. Short chain fatty acids (SCFA), products of bacterial fermentation of non-digestable carbohydrates, have been implicated in the regulation of colonic epithelial cell proliferation. SCFA are known to directly influence genes that regulate cell proliferation and cell cycle, and to directly or indirectly stimulate the release of growth factors and GI peptides that promote colonic growth. Antrectomy (eliminating endogenous gastrin) results in atrophy of the colonic mucosa, an effect that can be reversed by exogenously administered gastrin. Similarly, excess gastrin levels (either derived from the stomach or a gastrin-producing tumor) are associated with colorectal cancer. Therefore, gastrin may also play a role in the regulation of colonic growth.

WATER AND ELECTROLYTE TRANSPORT

Water Transport

The large intestine normally absorbs a smaller quantity of fluid than does the small intestine, but it does so much more efficiently. Approximately 1.5 to 2.0 liters of fluid enter the large intestine of the adult human each day. Over ninety percent of this fluid load is absorbed, leaving only 100 to 150 ml of water to be excreted in stools (Figure 8-2A). The maximum capacity of the human colon to absorb water may be as great as 5.0 liters per day. Provided that the rate of entry of fluid into the colon is less than the maximum absorptive capacity, net fluid secretion can occur in the small intestine, but diarrhea (increased fecal water excretion) will not oc-

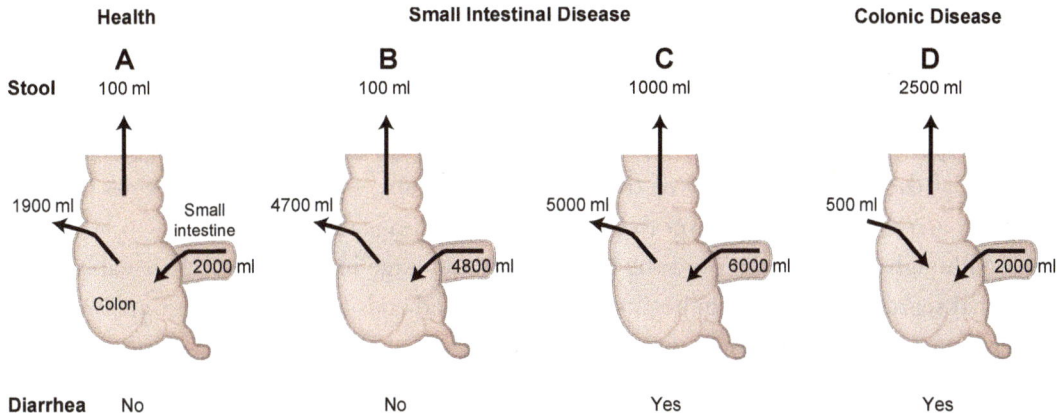

	Health	Small Intestinal Disease		Colonic Disease
	A	B	C	D
Stool	100 ml	100 ml	1000 ml	2500 ml

Small intestine, Colon labels; values: A: 1900 ml, 2000 ml; B: 4700 ml, 4800 ml; C: 5000 ml, 6000 ml; D: 500 ml, 2000 ml

Diarrhea	No	No	Yes	Yes

FIGURE 8-2: Interactions between ileocecal water flow, colonic water absorption, and stool water in health and disease. A. Normal dynamics; small bowel secretions well below colonic water absorption capacity. B. Enhanced mall bowel secretion that does not exceed colonic water absorption capacity (no diarrhea). C. Small bowel secretion greatly exceeds colonic absorption (diarrhea). D. Colonic fluid secretion adds to water entering from small bowel (diarrhea). [Modified from Binder HJ, Sandle GI. Electrolyte transport in the mammalian colon. In: Johnson LR (ed) Physiology of the Gastrointestinal Tract. 3rd ed. New York, Raven, 1994, pp 2133–2172.]

cur (Figure 8-2B). However, when ileocecal flow exceeds the absorptive capacity of the colon (Figure 8-2C), or if there is actual colonic fluid secretion (Figure 8-2D), diarrhea will ensue. Therefore, the capacity of the colon to absorb water is critical in determining whether or not there is diarrhea when either the small or large bowel is diseased.

Water absorption in the large bowel is coupled to, and dependent on, solute (electrolyte) absorption. The coupling of solute and water fluxes in the colon can be explained in terms of the standing osmotic gradient theory. An osmotic gradient is established between the lumen and lateral intercellular space by active transport of electrolytes into the latter compartment. The osmotic gradient draws water across the tight junctions and through the cell into the intercellular space (see Fig. 1-13 of Chapter 1). Unlike the small intestine and gallbladder, in which absorbed electrolytes are accompanied by an isosmotic equivalent of water, the large bowel generates a hypertonic absorbate. The human colon transports water against osmotic gradients of up to 50 mOsm per liter, indicating that a large osmotic gradient is needed to drive water through the colonic epithelial barrier. This observation is explained by the relatively high mucosal resistance (compared with jejunum and ileum) in the colon. The effective pore size of the colonic mucosa is approximately 2.3 Å (radius), which is significantly smaller than the 8 Å and 4 Å pore radii predicted for the jejunum and ileum, respectively. The relative tightness of colonic epithelia accounts for the ability of the colon to generate a hypertonic absorbate and to generate rather large concentration differences for electrolytes between lumen and plasma. This is explained by the fact that passive water movement into the intercellular spaces proceeds more slowly, owing to high resistance, than does electrolyte transport. Another functional consequence of the low permeability of the colonic epithelium is a larger spontaneous potential difference across the epithelial layer than those measured in jejunum and ileum.

Although we have employed the terms colon and large intestine interchangeably to describe that segment of bowel that begins at the cecum and ends at the anus, there are important differences in transport function among the different regions of the large intestine. Studies in humans and experimental animals indicate that the rate of water absorption is greater in the proximal large bowel (cecum to transverse colon) than in the distal segment (transverse colon to rectum). The cecum is the most permeable region of the large bowel (comparable to the ileum), whereas the rectal mucosa is the least permeable. Thus, the ability of the colon to generate a hypertonic absorbate increases in the aboral direction so that the final fecal water is hypotonic with respect to plasma.

Electrolyte Transport

Sodium

The large intestine is more efficient at conserving sodium than is the small intestine. Approximately 150 mEq of sodium enter the colon each day, yet only 1 to 5 mEq are lost in stools. The 95% or greater efficiency of the colon in conserving sodium compares with the 75% efficiency of the small bowel. Furthermore, the colon can absorb net sodium even when the luminal concentration is as low as 30 mEq per liter, whereas net sodium absorption in the jejunum does not occur when luminal sodium concentration falls below 130 mEq per liter. The differences in efficiency of sodium transport between the small and large bowels result from the relatively higher permeability of small bowel mucosa to sodium. The small intestine cannot absorb sodium against a concentration gradient because of a large permeability-associated plasma-to-lumen movement ("back-leak") of sodium.

The two major mechanisms of sodium absorption in the human colon are electroneutral NaCl absorption and electrogenic Na^+ absorption (Fig. 8-3, panel A). The electroneutral absorptive mechanism is dominant in the proximal colon, while electrogenic Na^+ absorption is more prominent in the distal colon. The electroneutral mechanism involves parallel Na^+-H^+ and Cl^--HCO_3^- transporters in the apical membrane of the colonic epithelial cells. The net effect of the coupled transporters is Na^+ and Cl^- uptake in exchange for H^+ and HCO_3^-. The

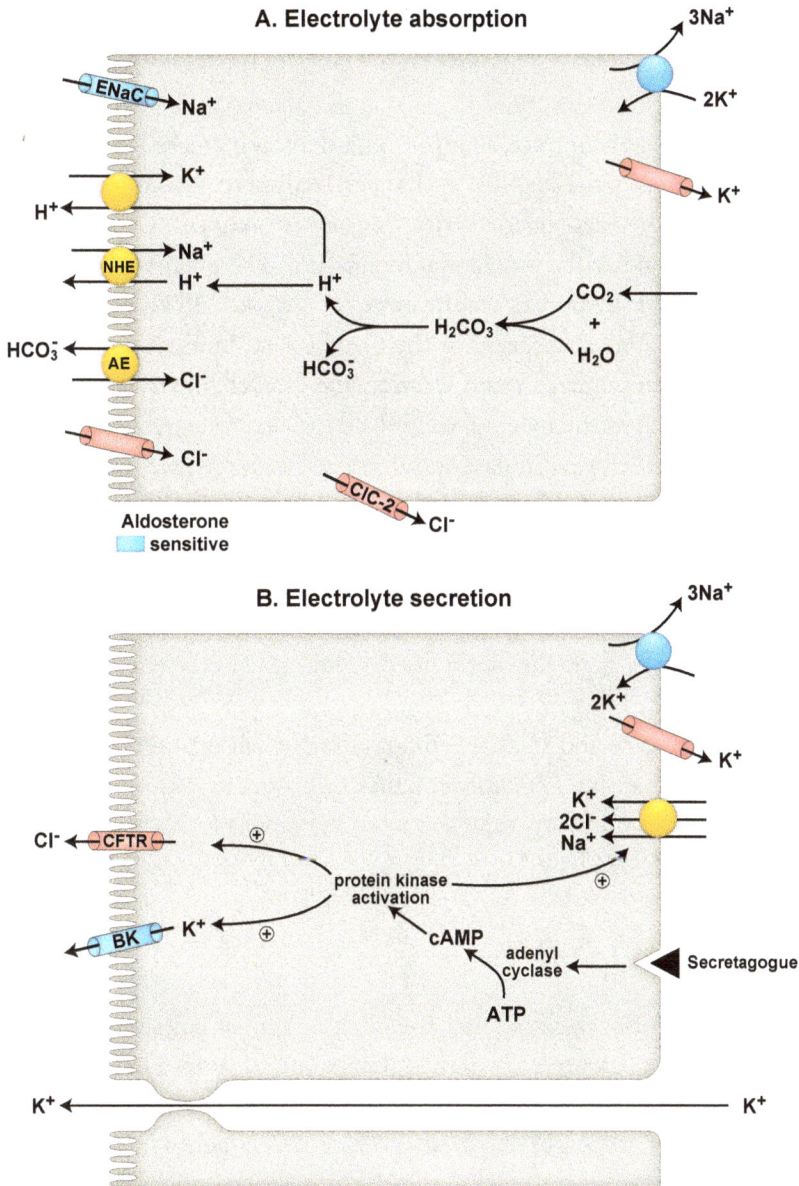

FIGURE 8-3: A. Mechanisms involved in the absorption of sodium, potassium, chloride, and secretion of bicarbonate ions by colonic epithelial cells. B. Transport mechanisms that underlie the ability of colon crypt epithelial cells to secrete chloride ions and water when stimulated bacterial toxins, neurotransmitters, and other secretagogues. Aldosterone enhances sodium absorption and stimulates potassium secretion at the designated pumps and channels.

H^+ and HCO_3^- used in this transport process are derived from the reaction of carbon dioxide (derived from blood and cellular metabolism) with water in the presence of carbonic anhydrase (which is present in high concentrations in the colonic mucosa). The two transporters are coupled through small changes in intracellular pH such that an increased Na^+ gradient stimulates Na^+ uptake and H^+ efflux via the Na^+-H^+ exchanger, leading to alkalinization of the cytoplasm and enhanced Cl^--HCO_3^- exchanger activity. Sodium is pumped out of the cell by the Na^+-K^+ exchange pump, located at the basolateral membrane. The energy required for the sodium pump is derived from ATP hydrolysis via the enzyme, Na^+, K^+-ATPase.

With electrogenic Na^+ absorption in the distal colon, luminal sodium enters the colonocytes at the apical membrane through electrogenic Na^+ channels (ENaC) that are highly specific for Na^+. Inasmuch as the mucosal cell interior is electronegative (−30 mv) with respect to the lumen (0 mv), and the intracellular sodium concentration is one tenth the lumen concentration, sodium moves into the cell down both electrical and concentration gradients. Sodium exit from the cell is directed against steep concentration (14 mM to 140 mM) and electrical (−30 mv to +20 mv) gradients. The Na^+, K^+-ATPase is responsible for the "uphill" transport of sodium out of the cell. The ability of the electrogenic transport mechanism to absorb Na^+ against large concentration gradients accounts for the highly efficient Na^+ conservation that occurs in the distal colon.

While luminal glucose and HCO_3^- do not alter Na^+ absorption in the colon, mineralocorticoids (e.g., aldosterone) can markedly enhance this process. Aldosterone increases sodium absorption in the distal colon by opening the ENaC channels to Na^+ ions and by stimulating the activity of the Na^+, K^+-ATPase.

Potassium

Approximately 10 mEq of potassium enter the healthy human large intestine each day, while 5 to 15 mEq K^+ are lost in the stools over the same period of time. The high stool K^+ concentration likely reflects the fact that net transepithelial K^+ movement in the colon is secretory. However, whether the colon absorbs or secretes potassium is largely dependent upon its concentration in the lumen; net secretion occurs when the luminal concentration is less than 20 mEq per liter, but higher lumen concentrations are associated with net potassium absorption. These observations are consistent with passive movement along electrochemical gradients. Dietary K^+ depletion enhances active electroneutral K^+ absorption in the distal colon via a process that involves the exchange of luminal K^+ for intracellular H^+ by an H^+-K^+ pump. The K^+ exits the cell via K+ channels on the basolateral membrane (Fig. 8-3, panel B).

Inasmuch as luminal potassium concentration is usually less than 20 mEq per liter, net secretion normally occurs (Fig. 8-3, panel B). The tight junctions between colonic epithelium are highly permeable to potassium ions. This high permeability, coupled to the transepithelial electrical potential difference (30-50 mv lumen negative with respect to plasma), allows potassium to diffuse readily from plasma to lumen via the paracellular pathway. Active K^+ secretion also occurs throughout the colon. In this process, K^+ is transported into the cell by the Na^+-K^+ pump and the $Na/K/2Cl$ cotransporter, which is energized by the low intracellular K^+ concentration produced by the Na^+-K^+ pump. Upon entering the cell at the basolateral membrane, K^+ is either recycled by moving across K^+-specific channels in the basolateral membrane back into the extracellular fluid or it crosses the apical membrane (secreted) via K^+ channels designated as Big K (BK) because of their large conductance. The BK channel is the rate-limiting determinant of active K^+ secretion and its activity is increased by stimuli such as aldosterone, VIP and cholera toxin, all of which enhance colonic K^+ secretion. Aldosterone also promotes K^+ secretion in the colon by increasing Na^+-K^+ pump activity. VIP and cholera toxin, both of which act via cAMP signaling to promote K+ secretion, do not alter Na^+-K^+ pump activity. Hence, these agents exert their prosecretory effects by exclusively targeting apical BK channel activity.

Chloride and Bicarbonate

The mammalian colon is highly efficient in absorbing chloride ions. Approximately 100 mEq of chloride enter the human colon each day, yet only 1 to 2 mEq are lost in stools. Net chloride absorption occurs even when the luminal chloride concentration is as low as 25 mEq per liter, indicating that it is actively absorbed. The principal route for Cl^- absorption in the proximal colon is the electroneutral NaCl absorptive mechanism described above for Na^+ (Fig. 8-3, panel A). Chloride ions enter the colonocyte in exchange for bicarbonate ions, while sodium is exchanged for hydrogen ions. The net effect of the Na^+-H^+ and Cl^--HCO_3^- transporters is Na^+ and Cl^- uptake in exchange for H^+ and HCO_3^-. The chloride ions then exit the cells via Cl^--specific (ClC-2) channels in the basolateral membrane. Chloride absorption in the colon can also occur via a purely passive process that is driven by an electrochemical gradient for chloride across the tight junctions (paracellular route) and/or the cell membrane (transcellular route). In the distal colon, electrogenic Na^+ absorption via the ENaC channels creates the lumen negative potential difference that drives the movement of chloride ions through the paracellular and/or intracellular routes. Enhanced electrogenic Na^+ transport creates a larger potential difference and yield a corresponding increase in passive chloride absorption.

Although electrolyte/water absorption normally occurs on the most luminal aspect of colonic mucosal epithelium (e.g., surface epithelial cells), epithelial cells of the crypt region are the source of electrolyte/water secretion. The physiologic benefit of water secretion from the crypts to the lumen is to maintain the hydration of the luminal contents. Water movement into the colonic crypts is largely driven by electrogenic chloride secretion. The cellular model for active chloride secretion in the colon is outlined in Fig. 8-3 (panel B). The major pathway for Cl^- secretion is the cystic fibrosis conductance regulator (CFTR) channel located at the apical membrane of intestinal crypt cells. Cl^- is taken up into the cell by a specific co-transporter in the basolateral membrane that allows for the uptake of Cl^-, K^+, and Na^+. Because this co-transporter moves 2 Cl^- (along with a Na^+ and K^+) into the cell, it provides adequate amounts of Cl^- for the secretory process. The energy required for this Na^+-K^+-$2Cl^-$ co-transporter is derived from the Na^+ gradient created by the Na^+-K^+ pump. The accumulating intracellular Cl^- then moves out of the cell via the CFTR channel. A small amount of Cl^- secretion occurs normally in the healthy colon; however, a number of secretagogues are known to markedly stimulate Cl^- secretion (Fig. 8-3, panel B). Secretagogues (e.g., bacterial toxins, neurotransmitters, activation products of immune cells) enhance Cl^- secretion by stimulating signaling pathways that involve the activation of enzymes such as adenyl cyclase, guanylyl cyclase, and phospholipase C. These enzymes generate mediators (cAMP, cGMP, IP_3) that activate protein kinases to stimulate pre-existing apical Cl^- channels (e.g., CFTR) and/or promote the insertion of more Cl^- channels into the apical membrane, thereby resulting in enhanced Cl^- secretion. While Cl^- exits the cell through apical channels, the resulting electrical gradient drives the movement of Na^+ through the paracellular channels, yielding an osmotic gradient that drives water movement into the lumen.

DIGESTION AND ABSORPTION OF FOOD RESIDUES

Although the human colon is incapable of absorbing the principal breakdown products of foodstuff (glucose, peptides, amino acids, long-chain fatty acids) in the form that they are delivered by the ileum, there are a large number of microbes in the colon that transform these and other food residues to substances that can be absorbed. An understanding of the ecosystem of microorganisms that live on the intestinal mucosal surface or within the gut lumen (commonly called the enteric microbiota) is necessary to appreciate its influence on digestion and transport.

Enteric Microbiota

Humans are born with a relatively sterile gastrointestinal tract that is successively colonized with microbial populations until a more diverse and stable adult-like microbial community is established at the end of the first year of life. In the adult human, the GI tract houses over 10^{14} microorganisms that represent over 1000 diverse species, most of which are bacteria. The density and diversity of microorganisms in the lumen increases progressively from the stomach to the colon (Fig. 8-4). The relatively low numbers of bacteria in the proximal GI tract is attributed to the influences of acid, bile, and pancreatic secretions on ingested microbes and the prevention of stable colonization of the lumen by phasic propulsive motor activity. The colon is the most densely populated and diverse bacterial habitat in the GI tract, with a concentration of 10^{11}–10^{12} living bacteria per gram of luminal contents. The presence of this bacteria-rich

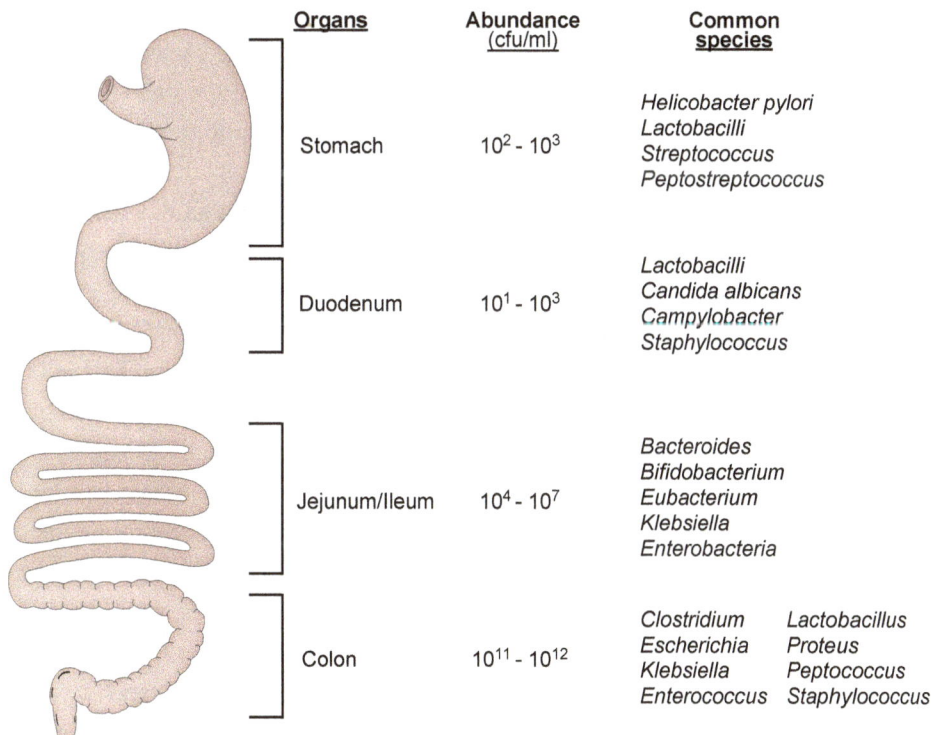

Organs	Abundance (cfu/ml)	Common species
Stomach	$10^2 - 10^3$	Helicobacter pylori Lactobacilli Streptococcus Peptostreptococcus
Duodenum	$10^1 - 10^3$	Lactobacilli Candida albicans Campylobacter Staphylococcus
Jejunum/Ileum	$10^4 - 10^7$	Bacteroides Bifidobacterium Eubacterium Klebsiella Enterobacteria
Colon	$10^{11} - 10^{12}$	Clostridium Lactobacillus Escherichia Proteus Klebsiella Peptococcus Enterococcus Staphylococcus

FIGURE 8-4: Density and distribution of bacteria in different regions of the gastrointestinal tract. The number and diversity of the bacteria increases from proximal to distal regions of the GI tract.

environment in the colon is generally attributed to slow transit time and the availability of fermentable substrates. The extremely large number of bacteria in the colon is exemplified by the fact that bacteria account for nearly 60% of solid colonic contents. Although the bacterial populations that reside in the proximal GI tract (stomach, jejunum) are aerobes and facultative anaerobes, anaerobic bacteria predominate in the colon, exceeding aerobic species by a factor of 10^2 to 10^4.

Although enteric bacteria are not essential for life, the microbiota does appear to exert a significant influence on human physiology. This is evidenced in germ free animals (born and raised in a sterile environment), which exhibit reductions in liver weight, intestinal wall thickness, gastrointestinal motility, and delayed development of the immune system. The beneficial physiological effects of enteric microbiota fall into three categories, i.e., metabolic, protective and trophic functions. The metabolic functions (discussed in greater detail below) relate to fermentation of nondigestible dietary substrates and endogenous mucus, as well as vitamin biosynthesis. Resident colonic bacteria also exert a protective function by resisting or restricting the colonization and growth of exogenous microbes or opportunistic bacteria that are present in the gut. Although the balance between species of resident bacteria provides stability in the microbial population, this equilibrium can be disrupted by antibiotic treatment that leads to the overgrowth of toxigenic bacteria, such as *Clostridium difficile*, which causes diarrhea. The trophic functions of the microbiota involve the control of epithelial cell proliferation and differentiation, as well as ensuring optimal development and maturation of the immune system.

Several gastrointestinal diseases, including inflammatory bowel disease, irritable bowel syndrome, and antibiotic-induced diarrhea, are associated with alterations in the gut microbiota (known as dysbiosis). Because a balanced gut microbial ecosystem is viewed as important for digestive and overall human health, the enteric microflora is now considered to be a viable therapeutic target for prevention and/or treatment of these diseases. Microorganisms (e.g., bifidobacterium and lactobacillus) that are administered in amounts adequate to confer a health benefit on the host are called *probiotics*. Food ingredients that are selectively fermented by beneficial enteric bacteria and elicit changes in the composition and/or activity of the GI microbiota that promote health are called *prebiotics*. For example, inulin-type fructans, which occur naturally in cereals such as wheat and barley, vegetables like onion and garlic, and fruits such as banana and tomato, stimulate the growth of "friendly bacteria" like bifidobacterium and lactobacillus to produce a healthy or balanced gut microbiota. Fecal transplantation, the transfer of stool from a healthy donor into a diseased recipient (most commonly achieved via colonoscopy), is another therapeutic strategy that is used to restore a balanced colonic microflora in patients who suffer from recurrent/refractory antibiotic-induced (Clostridium difficile-mediated) diarrhea.

Fermentation of Carbohydrates and Proteins

The majority of the bacteria in the colon are strict anaerobes that derive energy from fermentation. Substrates for fermentation that normally reach the human colon are listed in Table 8-1. These substrates typically include non-digestible carbohydrates, dietary fiber, the glycoprotein constituents of GI mucus (mucins), non-digested dietary proteins, other proteins of endogenous origin (enzymes, secretions, desquamated epithelial cells) and mucins that line the walls of the gastrointestinal tract. The amount of non-digestible carbohydrate available for colonic fermentation in individuals consuming a Western diet ranges between 20 and 80 g/day, while up to 50 g of protein are fermented in the human colon daily. Lipids that reach the colon are only present at influential levels in patients with severe pancreatic insufficiency.

TABLE 8-1: Fermentable Substrates that Reach the Human Colon*		
SUBSTRATE	**COMPONENT**	**AMOUNT (G/DAY)**
Carbohydrates		
	Resistant starch	5–35
	Non-digestible polysaccharides	10–25
	Oligosaccharides (e.g., fructo- or gluco-oligosaccharides, inulin)	2–8
	Monosaccharides (e.g., sugars, sugar alcohols)	2–5
	Mucins	3–5
	Synthetic carbohydrates (e.g., lactulose, polydextrose, modified cellulose)	(Variable)
Proteins		
	Dietary origin	1–12
	Endogenous origin (e.g., pancreatic enzymes and other secretions)	4–8
	Desquamated epithelial cells	30–50
Other		
	Non-protein nitrogen (e.g., urea, nitrate)	~0.5
	Organic acids, lipids, bacterial recycling	(Unknown)

* [From: Guarner F. *The Enteric Microbiota*. In: Colloquium Lectures on Integrated Systems Physiology: From Molecule to Function to Disease (Granger DN, Granger JP, eds), Vol. 29. Morgan and Claypool Publishers, 2012. doi: 10.4199/C00047ED1V01Y201110ISP029

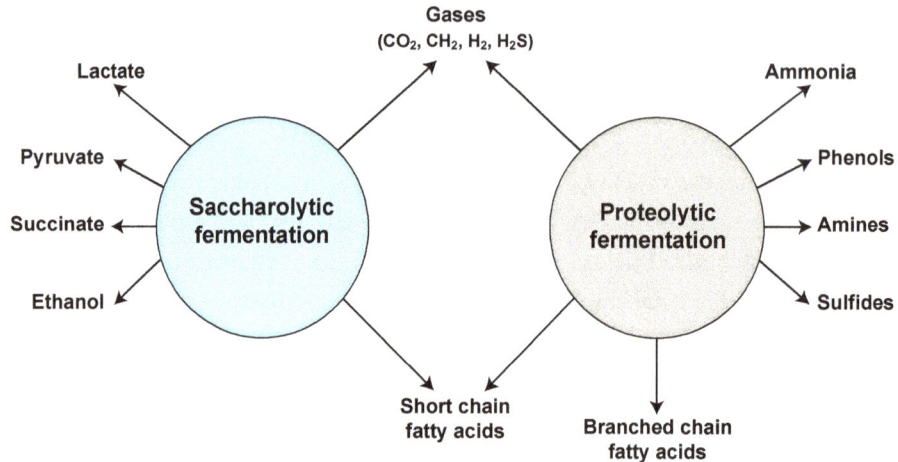

FIGURE 8-5: Products of carbohydrate (saccharolytic) and protein (proteolytic) fermentation in colon. Short chain fatty acids and gases are products shared by both fermentation processes. Saccharolytic fermentation is more robust in the proximal colon, while proteolytic fermentation occurs primarily in the distal colon.

The fermentation of carbohydrates (saccharolytic fermentation) in the colon leads to the generation of short chain fatty acids (acetic, propionic, and butyric acids), lactate, ethanol, succinate, pyruvate, and different gases (Fig. 8-5). The short-chain fatty acids (SCFA) produced by carbohydrate fermentation can achieve a luminal concentration of 100–150 mM and account for over 50% of the total anion content of human feces. Nearly all (>95%) of the SCFA produced in the colon is appropriated by this tissue, with very little excreted in feces. The luminal concentration of SCFA increases markedly approximately two hours after a meal. Luminal bicarbonate neutralizes a significant fraction of the acid load generated by these volatile short-chain fatty acids, resulting in the production of carbon dioxide and water. The human large intestine absorbs significant quantities of SCFA, by non-ionic diffusion and carrier-mediated transport via an SCFA-HCO_3^- exchanger. Once absorbed, the SCFA either are metabolized to ketone bodies by colonic epithelium or leave the cell by non-ionic diffusion or SCFA-HCO_3^- exchange on the basolateral membrane. Of the SCFA, butyrate is unique because it normally serves as the primary nutrient source for colonocytes. Upon exiting the colonocytes, the SCFA are transported via the portal vein to other tissues (e.g., liver) where they can be utilized as an energy source. Although it is estimated that absorbed SCFA provide 5% to 15% of the total caloric needs in man, some ruminants obtain up to 70% of their energy requirements from short-chain fatty acid production and absorption in the fore-stomach. A potentially important physiologic effect of SCFA in the human colon is augmentation of sodium, potassium, and water absorption. SCFA stimulate electroneutral Na^+ absorption (by activating the Na^+-H^+ exchanger) up to 5-fold in the human colon.

Saccharolytic fermentation largely occurs in the proximal colon because most enteric microorganisms preferentially ferment carbohydrates. Because carbohydrate availability diminishes in the distal colon, proteins and amino acids become important energy sources for bacteria in this region. Cellular desquamation is the principal (and least variable) source of proteins presented to the colon for fermentation (Table 8-1). Proteolytic fermentation, while yielding SCFA at much lower levels than produced from carbohydrates, is associated with the production of significant quantities of branched chain fatty acids such as isobutyrate and isovalerate, a variety of nitrogenous and sulphur-containing compounds, and different gases (Fig. 8-5). Unlike the products of carbohydrate fermentation that are believed to offer health benefits, some of the end products of proteolytic fermentation (e.g., ammonia, amines and phenolic compounds) may be toxic to the host. This may explain why colon cancer and chronic ulcerative colitis, diseases that generally affect the distal region of the large intestine, have been linked to proteolytic fermentation. Such an association between colonic disease and protein fermentation justify efforts to shift the gut fermentation towards saccharolytic activity by increasing the proportion on non-digestible carbohydrates in the diet.

Bile Acid Metabolism

Bile acids that are not absorbed in the ileum enter the colon, where they are extensively metabolized by the microflora. The number of bile acid metabolites produced by colonic bacteria is enormous. The degradative reactions include deconjugation, dehydroxylation, desulfation, oxidation, reduction, and even desaturation of the steroid ring. The products of these reactions are more lipid-soluble than the substrates, which facilitate passive bile acid absorption in the colon. It is estimated that 50 mg per day of bile acids are passively absorbed in the human colon, primarily as a result of microbial degradation, and returned to the liver via the portal vein. In addition to promoting the salvage of bile acids, the degradative reactions (e.g., deconjugation) mediated by colonic bacteria diminish the toxicity and carcinogenic potential of bile salts in the colon.

The biotransformed bile acids that are not passively absorbed in the colon are excreted in the stool. The normal daily fecal loss of bile acids ranges between 300 and 600 mg. Fecal loss of bile acids is influenced by a number of factors related to diet. An increased fat or dietary fiber load to the colon is associated with an increased fecal loss of bile acids. Both fat and fiber sequester bile acids in the lumen of the colon and thereby interfere with their absorption. The ideal diet for bile acid conservation in the colon is one that has no solid residue to act as a sequestrant.

Vitamin Production and Absorption

Colonic bacteria can synthesize a variety of water-soluble vitamins in amounts that are of nutritional value to the host and especially to local colonocytes. The colonic microbiota in humans produces considerable amounts of thiamine (vitamin B_1) and the thiamine thus generated is absorbed by colonocytes using specific thiamine transporters (THTR-1 and THTR-2). Other water-soluble vitamins that are synthesized by colonic bacteria and absorbed by colonocytes via specific carrier-mediated mechanisms are riboflavin, niacin (vitamin B_3), pyridoxine (vitamin B_6), biotin (vitamin B_7), and folate (vitamin B_9). In some instances (e.g., biotin, folate) the amount of vitamin produced by colonic microbiota is similar or greater that the amount consumed in the diet, and the amount of microbiota-generated vitamin is of nutritional value to the host (e.g., folate). In other instances, the amount of vitamin synthesized by the colonic microbiota is either too low to meaningfully contribute to human nutrition and/or the bioavailability of the vitamin is limited in the colon. Vitamin B_{12}, for example, is produced in significant quantities by colonic microbiota, however in the absence of intrinsic factor to form a complex with (and facilitate the absorption of) B_{12}, the vitamin cannot be absorbed by colonocytes. Similarly, while the amount of bioactive vitamin K (K_2, menaquinone) produced by colonic bacteria is sufficient to satisfy human vitamin K requirements for coagulation, poor bioavailability limits its absorption and potential to contribute to vitamin K requirements.

Ammonia Metabolism in the Colon

Approximately 25# of urea synthesized in the liver reaches the gastrointestinal tract by diffusion from the blood. Bacterial urease in the colon converts it to ammonia (NH_3), which diffuses readily across the colonic mucosa into portal venous blood and returns to the liver, where it is resynthesized to urea. A second source of NH_3 in the colon is proteolytic fermentation (discussed above) and the subsequent breakdown of amino acids. Colonic NH_3 generation by the combination of bacterial ureases and proteolytic fermentation accounts for approximately half of the NH_3 that is normally detected in plasma. Since the ammonium ion (NH_4^+) does not diffuse across the mucosa, some trapping of NH_3 in the colon occurs if its contents are acidified, allowing for its elimination in the feces. NH_3 is highly toxic to neuronal tissue. Consequently, the clinical importance of NH_3 generation by the colon is evidenced in acute liver failure or with portal-systemic shunting caused by portal hypertension (secondary to cirrhosis), conditions leading to toxic elevations in blood ammonia levels and encephalopathy. Since the colon accounts for most of plasma NH_3, therapeutic strategies for prevention treatment of hepatic encephalopathy include dietary changes (e.g., ingestion of lactulose) that favor the formation of SCFA, acidification of colonic contents, and the generation of NH_4^+, which is eliminated in the stool. Alternatively, these patients may receive non-absorbable antibiotics (e.g., rifaximin) to reduce the number of colonic bacteria that generate NH_3.

COLONIC GAS

The gastrointestinal tract of normal individuals contains approximately 200 ml of gas (Table 8-2). There are four major sources of gastrointestinal gas: (1) swallowed air, (2) carbon dioxide liberated from the reaction of hydrogen and bicarbonate ions in the gut lumen, (3) volatile metabolites produced by intestinal bacteria, and (4) diffusion of gas from blood to lumen. The latter two sources contribute significantly to gas production in the colon. A variety of bacteria in the colon produce hydrogen gas, methane, and carbon dioxide through saccharolytic and

TABLE 8-2: Composition of Gastrointestinal Gas			
	STOMACH (%)	INTESTINE (%)	FLATUS (%)
Nitrogen	79	64	61.2
Carbon dioxide	4	14	8.1
Hydrogen	0	19	19.8
Methane	0	8.8	7.3
Oxygen	17	0.7	3.6

[From Levitt, MD, Bond, JH, and Levitt, DG: Gastrointestinal gas. In Johnson, LR (ed): Physiology of the Gastrointestinal Tract, Vol. 2. New York, Raven Press, 1981.]

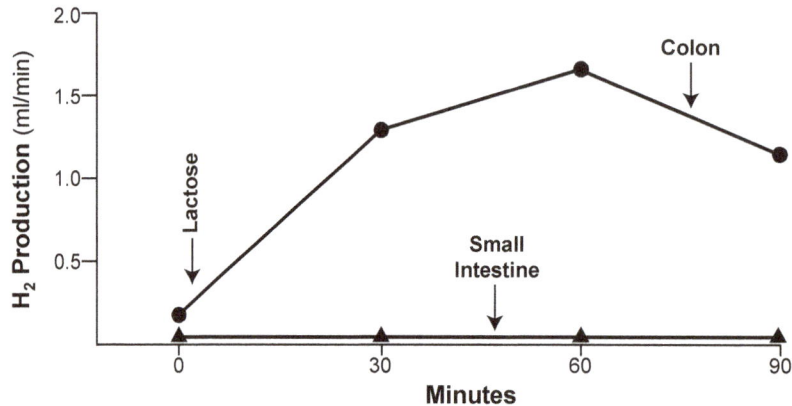

FIGURE 8-6: Hydrogen gas production in the small bowel and colon of normal subjects before and after intraluminal infusion of 2 grams of lactose. [Modified from Levitt, MD, Bond, JH, and Levitt, DG: Gastrointestinal gas. *In* Johnson, LR (ed): Physiology of the Gastrointestinal Tract. New York, Raven Press, 1981.]

proteolytic fermentation. Ingestion of fermentable substrates results in negligible hydrogen gas production in the small intestine (Fig. 8-6). In contrast, delivery of fermentable substrates to the colon leads to the production of appreciable volumes of hydrogen gas. This difference in the ability of the small and large bowel to produce H_2 largely reflects the greater number and types of bacteria in the large intestine. Ingestion of certain foods, particularly beans (e.g., soybean) that are rich in undigestable oligosaccharides (e.g., stachyose, raffinose), lead to massive increases in the H_2 content of flatus (air expelled through the anus). The rate of excretion of hydrogen gas in the breath can be used as a sensitive test for carbohydrate malabsorption, inasmuch as significant fractions (25–65%) of the H_2 produced in the colon is absorbed into the blood and eventually eliminated by the lungs.

Methane and carbon dioxide are also produced by colonic bacteria and eliminated in flatus. Methane gas that is produced in the colon becomes entrapped in feces, causing the stool to float in the toilet bowl (the floating or sinking tendency of a stool is determined almost entirely by its gas content). It is interesting to note that the ability of an individual to produce appreciable quantities of methane is familial. When both parents are CH_4 producers, there is a 95% incidence of production in the siblings, whereas the incidence is only 5% when neither parent is a CH_4 producer. Since children raised in close quarters (but with no common genetics) also have the same CH_4 producing status, the familial tendency may reflect an early environmental rather than hereditary cause.

Nitrogen is the predominant component of colonic gas in normal individuals (Table 8-2). The nitrogen gas found in the lumen of the colon (and flatus) is derived from diffusion of the gas down a partial pressure gradient from blood to lumen. A reduced partial pressure for nitrogen in the lumen results from bacterial production of CO_2, H_2, and CH_4 in the lumen.

It is estimated that 0.5 to 1.5 liters of flatus are passed in a 24-hour period, at a rate of about 14 times a day. The variation in flatus volume among individuals likely reflects differences in the quantities of fermentable foods that are ingested, and person-to-person variations in resident colonic microbiota. The passage of flatus normally accompanies the increased motor activity associated with food ingestion. Flatulence begins about one hour after the start of a meal and lasts for 20 minutes or more. The odor of flatus is largely determined by trace amounts of certain gases (e.g., hydrogen sulfide, ammonia) and volatile organic compounds (e.g., aliphatic amines, branched chain fatty acid).

MOTILITY

Intrinsic Influences

The contractile activity of the colon is dictated by the basal electrical rhythm or slow waves of its external muscle layers. The frequency of the slow waves determines the maximal number of contractions possible, while the number of spike potentials generated during the depolarization phase determines the strength of the contractions. The slow-wave frequency of colonic smooth muscle is highly variable and is not characterized by a proximal-to-distal frequency gradient. The dominant slow-wave frequency is 11 cycles per minute, and it occurs along a frequency plateau that extends from the proximal portion of the transverse colon to the descending colon. The slow-wave frequency in the ascending colon, cecum, and sigmoid colon is generally lower than that observed on the frequency plateau. The highest recorded slow-wave frequency in the entire gastrointestinal tract occurs at the level of the rectum (17 cycles per min). This pattern of slow-wave frequency is the basis for the erratic and slow transit of "chyme" through the colon.

Neural and Humoral Influences

The neurons of the myenteric plexus exert a net inhibitory influence on colonic smooth muscle cells. Their tonic inhibitory nature is exemplified by a condition in which the myenteric plexus is absent (Hirschsprung's disease). In Hirschsprung's disease, the aganglionic colonic segment is tonically constricted. The myenteric plexus of the colon includes excitatory neurons that release stimulatory neurotransmitters such as acetylcholine, substance P and neurokinin A, and inhibitory neurons that release nitric oxide and VIP upon activation.

Axons from both branches of the autonomic nervous system impinge on the neurons of the myenteric plexus. The parasympathetic supply to the proximal colon is by way of the vagus, whereas that to the distal colon is by way of the pelvic nerves. Although the parasympathetic nerves have an excitatory influence on colonic smooth muscle, different types of contractions are elicited during nerve stimulation. Vagal stimulation induces segmental contractions in the proximal colon; pelvic nerve stimulation produces propulsive contractions in the distal colon. The sympathetic supply to the proximal colon is derived from the splanchnic nerves that emerge from the superior mesenteric ganglia. The lumbar nerves, which originate from the inferior mesenteric ganglia, supply sympathetic nerves to the distal colon. Stimulation of either the splanchnic or lumbar nerves causes the colonic musculature to relax.

The extrinsic nerves are involved in several reflexes that alter colonic motility. The sympathetic branches mediate the reflex inhibition of colonic motility initiated by distension of the small intestine (intestino-colic reflex) or colon (colo-colic reflex). The parasympathetic (pelvic) nerves are involved in reflexes associated with defecation.

Various substances of gastrointestinal origin alter colonic motility. In general, their effects on colonic smooth muscle are similar to those observed in the small intestine (see Chapter 7).

Patterns of Contractions

The commonly observed motor patterns in the colon are either non-propagating or propagating (propulsive) contractions. Most colonic contractions are non-propagating. These are similar to the segmentation contractions described for the small bowel and serve to mix the contents of the large bowel and to enhance exposure of the fecal material to the mucosal surface, thereby facilitating fluid and solute absorption. Intrinsic slow wave activity regulates the maximum frequency and timing of the segmentation contractions. This motility pattern underlies the formation of sac-like segments called haustra. At any given locus, haustrations are formed and disappear in a random fashion owing to segmental contractions of the circular muscle. These contractions lead to haustral shuttling of the lumen contents that are simply displaced in both directions, and there is little or no movement. Extensions of this simple shuttling activity are haustral propulsion and retropulsion. Haustral propulsion occurs when adjacent haustra contract sequentially, so that the contents of a single haustrum are displaced into the adjacent segment and subsequently to another haustrum downstream. Haustral retropulsion occurs when the sequential haustral contractions move fecal material orally from one haustrum to another. Multihaustral propulsion is accomplished by the simultaneous contraction of several haustra that empty their contents into several adjacent haustra downstream. In general, all of the aforementioned segmental contractions, involving either a single haustrum or several haustra, occur with increasing frequency after a meal.

Propagative contractions occur much less frequently than the segmental type of motor activity. Peristalsis in the colon is a similar phenomenon to that observed in the small intestine, i.e., a propagated wave of contraction preceded by an area of relaxation that moves the fecal material. However, these high-amplitude propagated contractions (also called giant migrating contractions) can move colonic contents 18 to 20 cm, a distance ten times greater than that traveled by peristaltic waves in the small intestine. These mass movements, i.e., rapid propulsion of luminal contents over long segments of the gut, that are elicited by giant migrating contractions occur between 3 and 10 times per day in healthy humans. This motility pattern is not regulated by intrinsic slow wave activity. Mass movements usually begin in the transverse colon and can move chyme as far down as the rectum (Fig. 8-7). Mass propulsion or peristalsis is

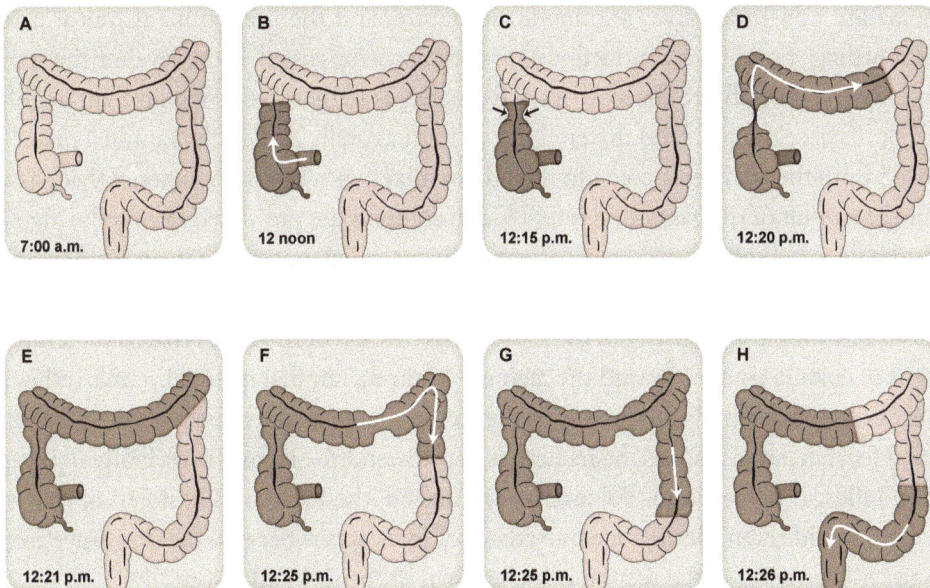

FIGURE 8-7: The movement of chyme in the large intestine at different times following food ingestion. The subject had a breakfast containing barium sulfate at 7 a.m. (*A*). At 12 noon, the barium is visible in the terminal ileum, cecum, and ascending colon, and a second meal is ingested (*B*). This results in the movement of more chyme from the ileum into the ascending and transverse colon (*C, D*). Haustral (segmental) contractions that mix chyme are evident in the transverse colon (*D, E*). About 5 minutes later the shadow becomes elongated, and then extends to the splenic flexure (*F*) and down the descending colon (*G*) to the level of the sigmoid colon (*H*).

rarely observed in the fasting individual but usually occurs soon after ingestion of a meal and often precede and accompany the act of defecation.

Effects of Feeding on Colonic Motility

The contractile activity of the colon increases after feeding (Fig. 8-7). The amount of food ingested does not appear to influence the response. However, the intensity of the postprandial contractile activity is directly related to the caloric content of the ingested food. Of the three major classes of nutrients, fats appear to be the most important determinant of the postprandial response. The postprandial motor response involves serotonin as well as a cholinergic component. The term *gastrocolic reflex* has been applied to the enhancement of colonic motor activity produced by ingestion of food and which is often associated with the urge to defecate.

The type of contractile activity induced by feeding is different in various segments of the colon. In the cecum and ascending colon, the major sites of fluid and electrolyte absorption, the movements are non-propulsive (segmental) in nature, and the predominant motor activity is haustral shuttling. Propulsive contractions are commonly observed at the level of the transverse colon following a meal. Since the slow wave frequency is higher in this region than in either the proximal or distal segments, fecal material moves both aborally and orally. Luminal material that is propelled from the transverse colon to the cecum mixes with the fresh material that passes into the cecum as a result of the gastroileal reflex. Cecal contents are prevented from refluxing into the ileum by the ileocecal sphincter. The net effect of postprandial motor activity in the proximal portion of the colon is to retain and mix the luminal contents, and to prolong the time available for absorption. The propulsive activity of the more distal portion of the transverse colon results in the movement of colonic content toward the rectum. At times, luminal material travels rather slowly. However, it can move rapidly all the way to the rectum if a mass movement is elicited. When the semi-dried fecal material reaches the rectum, it may distend the rectum sufficiently to initiate the defecation reflex. If defecation is voluntarily suppressed, the fecal mass is stored or moved back toward the sigmoid colon. Retrograde movement of feces occurs because the slow wave frequency (and contractile activity) in the rectum is higher than in the sigmoid colon.

The transit of fecal material in the large intestine is expressed in terms of days rather than hours. On a Western diet, 1.5 (average) to 4 (maximum) days are required for luminal contents to advance from the cecum to the rectum. The transit time is slightly faster in men with a corresponding increase in fecal volume. The fiber content of the diet is a major determinant of colonic transit time (Figure 8-8), as evidenced by the significant reduction in transit time when a large amount of bran (fiber) is added to the diet.

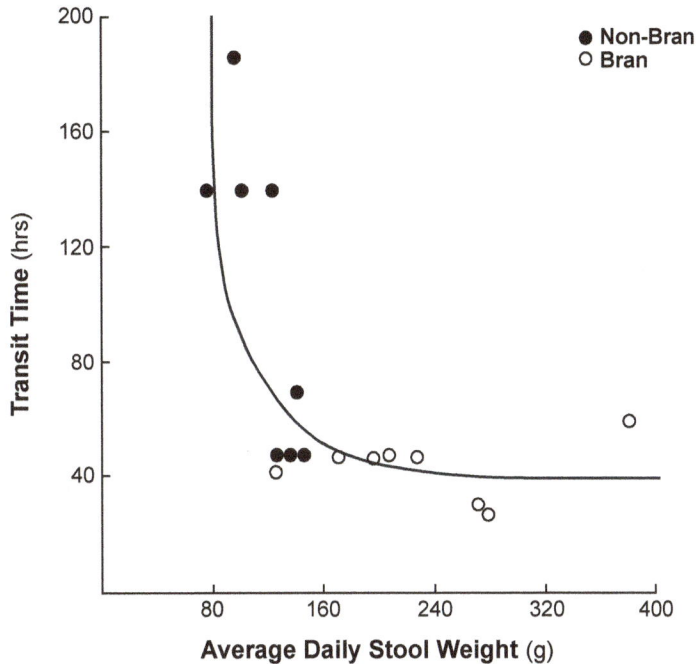

FIGURE 8-8: The effect of dietary fiber on colonic transit time and stool weight. The addition of fiber (bran) to the diet reduces stool transient time and increases stool weight. [Modified from Connell, A.M. *In* Reilly, RW, and Kirsner, JB (eds): Fiber Deficiency and Colonic Disorders. New York, Plenum Publishing Corp., 1975, p. 82.]

DEFECATION

Anatomic Considerations

The rectum and anus are particularly well suited for the controlled elimination of feces. The rectal mucosa contains crypts, which are deeper than those of the rest of the colon and are almost exclusively populated by goblet cells. The mucus secretion produced by these cells lubricates the rectal surface and eases the passage of semi-solid stools. The surface epithelium changes progressively from columnar epithelia to a stratified squamous epithelium, which, at the level of the anal orifice, becomes epidermis. The exit of fecal material is controlled by both smooth (internal anal sphincter) and skeletal (external anal sphincter and puborectalis) muscle components. The circular smooth muscle of the anal canal is three to four times thicker than that in the remainder of the colon and forms the internal anal sphincter. The longitudinal muscle layer becomes thinner over the internal sphincter and inserts into adjacent connective tissue. The external anal sphincter, which surrounds the internal sphincter, is an expansion of the striated levator ani muscle that comprises the pelvic floor. The levator ani includes the pubococcygeus, iliococcygeus, and puborectalis muscles. The puborectalis muscle forms a sling around the rectum and creates a sharp angle between the levator ani and external anal sphincter muscles, which (along with the tone of the internal and external sphincter muscles) helps to maintain fecal continence (Figure 8-9). Like the external anal sphincter, the puborectalis muscle is under voluntary control.

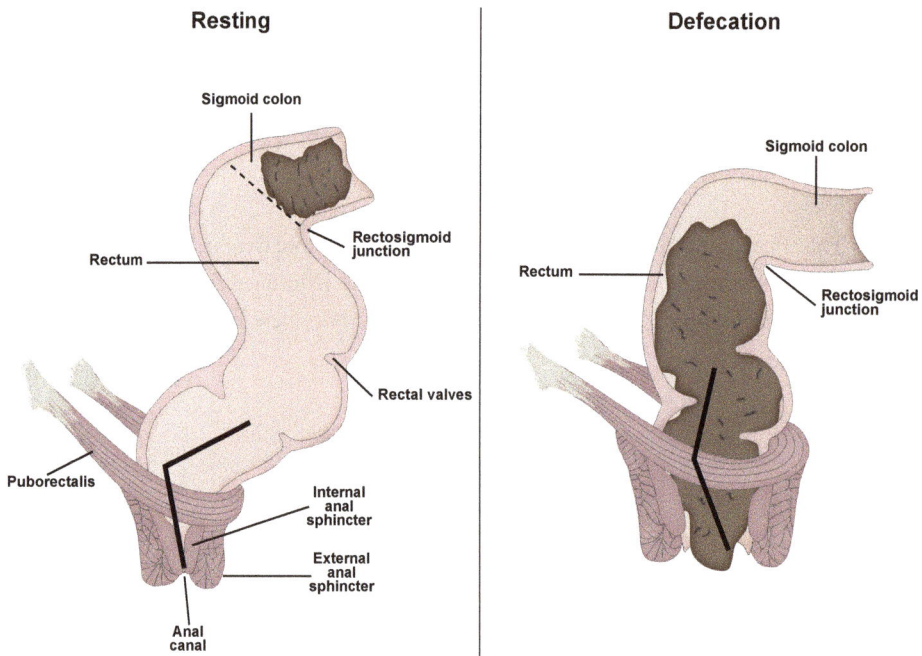

FIGURE 8-9: Anorectal responses during defecation. Under resting conditions (left panel), the rectum and anal canal are empty, the puborectalis muscle and the internal and external anal sphincters are constricted, and continence is maintained. Following a mass movement (right panel), fecal material enters and distends the rectum, which results in reflex relaxation of the internal anal sphincter. With voluntary relaxation of the external sphincter and puborectalis muscle, the increased pressure gradient between rectal lumen and anal canal will move the stool through the anal orifice.

Innervation of the Rectum and Anal Sphincters

The rectum and anal sphincters are innervated by both branches of the autonomic nervous system. The sympathetic fibers to the rectum are derived from L1–L3 spinal cord segments, while the parasympathetic fibers come from S3–S4 spinal cord segments. Afferent parasympathetic fibers in the nervi erigentes transmit the sensation of rectal distension, which is perceived via mechanoreceptors in the rectal wall and pelvic floor.

The parasympathetic supply to the internal anal sphincter is via the pelvic nerves, whereas sympathetic fibers are supplied by the hypogastric nerves. Activation of the parasympathetic nerves results in relaxation of the internal anal sphincter, a response mediated by the inhibitory neurons of the myenteric plexus. Sympathetic discharge tends to constrict the internal sphincter by activating alpha-adrenergic receptors on the smooth muscle.

The external anal sphincter, composed of striated muscle, receives somatic innervation via the pudendal nerves. The innervation of the external sphincter is excitatory, and the sphincter can be paralyzed by cutting the pudendal nerves. The external sphincter is primarily under voluntary control.

Reflex Activity of the Anal Sphincters

Both the internal and external anal sphincters are tonically constricted. However, the reflex responses of the two sphincters to rectal distention are different and provide the basis for the defecation reflex (Figure 8-10). Mild distention of the rectum results in a passive increase in rectal pressure. If the distention pressure is great enough, the passive increase in pressure is reinforced by an active contraction of the rectal muscles. The internal sphincter relaxes as the rectum is distended, via a local (intrinsic) reflex that is mediated by the release of NO and VIP from myenteric neurons. The external sphincter contracts in response to rectal distention. The contraction is mediated through an extrinsic somatic reflex activated by stretch receptors in the rectum.

Continence

As feces enter the rectum, either slowly during the interdigestive period or as a result of a mass propulsive wave after a meal, the rectal lumen distends. If the volume of the fecal mass is sufficiently large (when rectum is 25% filled), rectal contractions ensue, and the urge to defecate is felt. The intensity of this urge to defecate increases as the rectum continues to fill. Although the internal anal sphincter relaxes in response to rectal distenion, the external sphincter contracts

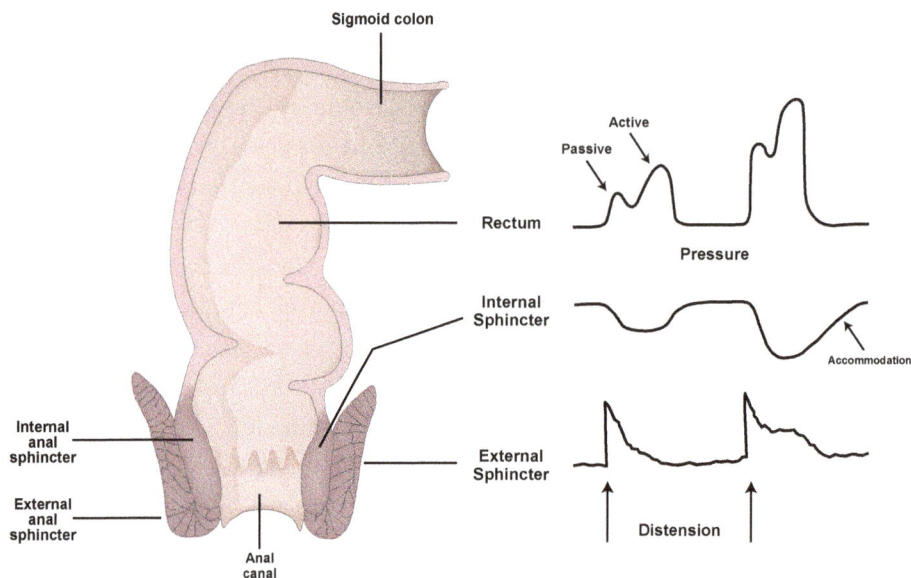

FIGURE 8-10: Responses of the rectum and anal sphincters to distension of the rectum. Mild distension of the rectum stretches its walls and causes a passive increase in pressure. Further distension reinforces the passive increases in pressure by stimulating contraction of the rectal walls. Each increment in rectal pressure is accompanied by a decrease in pressure at the internal sphincter and an increase in pressure at the external sphincter.

(Fig. 8-10). If the urge to defecate is opposed by voluntary contraction of the external anal sphincter, the rectal receptors eventually adapt, and the internal sphincter regains its tone. The rectum accommodates the fecal mass or propels it toward the sigmoid colon, and the urge to defecate subsides. This ability of the individual to override the urge to defecate by voluntary control of the external sphincter is referred to as continence. The continuous tonic contractions of the internal and external anal sphincter and puborectalis muscles are an important deterrent to involuntary defecation when stool has entered the rectum and the urge to defecate has passed. Similarly, these contractions are important in the prevention of involuntary defecation following sudden increases in intra-abdominal pressure, such as during sneezing, coughing, or lifting heavy objects.

Act of Defecation

If the rectum is appropriately distended at a convenient time either by the entrance of feces into the rectum or a voluntary increase in abdominal pressure, the defecation reflex will be voluntarily reinforced (Fig. 8-10). The internal sphincter will relax, but the external sphincter will only transiently constrict owing to voluntary removal of excitatory input to this sphincter. When both sphincters are relaxed, the increased pressure gradient between rectal lumen and anal canal will move the stool through the anal orifice. Voluntary relaxation of the external anal sphincter is also accompanied by relaxation (and lengthening) of the puborectalis muscle, and straightening of the anorectal angle. This basic defecation reflex is reinforced by extrinsic reflexes that are mediated by the pelvic nerves. Activation of these nerves by rectal receptors causes contractions of the descending and sigmoid colon and enhances the intrinsic inhibitory influence on the internal anal sphincter. Increased pelvic nerve activity also releases VIP, which, in turn, increases mucus secretion in the distal colon to facilitate the movement of feces. This extrinsic parasympathetic reflex can actually result in emptying of the descending and sigmoid colon as well as the rectum. The sympathetic nerves do not contribute significantly to the act of defecation.

Evacuation of the rectum is assisted by voluntary acts such as the Valsalva maneuver and by assuming a squatting position. The Valsalva maneuver consists of a forced expiration against a closed glottis, which increases intra-abdominal pressure, created by contraction of the abdominal muscles and the diaphragm. The increase in abdominal pressure forces the pelvic floor downward, straightening the anorectal angle and thereby reducing the resistance to the movement of feces. Flexion of the hips while assuming the squatting position also helps in aligning the anorectal angle.

Most people defecate once a day. However, defecation frequencies vary between 3 times per day to 3 times per week. The weight of the stool can vary considerably from one individual to another and from one bowel movement to another, determined largely by the fiber content of the diet (Fig. 8-8). The fecal water content, however, remains relatively constant. The amount of solids contained in the stool largely determines stool weight. Seventy percent of the total solids in stools consist of bacteria and fiber. The Bristol stool scale, an assessment tool that classifies human feces into seven categories, is used clinically to address stool consistency and frequency, to diagnose conditions such as irritable bowel syndrome, and to evaluate the effectiveness of treatments for various diseases of the bowel.

COLONIC CIRCULATION

Resting colonic blood flow in the adult human is approximately 250 ml per minute, which accounts for 5% of cardiac output. Blood flows as large as 1500 ml per minute (30% of resting cardiac output) have been measured in the colon of patients with ulcerative colitis. In humans, colonic blood flow (per unit mass of tissue) is approximately 40% and 60% of blood flow values measured in jejunum and ileum, respectively. The proximal colon receives a greater share of blood flow than the more distal portion.

Blood flow within the colon is unevenly distributed between the mucosa- submucosa (65%) and muscularis (35%) layers, presumably reflecting the different demands of the tissue layers for oxygen and other nutrients. Sympathetic nerve stimulation reduces total colonic blood flow while redistributing flow to the mucosa-submucosa layers, i.e., the vasoconstriction is more pronounced in the muscle layers. Pelvic (parasympathetic) nerve stimulation produces a biphasic (intense transient hyperemia followed by a smaller rhythmic hyperemia) increase in colonic blood flow, with the initial hyperemia confined to the superficial mucosal layer. The initial hyperemia is believed to be mediated by vasoactive intestinal polypeptide, which is released as a neurotransmitter from terminal fibers of the pelvic nerve. The sustained rhythmic hyperemia induced by parasympathetic activation is blocked by atropine.

Blood flow to the colon increases approximately 45 minutes following a meal. The postprandial hyperemia is confined to the mucosal layer, and the mechanism(s) mediating the vascular response have not been characterized. Volatile short-chain fatty acids may contribute to the hyperemia, since placement of SCFA in the colon lumen increases local blood flow. The postprandial hyperemia may also involve a local nervous reflex.

In contrast to the small intestine, net water absorption in the colon does not produce a rise in lymph flow. The absorbed fluid is removed from the colonic interstitium exclusively by the blood capillaries. The relative contributions of the blood and lymph microcirculations in transporting absorbed fluid are consistent with the distribution of these vessels in the colon. The initial lymphatics in the colonic mucosa are of a much smaller caliber than the lacteals of the small intestine, and they are located some 300 μm from the absorptive epithelium (the lacteals of the small intestine are about 50 μm from the enterocytes). Furthermore, the fenestrated capillaries of the colonic mucosa are much closer to the absorptive cells (1 μm) than their counterparts in the small intestine (2 μm), which is advantageous for the removal of absorbed fluid via the capillaries.

COLONIC PAIN

As in the small intestine, colonic pain arises from distension and stretching of the bowel wall. Nerve fibers conveying this sensation are found in the muscle layer. The pain is generally referred to the hypogastrium. Rectal pain is felt posteriorly, over the sacrum.

PATHOPHYSIOLOGY AND CLINICAL CORRELATIONS

Patients frequently seek medical care for symptoms such as constipation, diarrhea or colonic pain. These symptoms may result from either functional or structural changes in the large intestine. When compared with the small intestine, the colon has a higher propensity for development of chronic inflammatory disease and carcinoma. For example, *ulcerative colitis*, a chronic inflammatory condition that can extend from the rectum to the right colon, is characterized by mucosal abnormalities that include activation of resident and recruited immune cells, an imbalance between protective vs. aggressive commensal bacteria, increased epithelial membrane permeability, interstitial edema, leakage of protein-rich interstitial fluid with possible bleeding into the colon lumen. Patients with the more severe and prolonged forms of ulcerative colitis appear to be at increased risk for developing colorectal cancer. However, there are other risk factors that appear to predispose the colon to the development of carcinomas, including increasing age of a patient, and family history of colon cancer or polyps. Polyps are benign growths of the mucosa of the colon, but certain polyp types of larger size can proceed through a series of mutations to a carcinoma. Efforts at screening via occult blood testing or endoscopic evaluation by colonoscopy, with removal of polyps by a process called polypectomy, have led to a significant reduction in the number of patients diagnosed with colon carcinoma.

Another common finding in the colon of older patients is the presence of *diverticulosis*. Diverticulosis or *diverticular disease* is a condition associated with the formation of pouches (diverticula) of the colonic mucosa and submucosa through weakened areas of the colonic muscle layers. Diverticulosis, appears in patients over the age of 40, and is the most common cause of acute lower GI bleeding in this population. Whether fiber intake contributes or prevents the development of diverticuli is a source of controversy. The vast majority of patients are asymptomatic, but symptoms may develop when the diverticula are inflamed or bleed, which could require endoscopic intervention. Additionally, patients may develop a microabscess in a diverticula leading to pain, fever, and leukocytosis. Antibiotics are used to treat mild cases of diverticulitis, but patients may require radiologic drainage (percutaneous drainage using imaging guidance) for a larger abscess. Patients may also develop a perforation of the colon from a severe attack of diverticulitis, which would require surgical intervention. Fiber therapy has been found to soften stool and possibly prevent these complications after the development of diverticulosis.

Diarrhea may be induced in patients taking osmotic laxatives, such as magnesium citrate or polyethylene glycol, which extract and then retain fluid by osmosis into the colon lumen. Chronic intake of magnesium salt-containing antacids can also lead to diarrhea by this mechanism.

The laxative properties of magnesium citrate also account for its clinical use for bowel evacuation before diagnostic procedures (e.g., colonoscopy) or surgery of the colon. In lactase deficiency, the lactose that would normally be metabolized and absorbed in the small bowel enters the colon where it is metabolized to carbon dioxide, water and hydrogen by-products. The accumulated gases produce abdominal symptoms, while the remaining carbohydrates and fermentation products will draw water, by osmotic pull, into the colon to yield diarrhea.

Constipation refers to infrequent bowel movements (fewer than 3 per week) with hard or lumpy stools. Slow colonic transit time with increased frequency of segmental (non-propulsive) contractions will increase the time available for water absorption, with a resultant hardening of fecal material. Patients with slow transit constipation often exhibit either a blunted or no colonic motor response to eating (i.e., gastrocolic reflex). Dietary (e.g., fiber intake) and pharmacological (e.g., opiods) factors are common causes of constipation. In some patients, constipation is the result of a generalized depression of smooth muscle activity in the body. A good example of this generalized depression is the constipation of pregnancy, where high circulating levels of progesterone depress the tone of visceral smooth muscle. Aging is also associated with an increased incidence of constipation, with elderly women exhibiting a slower colonic transit time and are more likely to pass hard stools than their male counterparts.

REFERENCES

Binder HJ, Sandle GI. Electrolyte transport in the mammalian colon. In: Johnson LR (ed) Physiology of the Gastrointestinal Tract. 3rd ed. New York, Raven, 1994, pp 2133–2172.

Gearhart, S. Diverticular Disease and Common Anorectal Disorders. Chapter 297. Harrison's Principles of Internal Medicine. 18th Ed. Electronic Last Accessed: 08/25/2014.

Grundy D, Brookes S. *Neural Control of Gastrointestinal Function.* In: Colloquium Series on Integrated Systems Physiology: From Molecule to Function to Disease (Granger DN, Granger JP, eds), Vol. 30. Morgan and Claypool Publishers, 2012. doi: 10.4199/C0004 8ED1V01Y201111ISP030

Guarner F. *The Enteric Microbiota.* In: Colloquium Lectures on Integrated Systems Physiology: From Molecule to Function to Disease (Granger DN, Granger JP, eds), Vol. 29. Morgan and Claypool Publishers, 2012. doi: 10.4199/C00047ED1V01Y201110ISP029

Kvietys PR. *The Gastrointestinal Circulation.* In: Colloquium Series on Integrated Systems Physiology: From Molecule to Function to Disease (Granger DN, Granger JP, eds), Vol. 5. Morgan and Claypool Publishers, 2010. doi: 10.4199/C00009ED1V01Y201002ISP005

Levin B, Lieberman D, McFarland B, Andrews K, Brooks D, Bond J, Dash C, Giardiell F, Glick S, Johnson D, Johnson C, Levin T, Pickhardt P, Rex D, Smith R, Thorson A, Winawer S. Screening and Surveillance for the Early Detection of Colorectal Cancer and Adenomatous Polyps, 2008: A Joint Guideline from the American Cancer Society, the US Multi-Society Task Force on Colorectal Cancer, and the American College of Radiology. *Gastroenterology* 2008;134:1570–1595.

Rao MC, Sarathy J, Ao M. *Intestinal Water and Electrolyte Transport in Health and Disease*. In: Colloqium Series on Integrated Systems Physiology: From Molecule to Function to Disease (Granger DN, Granger JP, eds), Vol. 31. Morgan and Claypool Publishers, 2012. doi: 10.4199/C00049ED1V01Y201112ISP031

Rex D, Johnson D, Anderson J, Schoenfeld P, Burke C. Inadomi. American College of Gastroenterology guidelines for colorectal cancer screening. *Am J Gasterol* 2009;104: 739–150. doi: 10.1038/ajg.2009.104.

Said HM. Recent advances in transport of water-soluble vitamins in organs of the digestive system: a focus on the colon and pancreas. *Am J Physiol Gastrointest Liver Physiol* 305: G601–G610, 2013.

Sarna SK. *Colonic Motility: From Bench Side to Bedside*. In: Colloqium Series on Integrated Systems Physiology: From Molecule to Function to Disease (Granger DN, Granger JP, eds), Vol. 11. Morgan and Claypool Publishers, 2012. doi: 10.4199/C00020ED1V01Y20101 1ISP011